Silver and Gold Mining Camps
of the Old West

Also by Sandy Nestor

Indian Placenames in America
Volume 1: Cities, Towns and Villages;
Volume 2: Mountains, Canyons, Rivers, Lakes,
Creeks, Forests, and Other Natural Features
(McFarland, 2003 and 2005)

Silver and Gold Mining Camps of the Old West

A State by State American Encyclopedia

SANDY NESTOR

McFarland & Company, Inc., Publishers
Jefferson, North Carolina, and London

Library of Congress Cataloguing-in-Publication Data

Nestor, Sandy.
Silver and gold mining camps of the old West : a state by state American encyclopedia / Sandy Nestor.
p. cm.
Includes bibliographical references and index.

ISBN-13: 978-0-7864-2813-7
ISBN-10: 0-7864-2813-9
(illustrated case binding: 50# alkaline paper) ∞

1. Gold mines and mining — West (U.S.) — History — Encyclopedias.
2. Silver mines and mining — West (U.S.) — History — Encyclopedias.
3. Miners — United States — History — Encyclopedias.
4. Frontier and pioneer life — West (U.S.) — History — Encyclopedias.
5. West (U.S.) — History — Encyclopedias.
I. Title.
TN413.A5N47 2007 622'.3420978 — dc22 2006036399

British Library cataloguing data are available

On the cover: Butte County miners *(Butte County Historical Society, Oroville, California);* background ©2006 Photodisc

Manufactured in the United States of America

McFarland & Company, Inc., Publishers
Box 611, Jefferson, North Carolina 28640
www.mcfarlandpub.com

For Dad
Who passed away a few months
before this book was published.
You are deeply missed.

For my parents, who swore they would never return to that awful,
hot Mojave Desert after we crossed it moving from Pennsylvania
to California in the 1950s, but did. Our family ultimately became partners
in a uranium mine among all the gold in the region. Our mine was located
a stone's throw from the Tropico Mine at Rosemond, California.
Here is a view of the Nestor camp from the ore hopper
(the kitchen is in the middle):

Acknowledgments

Helen Giffen wrote, "No historian [or writer] is sufficient unto himself. He must rely upon the help of others to expand his own knowledge."

My admiration to all those hardy souls who wrote about the thousands of ghost towns, traveling to some of the most remote outposts where even four-wheelers could barely travel. Thanks to all the libraries, historical societies and museums, and individuals who aided me in my research.

To Susan Pritchard of Glendale, Arizona, whose late husband, Thomas, took wonderful photographs of numerous ghost towns and mining camps during their travels. Special thanks to you for giving me carte blanche to use them. This book is dedicated, in part, to his memory.

To my brother, David Nestor, a full-time RVer (you lucky stiff) who also shared his pictures with me.

And of course, Dad and Mom, who catapulted us into our wondrous times in the Mojave Desert, driving out to our camp every weekend at Rosamond, 80 miles from home in an old Willy's jeep over old Highway 99. You taught your kids how "not to fall" into mine shafts, how to take care of the desert, and to respect its wildlife. We will always have wonderful memories of those adventures that were some of the best years of our lives. Thanks, Mom, for assisting me with the final details of the manuscript.

My gratitude goes out to those individuals who graciously allowed me the use of their photographs: Robert Garrett, Arthur Alspector, Douglas Yetter, Stephen M. Murphy, John R. McNalley, Kevin Leach, Art Naffziger, Peter MacGregor, Bruno F. Smid, David Decareaux, Barbara Alger, Jeff Lotze, William Dilworth, Reeve Hennion, Channing Lowe, Chad Carter, and Dean LeMaster.

Table of Contents

From the Rockies to the Pacific, from the border of Mexico to the Klondike, hundreds upon hundreds of ghost towns dot the American landscape, silent reminders of an age when men sought their fortunes with picks and shovels. It took the peculiar nature of mining — its boom-or-bust fickleness — to cause a town to spring up overnight, flourish, then suddenly die.

—*Chuck Lawliss, 1985*

The miner has been California's heart. Indefatigable in everything he has undertaken appertaining to his vocation, with money or without, he has turned the river from its ancient bed and hung it for miles together in wooden boxes upon the mountain's side, or thrown it from hill to hill in aqueducts that tremble at their own airy height; or he has pumped a river dry and taken its golden bottom out. He has leveled *down* the hills, and by the same process, leveled *up* the valleys. No obstacle so great that he did not overcome it; "can't do it" has made no part of his vocabulary; and thus, by his perseverance and industry, were the golden millions sent rolling monthly from the mountains to the sea!

— Hutchings' California Magazine, *circa 1857*

Preface

In 19th century America, the only news that traveled faster than gossip and scandal was the cry of GOLD, GOLD, GOLD! Those fevered words forever changed the landscape of the West and the lives of those who came for the precious metal. The cry was heard around the world, bringing people from the East and Europe to find their fortunes, or their fates. After the initial furor over gold, silver came into prominence.

The discovery of gold at Coloma, California, spurred one of the greatest rushes the West has ever seen, something that will never occur again. Fortunes were made and lost, men were killed, and the land was desecrated by hydraulic mining. Some believe that the beginning of the rush was actually caused by an event that occurred near Mokelumne Hill. A man who was with Stevenson's regiment found a nugget along the Mokelumne River that weighed more than 24 pounds and assayed at over $5,000. The nugget was sent to Colonel E.F. Beale at Washington, D.C., by a Colonel Mason, who wrote there was gold in the West. Although gold had already been found at Coloma, initially that news didn't generate a lot of interest. The Mokelumne Hill nugget was shown at a public display, and may have sparked the gold rush to California.

Most famous of the gold-rush regions was the Mother Lode, named by the Mexicans who called it Veta Madre. This gold-yielding belt was defined from Mormon Bar on the south and Placerville on the north. The U.S. Geological Survey set the boundaries from just north of Bagby along the Merced River to the gold camp of Plymouth. The entire region yielded over $600 million in gold, and saw more than 120,000 miners searching for that glint that would forever change their lives, for better or worse.

Most of the gold was found in California, and a great deal was also located in Nevada, but the majority of ore mined there was silver. Gold and silver were found in Idaho, Oregon, and Washington (which also contained coal). Arizona was wealthy in copper, as was New Mexico (in addition to turquoise), though both yielded precious metals. Alaska produced a good amount of gold, although the majority of the ore in the region was located in the Klondike in Canada's Yukon Territory, which is not discussed here. An exception is Fortymile, since this was the place where the beginning of the great Klondike rush began, and from which the miners left Alaska. Utah also had a prominent coal industry. Wyoming was more notable for its coal and gas production, but did have a few gold mines.

This book focuses on mines and mining camps in the western states. Although other minerals were discovered throughout the United States (copper, iron, turquoise, uranium, manganese, etc.), only gold and silver camps are included here since those discoveries began the great rush to the west. People from the east headed to California with news of gold discoveries. Their migration created thousands of new settlements, many of which became permanent. Entries are arranged by state, and with the exception of Alaska, the modern county in which the camp is located accompanies each entry. (Alaska is divided into sixteen boroughs and one unorganized borough; the latter is made up of the area not encompassed by the organized boroughs and contains more than half of the land in the state.)

Camps were often analogous to mines. In many instances, there were one or two people who worked a mine. Next to their claim they made a "camp," which usually consisted of a tent or lean-to. However, many mining camps were centralized and became headquarters (home bases) where miners got their provisions; many passed into oblivion, while others survived long after mining ceased to central to their existence. Walla Walla, Washington, for example, began its life as a supply center for thousands of men headed to the Idaho gold fields, became firmly established, and evolved into a permanent community. Other camps flourished for a time, then began to decay as the ore played out and miners moved on to better prospects. Mining camps were places where prospectors could go to gamble or drink away their hard-earned money, take a bath, attend church, or partake of a boarding house's good meal. Numerous camps went on to become towns and cities because there were other reasons for their existence, though many ended up as ghost towns.

Locations of the camps and some of the mines that were worked near their respective sites are listed. Many of the obscure camps had little written about them, so identifying their exact sites was sometimes a best guess (even for the experts). When available, names of the miners who owned their claims, year of discovery, and ore values are also included. There will always be discrepancies on the total amounts of gold and silver. Many of the production figures were not recorded during early the mining days. Inaccurate records were the result of mine owners "cribbing" dollar amounts so they would not have to pay investor dividends against what was actually recovered from the mines. Some of the places were so small that records weren't kept. Additionally, documents were destroyed after fires visited many a mining camp. Also listed are those trading, supply, and milling centers that served the mining populations in their respective regions. Interesting events about the miners' lives are interspersed throughout the pages because their experiences were an integral part of what made up the wonderful flavor of the mining camps.

Women played important roles, and not just as pleasant recreation for lonely miners. Some women came to marry for love, some for money, and a few stated their own terms; as one woman wrote: "Her age is none of your business. She is neither handsome nor a fright, yet an old man need not apply, nor any who have not a little more education than she has, and a great deal more gold, for there must be $20,000 settled on her before she will bind herself to perform all the above." This included such mundane tasks such as feeding the animals, cutting wood, spinning weave, washing, cooking, sewing, sawing a plank and driving nails.

Chinese miners were also an important factor, but were not a sufficient threat to bring about a sustained effort to banish them. They were simply beaten, robbed or killed, depending on the mood of the day. Most of them came after the mines had been worked out and were content to glean what was left. Their hard work and resourcefulness netted them millions. A great number of Chinese who were imported to build the Transcontinental Railroad were left to their own devices after the railroad was completed, so they headed for the gold fields.

The fortitude of the prospectors and mining companies was amazing, for many of the ravines and mountains were all but impossible to navigate, not just for the men themselves, but the work animals. It took incredible determination to get machinery weighing tons into tight gulches and exceedingly steep mountains, hauled only by mule or ox teams. They lumbered up twisting, widow-making trails to heights of 10,000 feet or more in places. Some of the equipment was hoisted up by makeshift mechanisms that would boggle the mind today. Roads to the sites were almost nonexistent. During the winter access was impossible and supply trains were unable to reach the camps. Even when roads were being built, it still took years for their completion. Miners also had to contend with Indian attacks. But survive they did under the most dire of circumstances, undaunted by extremes, all for that shiny stuff.

A number of the camps were touted as having grand buildings and all the comfortable amenities. Contrary to that belief, except for the large mining towns, most were hastily established overnight with no thought to aesthetics, simply ramshackle cabins built haphazardly at whatever place was convenient, which were meager at best. The miners were much more interested in attaining riches than building warm, comfortable homes. And for the most part the places were temporary, for as soon as news arrived of riches elsewhere the men were off and running. Most welcome were those brave women who established boarding houses where the men could enjoy a good meal.

Prospecting claims were sacrosanct, or were supposed to be, but there were many claim-jumpings. One miner named Thomas Hall wrote on his claim: "Notis: To all and everybody. This is my claim, fifty feet on the gulch, cordin' to Clear Creek District Law, backed up by shotgun amendments." There was also much honor among the men, as Shinn noted: "The men dwelt together in no distrust of one another and left thousands of dollars' worth of gold-dust in their tents while they were absent digging." Miners were seldom hesitant to help one another. If one of them was broke, got discouraged, and wished to return to his former home, the men would collect enough cash to send him on his way. If a miner was killed and left a wife and children, again the men would take up a collection and care for the family until they could get back on their feet.

Many of the prospectors found riches and then sold their claims for a few thousand dollars, only to have the properties yield ore with values mounting into the millions. Prospectors knew through experience that without a moment's notice, the veins could pinch out, so they took what they could and skedaddled off to search for another strike.

Miners were an intrepid people, as described by Shinn in the 1880s: "They learned to wash and mend their own clothes, and bake their own bread.... They were hardy, generous, careless, brave; they risked their lives for each other, and made and lost fortunes.... They died lonely deaths, or perished by violence, pioneers to the last."

The Miners' Ten Commandments is included as an appendix which typifies the character of the people. There is also a glossary explaining some of the mining terms. References are arranged by state.

Sadly, so many of the mining camps and ghost towns have been desecrated by those who have no regard for history. Vandals have shot at the buildings, hammered them down, and otherwise destroyed them for incomprehensible reasons. Because of this, some ghost town writers have excluded certain places in order to keep them secret and preserve their pristine conditions.

ALASKA

ANCHORAGE The largest city in Alaska had its beginnings in 1887 with the discovery of gold in the region. Located in Cook Inlet, Anchorage was the operational center for construction of the Alaska Railroad in 1913 after President Woodrow Wilson authorized the funds. Its post office was established in 1914. The railroad brought in a population of more than 2,000. Originally named Ship Creek, in 1915 it was changed to Anchorage. When ore was found near Fairbanks during the early 1900s, Anchorage teemed with miners who were seeking their wealth, at which time it became headquarters for the Willow Creek Gold District and served as a supply center for all the outlying mining claims.

The 1964 earthquake had a devastating effect on Anchorage, since it was located only 80 miles from the epicenter, causing a loss that ran into hundreds of millions of dollars.

ANIAK Aniak is a Yup'ik word for "place where it comes out," designating the mouth of the river. It was once called Yellow River because of the water's color caused by silting. Although gold had been found in the area by a Russian trader in 1832, there was not much activity. Prospectors were reluctant to enter this unknown vicinity until 1900 when miners from Nome began panning for gold along the river. In 1914 the village of Aniak was established after a trading post was built by Tom L. Johnson. Gold found in the gravels of the river produced a little over $13 million.

BEAVER Located just south of the Arctic Circle along the Yukon River, Beaver was settled after gold was discovered in the Chandalar River during 1906. Before that time it was originally an Eskimo village and used as a river landing. Since U.S. congressman William Sulzer had an interest in a few mining claims there, he purchased a 4-stamp mill and had it shipped to Beaver, but it never arrived. Weighing almost 28 tons, only part of the mill made it because the rest of it was strung out along the trail. He was still trying to get the finished product delivered in 1927, and wrote to someone: "I suppose you will be very much disappointed when you learn that I haven't moved any of the machinery. It stormed 40 days in Alaska I never saw such a winter. I do all I could do. Hope you will look at this in the right manner."[1] During the 1970s travelers could still see parts of the Allis-Chalmers stamp mill.

A trail was built from town to the mines in 1911 and the post office opened in 1913, along with a trading post established by Frank Yasuda that served the surrounding mines. Beaver was predicted to be a booming town, but the gold proved not to be as rich as anticipated. About $620,000 was recovered from the nearby mines. The only major money made in the region was during 1961 when a mining company found an ore body and realized about $1 million in gold.

CHICHAGOF Named for Arctic explorer Admiral Vasili Chichagof, this site was founded in 1905 and its post office opened in 1909. Early attempts at mining in the district were conducted under American rule during 1871, but nothing was developed until John Newell and Ralph Young found gold float in 1905. They showed their samples to Judge Edward DeGroff of Sitka, who was impressed and grubstaked the two. He later purchased their claims and, along with a group of men, created a gold mining company. One of the most productive lode mines was the Hirst-Chichagof

Pedro dredge at Chicken. The dredge was transferred from Fairbanks (courtesy Robert Garrett).

that was discovered by Peter Romanoff, Andrew Dixon and Bernard Hirst, which produced about $2 million in gold. The Golden Gate Mine was later located by Alexander Pihl, Joe Simmons and W.P. Mills, and yielded them more than $13 million.

CHICKEN Chicken got its start when gold was discovered about 1896 by Bob Mathieson on Chicken Creek. Food was scarce in the region so prospectors ate the ptarmigan, a bird that resembles a chicken. When the post office was established in 1903, Ptarmigan was suggested, but nobody could spell it, so Chicken was selected. Mail was delivered to the post office by plane twice a week. When the Yukon gold rush began, Chicken thrived by serving the Fortymile Mining District. The Pedro dredge was moved from Fairbanks to Chicken Creek in 1959 where it retrieved more than 55,000 ounces of gold up until 1967. The town was so remote that it had no electricity or phones and residents depended on generators. Finally, the Taylor Highway was brought through in 1948, which connected the town with the Alcan Highway, creating modern transportation facilities.

Chicken has a year-round population between 17 and 20, except when the miners arrive in good weather and the population rises to about 100. The Chicken Creek Hotel that was built in 1906 has adapted over the years as the peoples' needs have changed, from a store, to a saloon, and then a school.

CHISTOCHINA Chistochina is derived from an Ahtna word meaning "marmot creek." It was established during the time when initial gold was found along the Chisna River in 1898 by miners named Meals and Hazelet. Originating as an Ahtna fish camp, it later became a stopping place for traders and trappers, then a placer mining camp. During 1924 hydraulic mining came into play when J.M. Elmer & Company brought in equipment to Slate Creek. Total placer production was nearly $3 million.

CIRCLE This village was founded along the Yukon River by L.N. McQuesten who established a trading post. Believing he was located close to the Arctic Circle, McQuesten named it Circle in 1887, which became a supply point where goods were shipped to the Yukon and other mining camps. In 1893 hot springs were discovered near Circle. During the winter when the rivers were frozen over and the miners couldn't pan for gold, they stayed there until the spring thaw. McQuesten told the newcomers how to build a good cabin to survive the winter: "Don't worry if it ain't watertight. Just get it built fast. There ain't much rain anyways and if a little summer air comes in, it don't do no harm. Come winter, the snow will pack the roof and if you feel a draft coming through the logs, splash a pail of water over the spot and the crack'll be froze airtight with a sheet of ice."[2]

In 1893 gold was found along Birch Creek, and miners from Circle had to pack in their supplies more than 80 miles away to the creek's mines. As a result, freighters were later hired to do the job. A year later, two prospectors discovered gold along Mastodon Creek. Circle was a bustling town in the 1890s, boasting dance halls, an opera house, hospital, church, and newspaper, followed by a post office in 1896. When gold was discovered in the Klondike, the town's population drastically decreased. After stabilizing from the effects of the big strike, Circle was left with a few stores and restaurants, church and school. It was the first community in the region to have a road built that reached the outside world, and today Circle is accessible by this road only during the summer months. In 1902 news of gold in the Fairbanks region took what miners were living at Circle. Total production was about $14 million in placer gold.

McQuesten devised an ingenious thermometer while living in Circle. He filled three bottles: one with kerosene, one with a liquid painkiller called Perry Davis, and the last with whiskey. If the kerosene froze, it was a warning not to venture too far from the cabin. If the painkiller froze, he knew to stay right near the fire, if not in it. The whiskey he probably drank.[3]

COLD FOOT Thousands of greenhorns in search of gold used Cold Foot as a supply center. Originally called Slate Creek, the name was changed to Cold Foot after the prospectors got cold feet and left in order to avoid the long, freezing winters. At the height of the gold rush Cold Foot had a wealth of gambling halls, roadhouses, saloons, plus ten prostitutes. When the post office was established, mail was delivered once a month by dogsled. The town decayed in 1912 after the gold rush era ended and the miners left. Construction of the Trans-Alaska pipeline with its accompanying highway brought Cold Foot back to life.

On old maps there was a place on the outskirts of town called "Hotfoot." According to court records, ladies of the midnight sun were consigned to Hotfoot when the more genteel ladies came to town. About 50 miles from Cold Foot, just before the Atigun Pass, stands a lone pine tree which is the farthest tree in Alaska. Next to it was a sign that warned, "Do not cut!"

For the hardy traveler, this area boasts nature at its best, great fishing and camping, and the opportunity to see the northern lights. Visitors can take a guided dogsled excursion. The Arctic Acres Inn welcomes visitors who can stay and enjoy the Fourth of July equinox at this "Northernmost Resort in the World." Contact: (907) 678-5201.

COUNCIL Located about 60 miles from Nome, Council started as a fish camp for the Fish River tribe. In 1897 Daniel B. Libby discovered gold in the vicinity, and within a year more than 50 buildings were housing the miners. The camp soon swarmed with more than 10,000 people, along with the requisite saloons with their connecting gambling houses and brothels. Serving as a supply center, Council began to lose its population when gold was discovered at Nome in 1900. A dredge was brought in to glean the rest of the ore as Robert Steiner wrote: "My father was engineer on the Blue Goose gold dredge operating near Council. However, by my time in 1917, the town had dwindled to a village of perhaps 150 people.... We had a narrow-gauge railroad at Council that ran from the scow unloading dock on the Niukluk River to the Wild Goose Mining Company camp, about 8 miles from town.... The Railroad was built to haul supplies to the mining site, but also took passengers."[4] The company ultimately recovered about $21 million in gold from the mines. The 1918 influenza epidemic followed by World War II caused the population to drop to only a few people. Council today has reverted once again to a fish camp.

CRIPPLE Cripple was established along the Innoko River about 1912 and served as a supply and trading point. It was named for the famous mining town of Cripple Creek, Colorado, and residents hoped it would be as prosperous. Gold was discovered between 1906–1908 along the river. Cripple became one of the busiest camps in the region even though it was remote, serving thousands of miners who traveled between the Innoko gold fields and Nome. The placer diggings produced nearly $10.5 million.

DOUGLAS This camp was located on Douglas Island, named for Dr. John Douglas, bishop of Salisbury, who edited explorer Captain James Cook's

Ruins of Dyea (photograph by Jet Lowe, Library of Congress, Prints and Photograph Division, Historic American Buildings Survey, AK-16-4).

journal. The site was originally named Edwardsville for an early miner. About 1880 gold was found by Pierre Erussard soon after the discoveries at Juneau. Two other mines, the Mexican and the Ready Bullion, were also located there. The following year John Treadwell purchased Erussard's claim. After the Alaska Mill and Mining Company purchased stock in the mine, a 5-stamp mill was erected. The Treadwell became one of the largest underground gold mines in the world.

John Muir wrote during his travels in 1890, "On Douglas Island there is a large mill of 240 stamps, all run by one small water-wheel, which, however, is acted on by water at enormous pressure. The forests around the mill are being rapidly nibbled away. Wind is here said to be very violent at times, blowing away people and houses and sweeping scud far up the mountainside.... We arrived at Douglas Island at five in the afternoon and went sight-seeing through the mill. Six hundred tons of low-grade quartz are crushed per day."[5]

Inadequate construction caused the mine to close in 1917. Its tunnels had pillars of unmined ore left as supports that, when removed to get more gold, were not replaced with wood supports. As a result the mine began to cave in, and high tides finally caused the whole thing to collapse. Then two fires leveled the town. By 1905 the mines had yielded more than $26 million in gold.

DYEA Dyea developed rapidly as a supply point with the onset of the Klondike gold rush and was on its way to rivaling Skagway by 1898. This is where many a prospector headed for the Klondike since Dyea was a port of entry to the Dyea Trail through the Chilcoot Pass. Construction of the White Pass and Yukon Railroad caused the town to decline. In 1907 Emile Klatt burned the hotel to make room for his cabbage patch. The Klatt family later took over the camp and developed a large farm.

EAGLE Located on the left bank of the Yukon, Eagle was the supply center for a large placer mining area. It was established about 1874 and first called Belle Isle when Moses Mercier opened the camp. It was later renamed for the abundant eagles in the region, with a post office in 1898. Eagle was important economically, as it was the entry point for miners headed to Fortymile River and its gold fields. The placers yielded about $800,000.

FAIRBANKS The city is called the "Golden Heart" of Alaska and was once a mining camp, founded in 1901 after E.T. Barnett built a trading post. It was named for U.S. vice president Charles W. Fairbanks, who served under Teddy Roosevelt. Felix Pedro discovered gold placers on Pedro Creek, in addition to locating the Cleary and Goldstream Mines. The first issue of *The Fairbanks Miner* reported that "Felix Pedro discovered the Fairbanks placer mines in July, 1902.... Pedro was known to many to be a careful and competent miner ... when his prospects on pedro and cleary grew to be a certainty he was nervously afraid these camp-followers would descend upon him and stake the creeks before he could get his friends located.... Accompanied by Ed. Quinn and Smallwood, Pedro set out for the creeks secretly and by night and taking with them plenty of supplies. They sank a hold to bed rock, and there lay the glittering gold."[6]

The news spread, but when the miners arrived they discovered most of the gold was buried 80–100 feet deep in the gravel, requiring machinery to get the ore. The rush at Fortymile and Circle also kept many of the men away. Eventually the ore was mined profitably and dredges were put into operation. Production continued until World War II. Total gold recovery of the Fairbanks district was more than $150 million.

Fairbanks had its red light district, as many of the camps did, plus an abundance of drinking palaces. One notable character was Oscar Nordale who was drinking at one of the houses. Whiskey and water was the usual order of the day. Whiskey would be poured in the glasses and the boys would dip water from a barrel at any increment they desired. Oscar had so much to drink one evening he couldn't make the outhouse to relieve himself. The madam noticed the men had quit dipping into the barrel and were drinking their whiskey straight, so she asked them, "What's the matter with you. Trying to get so drunk you can't do right by my gals?" "It ain't exactly that. It's Oscar," said a miner. "He had to go real bad, and the water bucket seemed just the likely place to him, and damn if we want to drink out of Oscar's urinal."[7]

Fairbanks celebrates Golden Days "Garters and Gold" during summer. Information: Fairbanks Chamber of Commerce, (907) 452-1105.

FLAT Bearing a descriptive name from nearby Flat Creek, this camp was established about 1910 along the Iditarod River. On Christmas Day, 1908, John Beaton and a prospector friend found gold that became the widest pay streak found in the state. Peter Miscovich and Lars Ostnes founded the camp and its post office opened in 1912. Flat was connected to Iditarod by a narrow-gauge railroad. Before that time, wagons were pulled along wooden rails by mules, which was very difficult because the land was sodden tundra. Two years later there were more than 6,000 miners living at Flat. Guggenheim interests brought in a dredge to work the creek, and within the first three months more than $400,000 in gold had been recovered. In all, nearly $30 million was recorded.

FORTYMILE It was from this point that word spread about the great Klondike strike, and Fortymile became the first significant strike during 1886, making it one of the oldest placers in the Yukon. Gold was initially found along the river by Arthur Harper in 1874, but no development occurred until 1886 when Howard Franklin and a group of men relocated the claim. Since the topography was mainly permafrost, fires had to be built to thaw out the frozen ground. The camp was well isolated, and for the most part the miners lived on fish until boats began to ply the Fortymile River. One of the prospectors left a recipe for hootch to keep the men's insides toasty: "Take an unlimited quantity of molasses sugar; add a small percentage of dried fruit or berries; ferment with sourdough; flavor to taste with anything handy, the higher flavored the better, old boots, discarded and unwashed foot rags and other delicacies of a similar nature; after fermentation, place in a rough still, for preference an empty coal oil tin and serve hot according to taste."[8]

Large-scale mining, dredges and hydraulics began in 1890, and a ditch was built to supply water for the operations. Summarizing the Fortymile: "Though no one struck it rich in the Fortymile, the region had its continuing appeal. Long after the boom at Circle and the stampede to the Klondike, the Fortymile region continued to draw prospectors and miners content with moderate return for hard work."[9] Total recorded production for the region was a little over $8 million.

GIRDWOOD Girdwood was founded as Glacier City along Turnagain Arm about the turn of the century after gold was discovered on Crow Creek, and served as a supply camp and trail stop. The town was renamed for merchant James Gird-

Girdwood Trading Company (courtesy USDA–Forest Service).

wood who discovered gold in 1896. In addition to living its early life as a mining camp, Girdwood was further developed with construction of a railroad. During World II all the mines were closed, and Girdwood almost became a ghost town until 1949, when the Seward Highway that would connect Seward to Anchorage was being built. The land dropped between 8 and 10 feet during the 1964 earthquake, and the town was moved to its present location more than two miles away.

HOPE Hope was founded about 1896 along the shore of Turnagain Arm. Its name was descriptive of the miners' hopes of great wealth. The first lode deposits were discovered in 1896, but although they were very rich the total tonnage was quite small. Placer mining was conducted along Resurrection and Bear creeks during this period. A small rush occurred after Alexander King was grubstaked by the Kenai Trading Post and returned with four bags of gold. Ore was also found along Sixmile Creek where a party of prospectors recovered about $40,000. Their discovery led to a stampede near Hope with more than 5,000 miners. Total production was about $2.5 million in placer and lode gold.

IDITAROD Iditarod is a well-known name associated with the famous dog-sled race, which commemorates the historic serum run back in 1925 when a diphtheria epidemic threatened the town of Nome. The village bears a Shageluk name meaning "clear water," or "distant place." Gold was discovered near the settlement in 1908 along Otter Creek, a tributary of the Iditarod River. By 1910 the placers had yielded about $500,000. Along Flat and Willow creeks gold in the amount of more than $88 million was recovered using dredges and hydraulics.

Someone wrote years later, "The number of old timers in Iditarod is rather large. Some of the newcomers are Chechakos, or tenderfeet, and are few. This is essentially a community of sourdoughs, pioneers who have blazed many a hard trail and whose residence dates back to 1898 or before."[10] Donald McDonald, a resident at Iditarod, went on to become Alaska's foremost road builder.

JUNEAU The capital of Alaska began its life as a mining town. News of gold strikes in the region during 1880 prompted Joe Juneau and Dick Harris to begin exploration. They discovered rich deposits at Gold Creek and before long the camp sprang up with more than 100 men. Its post office opened in 1882. The Alaska Juneau Mine was located and went on to become the largest lode gold mine in the state, operating until about 1944 when costs became prohibitive. Juneau sold his mining interests, made his fortune, spent it, and died in

Coast Range Mining Company Mill near Hope (courtesy USDA–Forest Service).

1899 at Dawson. The Alaska Juneau Mine yielded the bulk of gold produced in the Southeastern Alaska region. Placer diggings were conducted along Porcupine Creek. About $100,000 in gold per year was produced from 1900 to 1906. Total production was about $142 million.

Juneau holds its Gold Rush Days the last weekend of June. Information: Juneau Chamber of Commerce, (907) 463-3488.

KETCHIKAN Sandwiched between the mountains and sea with its unusual geographical character, Ketchikan began as a Tlingit fishing camp. It came to prominence about 1899 when gold was discovered, and developed into a mining and outfitting town. Its name is a Tlingit expression for "thundering wings of an eagle," or "spread wings of a prostrate eagle," descriptive of the rocks that divide the waters of Ketchikan Creek. Discoveries of gold made at Ketchikan and news of the Klondike encouraged many people to prospect in the area. By 1900 there was feverish activity in the district. Production amounted to over $1 million.

KIANA Kiana, once a central village of the Kowagmiut Inupiat Eskimos, was located about 55 miles east of Selewik. It became a supply depot and transfer point in 1909 for placer mining camps along the Squirrel River, and acquired a post office in 1915. Kiana is an Eskimo word for "a place where three rivers meet." When the gold rush began in the Kobuk valley it also brought traders, most of whom came for fur. The natives traded their pelts for metal knives, axes, and guns. The climate was severe, making transportation difficult, so development of major mining was hindered until advanced mining techniques were established.

KNIK Knik is situated about 17 miles northeast of Anchorage on Cook Inlet, and bears a Tanaina name meaning "fire." Furs and mining operations caused Knik to be founded when Russians began trade. After American occupation, the Alaska Commercial Company bought out some of the Russian posts. During the Klondike gold rush days, Knik was a distribution center and disembarkation point for the miners on their way to the gold fields in the Talkeetna Mountains. After the Iditarod Trail was built, the camp became a freight hauling station to the goldfields of Nome, and carried ore to

Present-day Ketchikan (courtesy Arthur Alspector).

the boat landings for shipment. The town declined in the middle 1900s when the gold gave out and Anchorage came into prominence.

Knik is called the Mushing Center of the World, since many dog mushers live there. The Sled Dog Mushers' Hall of Fame is also located there. It was created in 1967 to perpetuate the history of the Alaska sled dogs and dog mushing. Information: Knik Museum: (907) 376-2005.

KOYUKUK This village is situated near the junction of the Koyukuk and Yukon rivers, about 300 miles west of Fairbanks. It bears an Athabascan name for "village under a clay." Koyukuk became an outfitting and supply point for miners when placer gold was found between 1885 and 1890. A trading post was established to serve the growing number of mines and increasing river traffic. By 1898 nearly $4,000 in gold had been recovered. Eventually, prospectors left the area until galena was found and mining operations picked up again. The region yielded about $9 million in gold.

LIVENGOOD Ore had been found as early as 1892 but no real interest occurred until 1914. This camp was not named for its good living, but for prospector Jay Livengood. In a placer mining district along Livengood Creek, a tributary of the Tolovana River, Livengood and N.R. Hudson found gold. Hundreds of miners flocked to the region to seek their fortune. The new arrivals didn't find anything and accused Livengood of starting a stampede for his own personal gain. Drift mining was later attempted, but that failed because of a water shortage. The placers in the region ultimately produced about $7.5 million in gold.

MARSHALL Also known as Fortuna Ledge, this site was the farthest point south on a channel of the Yukon River. It was established about 1913 and named for U.S. vice president Thomas R. Marshall, with a post office in 1915. E.L. Mack and Joe Mills found gold along Wilson Creek where Marshall served as a landing point for boats transporting merchandise and mining equipment. It was also a terminus and supply center for prospectors embarking on trips up the Yukon River. After the first few years of near-bonanza placer production, activity slackened, then declined after World War II. Total recorded gold recoveries were a little over $2 million.

NABESNA Gold was discovered in the area during the 1890s, but no development occurred until 1929 after C.F. Withan with the Nabesna Mining Company located the Nabesna Mine, which was the only producing property in the district. Tradition says the gold was found after a bear exposed a moss-covered outcrop of the principal vein while trying to dig out a gopher. Within ten years the vein had pinched out. A total of about $1.3 million in gold was recovered.

Main street of Nome, 1906 (courtesy State of Alaska, Department of Commerce, Community and Economic Development Division of Community Advocacy, Community Photo Library).

NOME Located on the Seward Peninsula, Nome was first named Anvil City, taken from the stream that had a rock resembling a blacksmith's anvil. Gold was discovered in 1855 by telegraph engineer Baron Otto von Bendeleben, but nothing came of it until 1897. With the boom at the Klondike and Yukon, miners heading there stopped at Nome and worked the gravels of the nearby streams in addition to the sands of the beaches, which produced almost $1 million in gold. News of these discoveries led to the great rush in 1899. It was this year the name was changed, but a British naval cartographer misinterpreted a chart notation of "? Name." Water was needed for the placer diggings so Charles D. Lane, who owned the Wild Goose Mining Company, brought in water via 40 miles of ditches from the Nome River. From the first strike on Anvil Creek, mining efforts in the Nome area yielded over $136 million.

Nome was a wild and lawless town. With more than 20,000 people and little law enforcement, it was open season. One of the most famous characters was Soapy Smith, who had been kicked out of Creede and Leadville, Colorado, and later headed to Skagway where he was killed. At one time someone stole Nome's entire butter supply, which was later resold back to the stores at inflated prices. One absentee home owner returned to find thieves literally moving his building away. Because the town could not support the massive number of people, claim jumping became the order of the day.

The women who arrived at Nome in 1900 were appalled at the conditions: "We reached Nome, that human maelstrom, at night.... We proceeded through the main street, and if ever pandemonium raged, it raged here. The streets fairly swarmed with a heterogenous mass of people. Drunken gamblers groveled in the dust; women, shameless, scarlet women, clad in garments of velvet, silks, laces, of exceedingly grotesque character but universally decolette, reveled as recklessly as any of their tipsy companions.... Dust settled around about us like a heavy fog. We waded through rivers of it before we reached our hotel. There were thirty thousand inhabitants in Nome at that time, of nondescript character. Cultured, intelligent men hobnobbed with the uncultured and ignorant. The one touch of nature that made them all akin was the greed for gold."[11]

Numerous events are held at Nome, notably the Midnight Sun Festival, on the weekend closest to the summer solstice (June 21), and Poor Man's Paradise in July, Nome's Centennial event. Information: Nome Convention and Visitors Bureau at (907) 443-624.

RAMPART Gold was discovered along the Hess River and Minook Creek in 1882, but it was ten years before any serious development occurred. Rampart was founded about 1896 when gold was located by Minook Ivanov. The site was named for a rampart-like canyon near the Yukon River. Ivanov recovered about $3,000 in gold, which brought in more than 1,000 miners. The town also prospered during the winters because steamers were unable to navigate the river to Dawson and had to lay over at Rampart. The village was governed by a U.S. commissioner who spent more time searching for gold than tending to business. Reaching a peak before 1910, the region had yielded more than $1.7 million in placer gold.

As of 1972, the only person living at Rampart was a Mr. Weisner, who was a combination postmaster, miner, lawyer, doctor, cannery operator, bookkeeper, and store owner. He said, "The Government has been talking about putting a dam here ... and I sure as hell wish they would. They would have to buy me out before flooding my property, and then I could get the hell out of here."[12]

RUBY First discoveries of gold were along Ruby Creek in 1907, but they were quickly exhausted. After more gold was found at Long Creek during 1911, the town was established along the Yukon River and may have been named for ruby-colored stones along the stream. Its post office opened in 1912. More than 1,000 miners came to Ruby. A local newspaper wrote that this region would prove to be the biggest in terms of production in the state, but within a few years the ore was gone, although total production was estimated at nearly $8 million in gold. After the gold rush, Ruby's population rapidly declined. Most of the buildings were torn down and the wood was used by natives at nearby villages.

SHEEP CAMP Receiving its name for the mountain goats in the region, this site was a tent town with a restaurant and hotel, and served mainly as a forwarding depot for the miners on their way over the Chilkoot Pass. It was also a resting place before trekking the last, and worst, four miles up the pass. About 100 people from Sheep Camp were killed in an avalanche while climbing up the pass during 1898. When the gold ran out the place was deserted, and today it has one cabin used by hikers.

SHUNGNAK Located about 300 miles west of Fairbanks, Shungnak was founded in 1899 as a supply station for mining activities in the Cosmos Hills. Because the Kobuk River kept eroding the village with annual flooding, it was moved ten miles downstream. The name is derived from Eskimo, meaning "jade." By 1910 most of the gold had been taken out. The region was not that rich in ore, since only $200,000 was recovered between 1898 and 1930. A few years later other prospectors tried their luck, but the placers yielded only about 200 ounces of gold. In addition to the ore, jade was mined and cut along the Shungnak River.

SKAGWAY Skagway was founded in 1887 by settler William Moore, who predicted a gold rush years before its onset. He claimed 160 acres and built a small log cabin. In order to keep out the winds, Moore and his wife papered the walls with newspapers. When the gold rush did arrive, his land would become Skagway after Moore lost his homestead when his land claim was denied. Gold stampeders originally named the place Skaguay, but the post office changed the spelling to Skagway. A Tlingit name, it means "end of salt water." After gold was discovered in 1896 by a Tlingit named Skookum Jim, the rush was on and more then 20,000 people had arrived at Skagway by the late 1890s. Because of rough traveling conditions, many miners never made it to the gold fields, 500 miles further inland. The town served as a base of operations for thousands during the Klondike gold rush of 1897-98. The *Skaguay News* wrote in 1897 that "Skaguay may be properly termed a child of necessity. When the first startling reports of the unprecedented gold strike in the Klondike region reached civilization in July last there began the greatest rush to the Yukon country that the world has ever witnessed."[13]

Skagway was known as a rip-snortin,' shootem'-up town with more than 70 saloons controlled by gangsters and prone to regular shootings. It was also loaded with con men. One of the most famous was Soapy Smith, known for his many devious ways of relieving prospectors of their wealth. He had earlier been kicked out of Leadville and Creede, Colorado, for his soap scam, then headed to Nome. A reporter wrote that "Soapy Smith, a noted sport, has been working a very smooth game at Skaguay, whereby a number of men were robbed of what money they had about them."[14] He did attempt some semblance of respectability, though. If his henchmen robbed a preacher, Soapy would donate money so the preacher could build a church. Or if one of his victims was killed, he would provide for the man's widow and children. The town finally got fed up with Soapy and the reputation Skagway

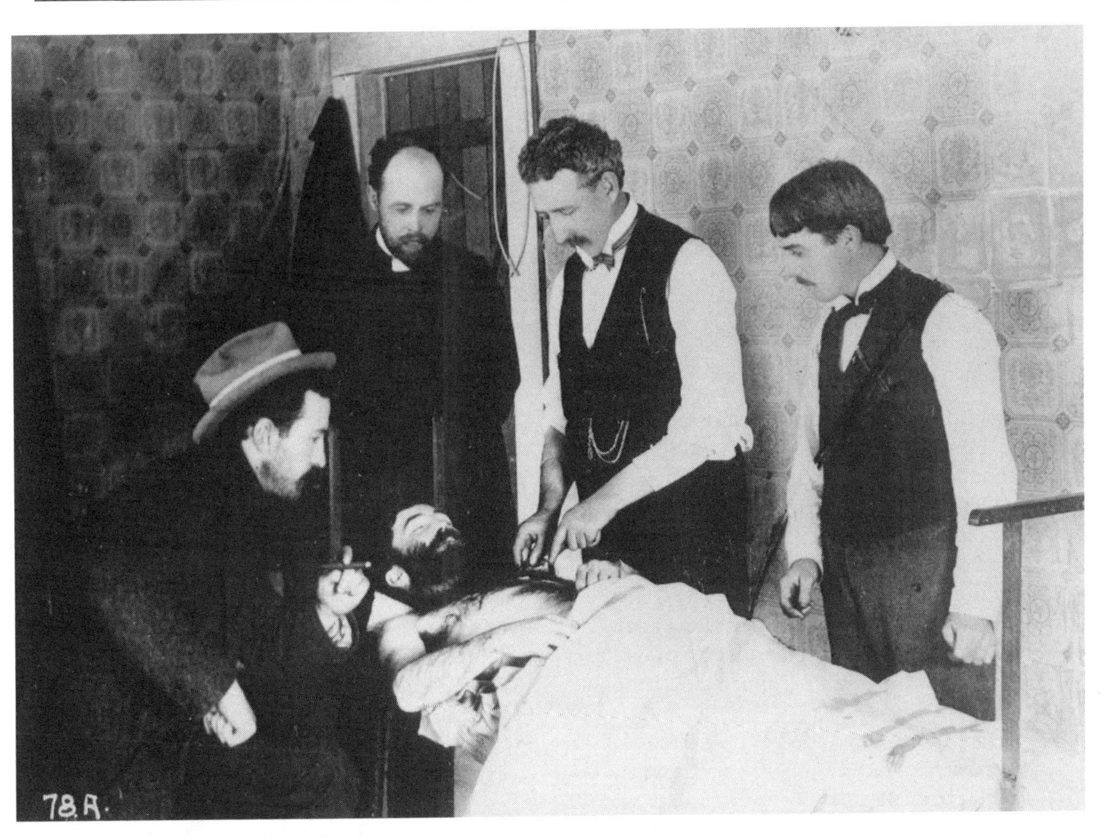

Death of Jefferson Randolph Smith, better known as Soapy Smith, Skagway, 1898 (University of Washington Libraries, Special Collections, Neg. Hegg17A).

was acquiring, so a vigilante committee was formed. Soapy met his end in a shoot-out at the hands of Frank Reid, the town surveyor and one of the vigilantes.

The area would have become a ghost town when the gold rush ended in 1900 had it not been for construction of the White Pass and Yukon Railroad, which served Whitehorse and the Anvil Mine with supplies and transportation. During World War II Army troops were stationed there to supply equipment for construction projects, in addition to materials for building the Alcan Highway.

Harriet Smith Pullen, better known as "Ma Pullen," was the best loved women in town. After her husband deserted her, she and her four children came to Skagway where she baked pies in her spare time while working as a cook. The pies became so popular with the miners that she acquired enough money to purchase horses so she could run a pack train to the pass where supplies were transferred and hauled to the mining camps. She later built a hotel called the Pullen House.[15]

William Moore's cabin is still intact, and in 2005 the National Park Service, in conjunction with the Klondike Gold Rush National Historical Park, decided to try to preserve the newspapers that lined its walls. They discovered that the papers were put up in a somewhat thematic nature, all dating from the 1880s and 1890s. One of the major stories in 1895 from the *San Francisco Examiner* was headlined "Where the Real Boundary Lies," in reference to England's contention over the Alaskan border. The papers are being retained by the Park Service until more funding is available for further preservation.[16]

Since the decline of gold and completion of the Alcan Highway, Skagway is supported mainly by tourism. The Historic District contains buildings and exhibits that depict the Gold Rush Days, a grand reminder of days past. Stalwart trekkers can take the Chilkoot Trail hike, which takes 3–5 days, giving host to historic ruins and artifacts along the trail. For the less hardy souls, visit the Klondike Gold Rush National Historic Park, a memorial for the many miners who lost their lives over the Chilkoot Pass during the 1898 gold rush. The gold seekers used two trails that went to the headwaters

Newspaper-lined wall, William Moore's Cabin, Skagway (Library of Congress, Prints and Photograph Division, Historic American Buildings Survey, AK-15-32).

of the Yukon: the Chilkoot Trail, 33 miles long originating from the little town of Dyea, and the 40 mile White Pass Trail out of Skagway. Many of the prospectors had little knowledge of how to survive in this region. Most of the equipment they were hauling was abandoned because they couldn't carry it over the passes. In 1898 an avalanche buried more than 100 men and women who had left Sheep Camp and were on their way up the summit of Chilkoot Pass. The men who were still at camp went up to rescue the prospectors, but many of the bodies were never found. Information for the following may be obtained from the Skagway Chamber of Commerce: (907) 983-1898.

SOLOMON Settled by Eskimos in the early 1800s, this camp was once known as Erok and was initially located between the Solomon and Bonanza rivers on a sand spit. During the 1880s an Eskimo named Tom Guarick found gold along Ophir Creek, which brought in thousands of miners. During 1898 Solomon switched from a native settlement to a mining camp with the discovery of more gold along the Casadepaga River, a tributary of the Solomon River. It was also a terminus of a narrow-gauge railroad that was eventually washed out by numerous storms. In 1899 placer gold was found along the Solomon River, but was quickly depleted. About a year later more extensive placers were discovered along Daniel Creek, and were worked by dredges and hydraulic methods. To bring water to the mines, numerous ditches were built to accommodate operations. In 1913 a horrendous storm wiped out the camp, so it was relocated. During the 1930s a dredge was brought in by Richard and Charlie Lee, who recovered about $400,000 in gold during its operation. A total of $5 million came from the region.

SULLIVAN Located at the confluence of Alder Creek and Gold Run, this small mining camp was settled with the rush of miners in 1899 and acquired a post office in 1902. So little gold was found that the prospectors went elsewhere.

TALKEETNA Before development as headquarters for the Alaska Engineering Commission during construction of the Alaska Railroad, Talkeetna was a supply station during the 1890s as gold miners headed for the Susitna River to realize their dreams. The site is located at the confluence of the Susitna, Talkeetna and Chultina rivers, and was once a Tanaina Indian village. It is a Den'ina word for "river of plenty." Talkeetna later became a riverboat steamer station, but continued to supply the miners until about 1940. Today it is a non-native community and serves as a year-round headquarters for hunters and fishermen. In 1993 Talkeetna was placed on the National Register of Historical Places.

The town celebrates its Annual Moose Dropping Festival traditionally held the second weekend in July. It also has a walking tour of the miners' cabins, the first general store, and other outbuildings. Contact the Visitor Center for dates: 1-800-660-2688 or (907) 773-2688.

TANANA Tanana is an Athabascan village that started out as a trading post established by the Alaska Commercial Company. Located at the junction of the Tanana and Yukon rivers, it was originally called Nuchalawoya, an Athabascan word that means "the place where the two rivers meet." The name was changed to Tanana, interpreted as "river trail." The village served as a trade center during the gold rush, and was well populated during the early part of the Fairbanks rush. Most of the deposits were shallow and by 1906 the prospectors had left for better opportunities. Total gold production was nearly $1 million.

TENDERFOOT CREEK The Tenderfoot and Richardson area was founded about 1905 after discovery of gold and silver. Most of the supplies were furnished by a steamer out of Fairbanks. Placer miners had a ditch built in order to provide adequate water. They worked their diggings for about four years, and recovered between $300,000–$400,000 annually. Total gold and silver mined in the district was about $17 million.

TYONEK Forty-three miles southwest of Anchorage lies the Tanaina village of Tyonek, a Tanaina word for "little chief." The Russian-American Company had its trading post there until the late 1800s, when the Alaska Commercial Company took it over. Tyonek became a supply center and debarkation point when gold was discovered at Resurrection Creek during the 1880s. Orville G. Nerning, who was employed by the Klondike & Boston Gold Mining & Manufacturing Company, took control over most of the placer diggings in 1898. The region was very productive, yielding more than $10 million in placer and lode gold. Latecomers to the gold fields packed in expensive equipment and supplies at Tyonek to be transferred to their claims, but the ventures were useless and most of the machinery was abandoned. The region was flooded about 1930, which precipitated the village's move to the top of a bluff a few years later.

UNGA Located on Unga Island, which is part of the Shumagin Islands of the Aleutian Chain, the village ceased to exist in 1925 after the school closed. Unga is Aleut for "south." During the 1830s the village was a Russian sea otter station called Delarov until 1836 when it was renamed Ougnagok. When the post office was established, the name was changed to Unga. Prospectors arrived in 1891 after a vein of gold-bearing quartz was discovered at the Apollo Mine which operated until 1912. Between $2 and $3 million in gold was recovered from the Apollo by the turn of the century.

VALDEZ Named for the minister of Spain, Antonio Valdes y Basan, the town was founded about 1898 when news of the Klondike rush got out. It became a center of commerce and debarkation point for the miners headed to the gold fields. When they arrived at the Klondike, the men found that most of the claims had been taken, so they returned to the Valdez area, where they discovered placer gold at Prince William Sound. One of the prospectors found a gold ledge that yielded him about $40,000. By the 1930s the vein had pinched out, but not before yielding about $2.8 million in gold.

WASILLA Wasilla was established about 50 miles north of Anchorage in 1917 during construction of the Alaska Railroad, and became a supply center for prospectors searching for gold during the 1940s. Wasilla is an Athabascan word meaning "breath of air." During the early 1900s Robert Hatcher discovered gold, and the Independence Mine was located in the Talkeetna Mountains near town. Total gold production was mainly from lode mines, estimated at about $13 million. Wasilla experienced tremendous growth during the 1970s when the Trans-Alaska pipeline was being constructed.

WRANGELL Wrangell was named for Vice Admiral Baron Ferdinand von Wrangell of the Imperial Russian Navy and manager of the Russian-American Company. Before the mining era, Russians built a stockade there about 1811 and conducted fur trading with the Tlingits. During the 1860s and 1870s the town grew as an outfitter for gold prospectors, filled with gambling dens and dance halls, and flourished with the thousands of miners traveling up the Stikine River to British Columbia in 1874, then to the Klondike in 1897.

In 1890 John Muir wrote of his travels to Wrangell:

> The Wrangell village was a round place. No mining hamlet in the placer gulches of California, nor any backwoods village I ever saw, approached it in picturesque, devil-may-care abandon. It was a lawless draggle of wooden huts and houses, built in crooked lines, wrangling around the boggy shore of the island for a mile or so in the general form of the letter S, without the slightest subordination to the points of the compass or to the building laws of any kind.... Most of the permanent residents of Wrangell were engaged in trade. Some little trade was carried on in fish and furs, but most of the quickening business of the place was derived from the Cassiar gold-mines, some two hundred and fifty or three hundred miles inland, by way of the Stickeen River and Dease Lake.... The fort was a quadrangular stockade.... It was built by the Government shortly after the purchase of Alaska, and was abandoned in 1872, reoccupied by the military in 1875, and finally abandoned and sold to private parties in 1877.[17]

ARIZONA

ALAMO CROSSING (Mohave) Tom Rodgers established this small camp in the late 1800s, and its post office opened in 1899. The place was very small, containing only a post office, store, and a small stamp mill that was built to serve the nearby mines. But the ore was quickly exhausted and Alamo Crossing declined. It saw a small resurgence during the 1950s when manganese was discovered. The site is now under a lake in Alamo Lake State Park.

ALEXANDRA (Yuvapai) Edmund G. Peck and his partners discovered ore while prospecting in the Bradshaw Mountains during 1875. Peck espied an interesting rock which turned out to be silver, and the camp was formed shortly thereafter about 20 miles from the city of Prescott. It was named for the first woman who moved there, Mrs. T.M. Alexander. The only major mine was the Peck, which produced almost $1.5 million. Alexandra was located in such a remote area that supplies were at a premium, and the ore had to be hauled out by pack trains over mountains and rocky terrain more than 30 miles away to the mills, an expensive venture. As a result, a 10-stamp mill was built in 1877 to accommodate the Peck. Placers were also discovered, which yielded about $300,000.

ALTO (Santa Cruz) Alto is Spanish for "high," an appropriate name for gold seekers, who had high aspirations. Beginning in 1687 the region was mined by Mexicans, until about 1857, when the Apache Indians attacked and drove them away. A prospector named Mark Lully discovered the old mines and reopened one in 1875, calling it the Goldtree, honoring a Joseph Goldtree. But Lully had no success, the mine closed, and Alto turned to dust.

AMERICAN FLAG MINE (Mohave) Located in the Hualapai Mountains, this silver mine was discovered in 1874 by a prospector named Shoulters. A mill at Mineral Park processed the ore valued between $300 and $1,000 of silver per ton. A post office named Old Glory was established in 1896 for the mine, with John H. Alexander as postmaster. Shoulters later sold his claim to W.M. Richards, John Covin and other partners. By 1884 the population had dropped drastically, and little was heard of the site after that.

ARIVACA (Pima) This region was occupied by the Pima and Papago Indians until the Piman Revolt in 1751. Arivaca is a Pima word for "small springs." Mining was conducted by Spaniards between 1751 and 1767 until attacks by Apache Indians caused them to leave. The land was part of a grant purchased by Agustin Ortiz in 1812, and Charles Poston bought the land in 1846. When silver was discovered in the Cerro Colorado Mountains near Arivaca, Poston and Herman Ehrenberg built a reduction plant for the mines. The Heintzelman Mine was discovered in the 1850s, and went on to become one of the most important properties in the mountains. Another claim that yielded well was the Cerro Colorado Mine, which recovered nearly $2 million. During the 1860s Apaches again waged war against the white miners, and for a period of time the place was abandoned. During the same time period, J. Ross Browne was passing through and wrote: "At the time of our visit it was silent and desolate — a picture of utter abandonment."[18] Arivaca is home to the oldest intact adobe schoolhouse in the state.

BAGDAD (Yavapai) Gold was discovered in the

Old dwelling at Arivaca (photograph by F.D. Nichols, Library of Congress, Prints and Photograph Division, Historic American Buildings Survey, 10-ARIV).

vicinity about 1880, followed by the establishment of Bagdad, an Iranian word for "God's gift." It received its name because settler William Lawler had read the book *Arabian Nights*. The post office opened in 1910 with Henry A. Geisendorfer as postmaster. The Hillside Mine was the biggest producer, with some smaller properties also yielding good results. Total output of the area was a little over $1 million in gold, mainly from the Hillside.

BIG BUG (Yavapai) Taking its name from the creek, Big Bug was certainly appropriate, since miners found large beetles about the size of walnuts in the stream. The first settler at Big Bug, in 1862, was prospector Theodore Boggs. Discovery of gold in the nearby Bradshaw Mountains by William D. Bradshaw brought in a surge of prospectors, and the Bradshaw Mining District was formed in 1863. Big Bug's post office opened in 1879. Gold placers were highly productive during the 1880s and from 1933 through 1942, after which time they declined in importance. Total gold production from 1867 until mining ended amounted to about $12 million.

BUENO (Yavapai) About 1863 Bob Grooms discovered a ledge rich in ore along Turkey Creek, which he called the Bully Bueno. Apache Indian attacks hampered development, but the prospectors were determined and formed the camp, which bears a Spanish name meaning "good." The site was very active with 250 people and two stamp mills. According to Lindgren in 1926, "Placers have been worked at several places.... A few years ago a Portugese [sic] is said to have taken out $20,000 near the old stone cabin, one mile below Howard. There are also small placer deposits near Turkey Creek station, and every year more or less dry washing is done by Mexicans in this locality."[19] Placer production came to a little over $25,000.

BUMBLEBEE (Yavapai) Originally a stage stop on the Phoenix to Preston line, Bumblebee was settled about 1879 and supposedly named for a bumblebee's nest found by prospectors in 1863. Another story says the name was selected because the Indians were thick as bumblebees. William D. Powell became postmaster when the office opened in 1879. First locations of gold were made as early as 1873, but no records were documented until 1904. Although the gold was rich, the placers were superficial, but they did produce a little under $1 million.

Bumblebee later became a trading center for ranchers.

BURCH (Gila) Burch was a small trading post along the banks of Pinal Creek. Named for Mr. Kenyon Burch, it served the small silver and gold mines that were found in the Sleeping Beauty Mountains about 1876. The claims quickly played out and Burch was deserted.

CASTLE DOME (Yuma) This site was established at the foot of Mt. Ophir. In 1863 Jacob Snively and his partner, a Mr. Conner, discovered old mines nearby. Snively started out as the district recorder, his office an old box situated in an arroyo out in the open with his papers stacked up on the ground, and a sign designated that his little hole in the wall was the recorder's office. The camp took its name from a nearby peak that resembled the dome of the nation's capitol, and a post office was established in 1875.

Miners who came to the prospects were disappointed when the ore proved to be more lead than silver and they left. One of the mines had silver ore that assayed at a little more than $100 a ton, which brought in San Francisco investors, but it was too expensive to conduct large-scale mining and transportation was undependable. Activities were sporadic until gold was found in 1884, although there was not much interest until about 1912. Accurate records of the main gold placers in the mountains are not known, but estimates were between $75,000 and $100,000.

CERBAT (Mojave) This mining camp was founded about 1870 with a post office in 1872. It was located along a branch of the Atchison, Topeka & Santa Fe Railroad. Cerbat was the county seat until it was later moved to Mineral Park. Ore rich in silver and gold was found in the nearby Cerbat Mountains. Because there was no way to process the ore there, a 5-stamp mill was built at Mineral Park in 1876. One of the silver mines was yielding ore worth about $140 to $500 a ton. Gold came into prominence when the price of silver dropped. After 1904 the mines of the district produced about $2.5 million.

Castle Dome remains (courtesy Douglas Yetter).

CHARLESTON (Cochise) Charleston was established in 1879 along the San Pedro River as a mill site for Tombstone's mines, which did not have enough water for its stamp mills. As a result, the Tombstone Mining and Milling Company was formed. Frederick Brunckow, originally from Germany, was a scholar and scientist who developed the Brunckow Mine a few miles from Charleston. Unfortunately the mine did not yield any appreciable ore, then Brunckow was killed by Mexican miners. During the late 1880s the mines at Tombstone began flooding, leading to the demise of Charleston. Its death knell came on May 3, 1887, when an earthquake hit, reducing the old adobe buildings to rubble.

A bunch of ruffians called the Galeyville gang arrived at Charleston one day. On a Sunday, they "attended" church, donated quite a bit of money when the plate was passed, but got into a snit when the preacher cut his sermon short. One of the gang members told the preacher, "We've paid our dough and we want a full length spiel and throw in a prayer too." A resident tried to calm the men down while the preacher continued his ministry. After hearing of the confrontation, the sheriff fined the resident for his actions. The resident said he was only trying to keep the peace. "Then I fine you for being in bad company on Sunday," was the sheriff's reply.[20]

CHEMEHUEVI VALLEY (Mohave) Placers were discovered in the foothills of the Chemehuevi Mountains, which bear a Yuma name meaning "Paiute Indian." Located about 18 miles from Topock, the valley was occupied by more than 30 men working the diggings during 1932. The largest production near the mountains was at the Spanish Diggings, which came to a total of about $7,000, most coming from the Chief claim.

CHLORIDE (Mohave) Located in the Cerbat Range, the mining camp of Chloride was established about 1863. Silver mining had been conducted during the 1840s, but the Hualapai Indians killed four of the prospectors, keeping many of the rest away. After a treaty was signed with the Indians in 1870, the prospectors returned. Chloride received its name from the exposed ores that carried heavy silver chloride. Its post office opened in 1873 with Robert H. Choate as postmaster. The 1870s were the most productive years, after which the price of silver began to decline. The value of the metals from the district came to about $5 million.

CLEATOR (Yavapai) This site was first called Turkey Creek, for the turkeys that lived along the stream. Gold was discovered nearby about 1864. After the Turkey Creek Mining District was organized, the camp was formed around a stage station and received a post office in 1869. It served as a supply depot for the Crown King Mine. When the Prescott and Eastern Railroad was built to the Crown King in 1904, Cleator experienced moderate growth. It was renamed for landowner James P. Cleater who moved there after making about $10,000 on a claim he held in California. By the early 1900s most of the deposits had been worked out.

CLIP (La Paz) Clip was established along the Colorado River as a supply center and shipping point for the Clip Silver Mine. Discovered in the Trigo Mountains by Anthony G. Hubbard and George Bowers in the 1880s, the claim produced more than $1 million. Hubbard was also postmaster when the office opened in 1884. A 10-stamp mill was erected to crush the ore, but when the price of silver dropped the mine closed. After 1900 the tailings were reworked.

CONGRESS (Yavapai) This camp was established in 1884 shortly after Dennis May located the Congress Mine. Three years later Diamond Joe Reynolds purchased the claim. He died in 1891 and the property became inactive for about three years when E.B. Gage, Frank Murphy and Wallace Fairbank bought it. Charles A. Randall served as postmaster when the office opened in 1889. The Santa Fe, Prescott & Phoenix Railroad was later built a few miles from the mine. Since the gold was not free-milling, property owners had to have a cyanide plant built to treat the ore. By 1910 most of the ore had been recovered. The Congress produced more than $8 million in gold.

After Reynolds died, the *Arizona Enterprise* wrote: "Diamond Joe died in one of his houses.... After his death we had him packed in ice, and the coffin padded all around inside. No better attention could be given to anybody during sickness as five men were continually waiting on him, and all his wants were fully attended to."[21]

CONTENTION CITY (Cochise) This town was established about 1879 along the San Pedro River and became one of the milling centers for the Tombstone mines. Prospector Hank Williams discovered gold on the halter of his mule and staked his claim, which was contested by another miner named Dirk Gird. Ed Schieffelin and Gird ended up purchasing the Contention claim, from which the camp took its name. The post office opened in 1880 with John McDermoot as postmaster.

Author William H. Bishop wrote of his visit in 1882, "We changed horses and lunched at Contention City. One naturally expected a certain belligerency in such a place, but none appeared on the surface during our stay. There were plenty of saloons — the 'Dew-drop'; the 'Head-light,' and others — and at the door of one of them a Spanish señorita smoked a cigarette and showed her white teeth. Contention City is the seat of stamp-mills for crushing ore, which is brought to it from Tombstone."[22] When the mines at Tombstone flooded, that was the end of Contention.

COPPER BASIN (Yavapai) This name belies the fact that gold was discovered between Skull Valley and Sierra Prieta. The ore was worked only intermittently until about 1929 when the Depression brought renewed interest. During 1932 three companies moved to the camp and began conducting large-scale placer operations. Water for the diggings was pumped from nearby shallow wells and reused whenever possible. The companies recovered about $150,000 in gold.

CROWN KING (Yavapai) A very active mining camp, Crown King was established in the 1870s with a post office in 1888. It differed from the usual mining camps since there was no drinking allowed in town. The Crown King Mine was located in the 1870s, in the middle of the Bradshaw Mountains. News of its discovery prompted others to move there to begin their own placer mining, but the diggings were shallow and soon exhausted. This was followed by the discovery of other lode mines which brought in a stampede, especially when rumors got out that the ore was running $180,000 a ton and touted as the richest strike in the state. But no one mentioned that there was a great deal of difficulty with the extraction process. The mills refused most of the ore because it proved so unyielding, demanding less refractory material.

In 1902 what was called the Impossible Railroad was brought to the mines by Frank Murphy, it was an appropriate name given the steepness of the grade and instability of the surface. The Crown King was the most productive lode in the region. Total gold recovered was estimated at about $2.5 million.

DENOON (Pinal) This camp was founded when James DeNoon Reymert, an attorney, opened his office; he also located the Reymert silver mine. During 1889 the camp's population grew dramatically and it was anticipated the town would boom. A post office was established, but closed within a year along with Denoon because of disappointing prospects.

DOS CABEZAS (Cochise) Dos Cabezas bears a Spanish name meaning "two heads," descriptive of two nearby bald summits. Located in the foothills of the Cabezas Mountains, the camp was established during the 1870s with a post office in 1879. The Casey brothers were prospecting nearby and made their strike in 1878. They were approached by an attorney from Tombstone who tried to purchase their property for $40,000, which they refused. The mines did not produce, but Dos Cabezas survived by serving as a supply center for the neighboring mines and cattle ranchers. Placer mining by Mexican prospectors began about 1901. Four years later a number of companies came in but they only recovered a small amount of gold, roughly $2,500. The lodes yielded nearly $183,000.

DUQUESNE (Santa Cruz) Duquesne was settled in the late 1800s along the Patagonia Mountains with a post office established in 1904. It was named by Westinghouse executives, and enjoyed a population of about 1,000. One of the major mines was the Westinghouse which produced about $4 million in gold. In 1890 the Duquesne Mining and Reduction Company of Pittsburgh purchased the property and built a reduction plant nearby. By 1920 the place was on the decline.

EHRENBERG (Yuma) This town was named for German mining engineer Herman Ehrenberg, who surveyed the site in 1863. He also aided in construction of a reduction plant at Arivaca. He was later killed in the Mojave Desert. Ehrenberg was settled along the Colorado River and served as a shipping point for Prescott and the nearby mining camps. Michael Goldwater operated a ferry and was proprietor of the first general store. Its post office was established in 1896 with Joseph Goldwater serving as postmaster. Martha Summerhayes wrote in 1873, "Of all the dreary, miserable looking settlements that one could possibly imagine, it was the worst. It was an important shipping point for freight to the interior and there always was an army officer stationed there."[23] But Ehrenberg thrived until all the gold had been recovered, about $7 million, mainly in placers.

FAIRBANK (Cochise) Fairbank, established about 1880, is located along the San Pedro River near the El Paso & Southwestern Railroad. It was named for a mining entrepreneur from Chicago, N.K. Fairbank, who was also a railroad stockholder and organizer of the Grand Central Mining Company. Fairbank served as a supply and shipping point for the ore at Tombstone that was hauled by freighters to the mills at Contention. Today the land that encompasses the site is part of the San Pedro Riparian National Conservation Area, owned by the Bureau of Land Management.

In 1900 a gang of outlaws robbed the train containing a Wells Fargo express car that had made a stop at Fairbank. Lawman Jeff Milton was guarding the strongbox when the holdup occurred. He killed one of the robbers, but not before getting shot in the arm. The rest of the hoodlums quickly departed. Milton was taken to the doctor, who said the arm had to come off. But Milton wasn't having any of it and told the doctor, "The man who cuts off my arm will be a dead man."[24] No amputation was performed and Milton regained partial use of his arm.

FORTUNA (Yuma) Gold was discovered by Charles Thomas, William Holbert and other partners in 1893 who named their claim the La Fortuna, and had a 20-stamp mill erected. Fortuna was founded near the Gila Mountains about 20 miles from Yuma, and its post office was established in 1896. Other mines were located, but the La Fortuna was the only profitable one. Because of lack of water, a ditch was built to pump in water from 20 miles away. The mine closed in 1901 after several fruitless attempts to relocate the vein. Total gold production of the district was about $2.5 million.

GALEYVILLE (Cochise) Situated in a canyon in the Chiricahua Mountains, the camp was established in 1880 and named for John H. Galey, who erected the Texas Mine and Smelter. He came from Pennsylvania and was tempted by the high price of silver, but left town a poor man. Folklore says that miners smelted Mexican money and declared it came from a mine, precipitating a small stampede. Isolated and with access by only a few small trails, extraction of the gold was unprofitable because of high production costs. Since the camp was settled along a trail between the U.S. and Mexico, it later became a haven for cattle rustlers and outlaws. It was from here that a group of rabble rousers went to Charleston and stirred up trouble.

There is a sign posted outside of Galeyville that reads: "John H. Galey opened the Texas Mine and Smelter in 1881. Mining boomed and died and the town became a haven for lawless men like Curly Bill and John Ringo who used the nearby gulches to hold stolen cattle while brands were being altered. In 1888 the San Simon Cattle Company forced out squatters and the remains of Galeyville were carried away to construct houses in nearby Paradise."

GERMA (Mohave) This camp was founded about 1896 when a gold deposit was discovered nearby. The Germa Mine was worked for four years, then sold to the German-American Mine Company. Lack of water caused the claim to be shut down and the town folded by 1906. The mine produced about $40,000 in gold.

GILA CITY (Yuma) Established as a stagecoach station during 1858, Gila City became an outfitting point after the discovery of placer gold in the Gila Mountains, about 25 miles from Yuma. A party of prospectors worked the rich gravels of Monitor Gulch and benches near the newly founded settlement in 1860, and made the local news: "Yesterday, I saw 152 ounces of gold from the Gila mines deposited with the agent of Wells-Fargo. Very beautiful it was; and there is more in the same country and 400 men digging for it."[25] In 1870, Raymond wrote, "At Gila City a San Francisco company has during the last year erected works to pump water from the Gila up into a large reservoir on top of the highest foot-hills in order to work the placers of the vicinity by hydraulic power."[26] Hinton wrote in 1878, "Within three months of their discovery, over a thousand men were at work prospecting the gulches and canyons in this vicinity. The earth was turned inside out. Enterprising men hurried to the spot with barrels of whiskey and billiard tables.... There was everything in Gila City within a few months but a church and a jail."[27]

The ore was not as rich as boasted, but about $500,000 was recovered, the bulk of which was mined before 1865. The entire town was nearly destroyed in 1862 by a large flood. Prospectors moved on to La Paz when gold was discovered there the same year. By 1865 most of the placers had been worked out. G.H. Mears attempted small-scale hydraulic operations in Monitor Gulch during 1931, but less than $1 per man was being recovered per day.

GLEESON (Cochise) This site was named during the 1880s for John Gleeson, cattleman and

Goldfield (courtesy Stephen M. Murphy).

prospector. Before gold was discovered, the Apache Indians were recovering turquoise. Spaniards enslaved the Indians and tortured them in order to get the location of the minerals, but the Apaches never revealed the source. Gleeson was located along a branch of the El Paso & Southwestern Railroad, and received its post office in 1900. Placer gold was located on the east side of the Dragoon Mountains about 18 miles north of Bisbee. Although the camp experienced only moderate growth, the nearby claims yielded a little over $1 million in gold. Some mining was conducted in the 1930s, but only $200 was realized.

GOLD BASIN (Mohave) Gold-bearing veins discovered in the 1870s brought about the establishment of Gold Basin, created to serve as a milling community. The post office opened in 1890 with Michael Scannion as postmaster. The remoteness of the area made it difficult to obtain water and supplies, so mining activities were hindered. Before 1900, however, the district yielded gold worth between $50,000 and $100,000, most of which came from the Eldorado Mine. Placer gold was found in 1932 by W.E. Dunlop, and a year later S.C. Searles installed a large-scale treatment plant. Value of the recovered placers came to about $14,500.

GOLDFIELD (Pinal) Goldfield was established near the Superstition Mountains in 1893, a year after gold was discovered. Most of the ore was marginal and the camp almost sank into oblivion. But during the early 1900s George U. Young brought in new mining methods to recover the low-grade ore. Unfortunately, production was low and the town finally folded. Young went on to become Arizona's secretary of the territory.

GOLDROAD (Mohave) Gold was found by John Moss about 1864, but with discovery of silver in the Cerbat Mountains the place was temporarily deserted. About 1900 Jose Jerez was grubstaked by store owner Henry Lovin and the camp was formed with a post office in 1903. Tradition says that while Jerez was searching for his burro he tripped over a quartz rock and discovered what would become the Gold Road Mine. Jerez and Lovin sold the claim for about $50,000 to a mining interest. It was later purchased for $275,000. By 1907 the new owners had recovered about $2.5 million. The old site of Goldroad was razed in 1949 to avoid taxes.

GREATERVILLE (Pima) Prospector A. Smith was the first to make a strike in 1874, which attracted other gold-seekers. Primarily a placer district, many of the nuggets found during 1876 were valued between $35 and $50 apiece. The largest nugget ever reported from the camp weighed 37 ounces. First known as Santa Rita, the camp was settled during 1874. It was renamed for a resident named Greater when the post office opened in 1879. Since the closest stream was about five miles away in Gardner Canyon, the miners had to bring in water using goatskin bags. By 1881 most of the rich gravels had been recovered and the camp began losing its residents. Erection of a hydraulic plant at Kentucky Gulch brought Greaterville back to life for a short time. But again, water was not plentiful enough and the plant closed down. Total production was close to $7 million.

During its heyday, Greaterville was a lively place, and dances were the order of the day. One evening in 1880, a group of wild cowboys came into town quite heady with drink. When they tried to enter the dance, the residents wouldn't let them in. To get even, one of the men climbed up the adobe house where the dance was being held and dropped his gun cartridges down the chimney into the hot fire, which promptly ended the festivities.

GREENWOOD (Mohave) Named for the numerous paloverde trees, Greenwood came into existence during the 1870s as a milling camp with the discovery of gold. It never had a post office, so residents had to pay about $1,000 a year to get their mail from private carriers. A 10-stamp mill was erected, but the ore was too low grade for any profit. A larger stamp mill that was built at Virginia City further lent to Greenwood's demise.

GUNSIGHT (Pima) During the 1880s gold and silver rushes, General John B. Allen visited a number of sites for a good place to build a hotel, and selected this camp which he named for himself. He picked a spot on a side of the Ben Nevis Mountains and built his enterprise. Allen's hotel served a number of transient miners for only a short time. The post office opened in 1882, and closed four years later. The name was changed to reflect the nearby mountain, for a ridge that resembled a rifle's gunsight. The Gunsight Mine discovered in 1878 was sold in 1883 and a smelter was built, but because of bad construction it was soon shut down. The Silver Gert Mining Company took over operations in 1892, and brought out more than $175,000 in silver from the property.

HACKBERRY (Mohave) W.B. Ridenour and Samuel Crozier were prospecting near Peacock Mountain about 1874 when they discovered the Hackberry Mine. The camp was established, taking its name from the hackberry trees in the area, and acquired a post office in 1878. A nearby spring ran the boilers of the mill that was constructed to treat the property's ore. Hackberry prospered until the mine closed after recovering about $300,000 in silver. A little diner and a few nearby farms survive today.

HARCUVAR MOUNTAIN (La Paz) The mountain rises to more than 5,100 feet high in the Harcuvar Wilderness, and bears a Mojave name for "little water," or "sweet water." During the 1920s the Hargus Ranch was established in a canyon at the foot of the mountain and operated as a sheepherding station until about 1940 when prospector Shorty Alger accidentally discovered a glory hole there. Climbing up a hill and using his pick for balance, he struck a gold nugget that weighed a half pound. His discovery brought out more than $100,000.

HARDYVILLE (Mohave) Located at the head of navigation along the Colorado River, this camp was settled by ferry operator Captain William H. Hardy in 1860. It served as a shipping point and trading center for the mines in the Cerbat Mountains and received a post office in 1865. Hardy was known for his generosity; if someone could not afford to pay for the ferry crossing, he allowed them free passage. He also invented riveted mail sacks. During the Pony Express days the thread used on the sacks frayed too easily as it wore on the horse's saddle. Fire visited Hardyville in 1872, followed a year later by another that just about destroyed it. The town saw its demise when a railroad crossing was built at Needles, California.

HARQUAHALA (Yuma) Rising about 5,690 feet high, the mountain was given a Mohave name meaning "running water," or "running water high up," descriptive of a nearby spring. The camp was established in the 1880s. Gold had been found by the Spaniards during the 1760s, and later by the Pima Indians in 1869. While Harry Watton, Robert Stein and Mike Sullivan were prospecting in 1888, Watton climbed up a hill and found a nugget that was valued at $2,000. He told his partners, who found even more nuggets at a spot that would be developed as the Harquahala Mine. Stein later wrote a letter to a newspaper: "The new strike has never been equaled in Arizona.... Boulders of float weighing from one to five tons are worth $50 to

Hackberry General Store (courtesy John R. McNally).

$1000 a ton.... For the last two weeks we have taken out every day very rich ore worth hundreds per ton."[28] And it was rich, so much that mine owners were fearful it would be stolen, so they had the gold cast into heavy bars that weight nearly 200 pounds. The region produced more than $2 million, half of which came from the Harquahala.

HARRISBURG (La Paz) This small camp served as a milling center for the mines at Socorro and other neighboring properties. It was named for founder Captain Charles Harris, who moved there from Canada. He and his partner, Frederick A. Tritle, had a 5-stamp mill brought in to process the ore. The post office opened in 1887, with William Beard as postmaster.

HARSHAW (Santa Cruz) When Indian agent Thomas Jeffords asked cattleman David Harshaw to remove his cattle from the Chiricahua Indian land, he complied. As Harshaw was moving on he found gold which resulted in the formation of the camp in 1880 along Harshaw Creek, with a post office established the same year. It was first called Peach for the peach trees planted by a Spanish padre at one time. The Hermosa Mine was located and went on to become a good producer, yielding in a four-month period about $365,000. In all it was thought to have recovered about $700,000. A 20-stamp mill was later built by the Hermosa Mining Company. Harshaw was also a service center for the mines in the Patagonia Mountains. By 1909 the camp was almost gone, then it was finished off by a fire.

HASSAYAMPA (Maricopa) The waters of the Hassayampa flow beneath a sandy stream bed through part of its course, and intermittently disappear under the sand. It bears a descriptive Apache word meaning "river that runs upside down," or "water that is hidden." The name was also thought to be Mojave for "water place of big rocks." Indian legend that says if you drink the water from this river above the trail you will always tell the truth, but if you drink below it you will never know the truth. During 1864 gold and silver were discovered along the river, and a camp was established near the Bradshaw Mountains with a post office in 1881. Placers were worked between 1885 and 1890; thereafter operations were carried out on a small scale. Total gold production was about $2.6 million.

Harshaw (courtesy Thomas Pritchard).

HUMBOLDT (Yavapai) Named for the German scientist and explorer Baron Alexander von Humboldt, this site was settled as a smelter town during the 1860s. Humboldt predicted "the riches of the world would be found" in the center of the Great Peck Mining Company area.[29] The camp was situated along a branch of the Atchison, Topeka & Santa Fe Railroad from where the smelted ore was shipped out to various destinations. The smelter operated until the 1920s.

KATHERINE (Mohave) This camp was established at Katherine Landing along the Colorado River in the 1900s. Although gold was discovered nearby during the 1880s, it was about 1901 before high-grade ore was located in outcrops of the Gold Road vein, and a rush was on. The Katherine Mine was discovered about 1920, named for the sister of its discoverer, J.S. Bagg, as was the town. The post office opened in 1921 with Alva C. Lambert as postmaster. The Katherine yielded over $2 million in gold. Total production for the district was about $42 million.

KLONDYKE (Graham) Klondyke was named by prospectors who had been mining in the Yukon Territory. The camp was established along Aravaipa Creek, and bore an Athabascan name for "hammer water." A post office began operations in 1907. While searching for gold the prospectors also came across some silver. Although there were more than 400 miners residing there, the ore was insufficient. Klondyke managed to survive for a time as a trading center for ranchers.

KOFA (Yuma) Kofa's mines were discovered on the southwest slope of the Kofa Mountains about 1899. Nearly the entire gold revenue came from the King of Arizona Mine. Felix Mayhew was the locator of the North Star, and tradition says he found it while searching for a good place to dig a well. Mayhew later sold his claim to New York capitalists for more than $100,000. The new owners proceeded to have a 50-ton mill erected, but their efforts proved to be disappointing since less than $10,000 was recovered. Charles Eichelberger found the rich King of Arizona Mine during the early 1900s while looking for drinking water. He worked his claim until he recovered about $500,000 in gold, then sold the property to Eugene Ives for $250,000. A stamp mill was built to crush the ore, and the first two tons were valued at more than $1,200. The King of Arizona operated until about 1910. Total gold production of the region was about $4.8 million.

LA PAZ (Yuma) A stampede to the area began about 1862 with the discovery of gold placers, and a camp sprang up with a few hundred miners. It was given the Spanish name meaning "the peace." Located along the Colorado River, La Paz prospered not only as a mining camp but as a river port, and grew to a population of about 5,000. Its post office was established in 1865.

A. McKay, who was a member of the territorial legislature, wrote: "Of the yield of these placers, anything like an approximation to the average daily amount of what was taken out per man would only be guess work.... Of the total amount of gold taken from these mines, I am at a loss to say what it has been ... I have failed to find any pioneer whose opinion is that less than $1,000,000 were taken from these diggings within the first year, and in all probability as much was taken out in the following years."[30]

Much of the gold was used for gambling in La Paz, and a large portion was sent to Mexico by the Mexican miners. There was some lode mining in addition to the placers. Store owner Manuel Ravenna purchased a lode claim called the Conquest. He worked the claim for a while, then sold it to an English firm for about $2 million. By 1864 the high-grade placers had been exhausted.

Five years later the Colorado River changed its course and La Paz lost its status as a river port. Placer gold production came to between $2 and $8 million. In 1910 plans were made to use hydraulics to recover the ore, but most of the land was on an Indian reservation, so the idea was abandoned.

LOCHIEL (Santa Cruz) Lochiel was named by cattlemen Colin and Brewster Cameron in honor of their ancestral home in Scotland. Since the camp was so close to the Mexican border, it prospered from goods that were smuggled across. The camp was once called Luttrell, for Dr. J.M. Luttrell who built the Holland Company Smelting Works that served the surrounding mines.

LOST BASIN (Mohave) Established along the Colorado River, Lost Basin received its name because it was isolated and surrounded by small hills. The post office opened in 1882 with Michael Scanlon as postmaster; he also served as a ferry operator. The only gold found before 1930 amounted to a mere $700. The nearby King Tut placers were discovered during the 1930s by W.E. Dunlop on land owned by the Santa Fe Railroad. The Duncan Ranch also had title to part of the property on which an ore concentrator was built. Water for the diggings had to be hauled in from a well five miles away. The placers yielded only about $23,500.

MCCABE (Yavapai) An early miner to the area during the 1860s was John H. Marion, who worked his claim only a short time because he was chased out by hostile Indians. About 1883 Frank McCabe relocated Marion's old claim, which started a stampede and the camp was formed. It was officially established with a post office in 1897, and Marion C. Behn served as postmaster. The camp even had a telephone line that connected it to Prescott, and a small hospital was built by Dr. Robert N. Looney. The McCabe mine shut down in 1913, causing the town to decline.

During the rainy season in 1937 a horrendous storm occurred. Water rushed down the Bradshaw Mountains, carrying with it rocks and timber, creating a dam. The pressure built up to a point that the dam broke, washing out all the buildings. There was some small-scale mining into the 1980s, then copper was discovered and those operations continued until about 1993.

MCMILLANVILLE (Gila) Charles McMillan and his prospector friend, Theodore Harris, were the first to discover silver ore in 1874. While taking a break from their prospecting, Harris idly picked at the rocks and found a chunk of quartz. This would be the beginning of the Stonewall Jackson Mine and a camp with more than 1,000 miners. A 5-stamp mill was later built for the nearby claims. Unknown to some, the site was located on the San Carlos Indian Reservation. During 1882 Apache Indians were angered that the white men were tearing up their hills and began attacking. Town residents took refuge in the Stonewall Jackson tunnel. After the foray, the men returned to their claims and continued working. Congress later severed that part of the land from the reservation. McMillan and Harris recovered about $60,000 and then sold their claim for $160,000; it ultimately yielded another $2 million. By 1884 most of the ore had been mined and the people left.

MILLTOWN (Mohave) This camp was established as a milling center for the area's mines. In 1903 a 40-stamp mill was built by the Mohawk Gold Mining Company, and the ore was shipped via a narrow-gauge railroad to the town of Needles, California. Milltown was located along the Colorado River, but after only a few years of prosperity the river flooded and washed out most of the train tracks.

MINERAL PARK (Mohave) Twenty miles from the town of Kingman, this site was established in 1871 after the discovery of rich-paying ore near Ithaca Peak. The town served as a trading and supply point, in addition to treating some of the ore with a 5-stamp mill. Harris Solomon was a freighter who brought supplies to the camp, but at a high price. When the Atlantic and Pacific Railroad completed its line to the region, it enabled an economical means of transportation. Mineral Park became the county seat in 1873, but lost out to the town of Kingman in 1887. That same year the mines started to close.

MOWRY (Santa Cruz) Mowry is one of the oldest mining camps in the state, having been established in the 1850s. The mine was first worked by a Mexican herder during that time. He later sold it to R.S. Ewell and J.W. Douglas for a pony and animal traps. In turn, they sold it for about $20,000 to an Army officer, Lieutenant Sylvester Mowry from Fort Crittendon, in 1858. The Mowry was located near the Patagonia Mountains, and yielded more than $1 million in silver. The lieutenant was later imprisoned for six months as a Southern sympathizer, and as a result lost title to his claim.

MUGGINS MOUNTAINS (Yuma) Named for a prospector's burro, the site was located about 25 miles from the town of Yuma. Gold placers were not worked extensively until the 1930s. Between 20 and 25 men were working the diggings, and had to haul the water in from wells miles away. Their hard work yielded them barely $1 a day per man. The only recorded production was a little over $6,000.

OATMAN (Mohave) Soldiers from Camp Mohave discovered gold about 1863. Although some rich ore was taken, development was slow and the region was not active until 1902 when Ben Taddock espied a shiny gold rock and thereby laid his claim. The following year he sold his property to Judge E. Ross and Thomas Equing, who worked the mine until 1905, when they sold it to the Vivian Mining Company which began large-scale development. A narrow-gauge railroad was built to the mines and the ore was shipped to the Colorado River. From that point a ferry unloaded supplies from Needles, California. The company recovered about $3 million in gold within a two-year period. In 1903 the Tom Reed Mine was discovered, thought to have yielded more than $13 million before it shut down in the 1930s.

The Gold Road Mine Tours out of Oatman offers excursions of the Gold Road Mine at the old camp of Gold Road. Reservations can be made by calling the company at (928) 768-1600.

OCTAVE (Yavapai) During the 1860s Indians traveled to the town of La Paz and told the white men of gold deposits they had found. Their discovery sparked the interest of Abraham H. Peeples, who organized a group of gold seekers led by explorer Pauline Weaver to a place called Rich Hill. Tradition says after a pack animal strayed up into the mountains, and while retrieving the mule one of the men found a gold nugget. Peeples dug up about $7,000 in gold in one morning.

Octave was established at the head of Weaver Creek near the Weaver Mountains. The Octave Mine recovered between $2 and $8 million in silver and gold, and operated until 1930. The property was shut down until 1934 when the American Smelting and Refining Company took it over. Placer deposits were credited with more than $6 million. Much of the lode gold came from the Octave Mine. After World War II the camp was razed to reduce property taxes.

A few miles from Octave were the Old Hat placers, where about 30 men worked their diggings during the 1930s. Most of the claims yielded very little, although a few miners were rewarded with some $5 nuggets, and occasionally one would pop up worth $25. Tradition says a couple of miners found a nugget that weight 16 pounds, but they were found murdered and the nugget disappeared.

OLD TRAILS (Mohave) This long-abandoned site was named for the old trails that linked nearby settlements. During 1915 it was a busy place and there were high hopes that it would become a large town. The camp boasted a hospital, electric lights, telegraph, and phones, but the nearby claims proved to be inadequate and the camp was deserted.

OLIVE (Pima) Olive was established as a result of silver discoveries during the 1880s. The camp was named for Olive S. Brown, whose husband, James, was part owner in the Olive Mine. The miners were very fond of Mrs. Brown, since she treated the men who worked at the mine to free chicken dinners every Sunday. The family later sold the property. Olive began its decline when the silver was depleted, and prospectors turned to copper mining.

ORACLE (Pinal) This site was named for a ship that sank on its way to San Francisco. Albert Weldon was a passenger on the *Oracle* and headed for Arizona about 1876. After locating a claim which he named the Oracle, Weldon established a small camp. His mine didn't pan out but he continued searching for more wealth. In 1880 a post office opened, businesses were established, and a freighter named Curly Bill Neal began delivery supplies to the camp. A year later Weldon and a friend, Alexander McKay, found another claim they christened the Christmas Mine. Other small deposits were found during the 1890s. After the mining era, people began to move into the region because of its dry climate, which was conducive to people suffering from tuberculosis, so doctors sent their patients there to recover.

ORO BLANCO (Santa Cruz) Bearing a Spanish name meaning "white gold," Oro Blanco was occupied by Spanish prospectors about 1740. They worked a few shallow mines, but it is not known how much gold they recovered. American miners arrived on the scene during 1873. Discovery of the Warsaw Mine resulted in the building of a 10-stamp mill which processed about $25,000 from the claim. Oro Blanco is about 85 miles south of Tucson. Its post office was established in 1897 with William J. Ross as postmaster. Between 1896 and 1904, only about $2,000 was realized from the placers. Blake wrote in 1899, "Most of the placer mining is carried on in a desultory way, often with a small and wholly inadequate water supply.... The returns are small."[31] Gold lodes from the district produced about $2.5 million.

PEARCE (Cochise) Tombstone miners James and William Pearce purchased a ranch during the 1890s. While riding the range one day in 1894, James found gold and named his newly found claim the Commonwealth. Formation of a camp soon followed with a post office in 1896. Pearce was a fairly wild place and also a hangout for outlaws. Bullion theft became a problem, so to circumvent any more robberies the ore was shaped into very large bars and transported to the town of Cochise in farmers' wagons. James later sold his interest for about $250,000, not realizing it would produce more than $15 million. The property was temporarily shut down in 1904 when part of the mine caved in. The following year a cyanide plant was built to extract gold from the tailings.

PICK-EM-UP (Cochise) This interesting name came about because of a would-be lynching. Gambler Johnny O'Rourke shot a mining engineer named Henry Schneider. Miners rushed in to lynch Johnny but the sheriff stopped them, grabbed the killer, and fled from the irate citizens. Mr. McCann, the owner of the First Chance Saloon, had a racing mare. He heard the sheriff holler that he was trying to save Johnny from the lynch mob, and told the lawman to "pick em up" (get on the horse). The sheriff got his prisoner safe and sound to Tombstone. This camp was located between Charleston and Tombstone and inhabited during the 1890s. The First Chance Saloon got its name because it was the first chance for miners to get a drink after leaving Charleston on the way to Tombstone.

PINAL (Pinal) This defunct camp was established during the 1870s after the nearby Silver King Mine needed a smelter to refine its ore. Mule teams hauled the silver to the mill which was producing up to $20,000 a week. The camp was also an important stage station on the road to the town of Globe. With the riches from the Silver King, Pinal grew rapidly with churches, schools, hotels, and of course the usual saloons, and accommodated a population of about 3,000. A drop in production caused a drop in Pinal's population, and by 1890 no one was left.

PLACERITA (Yavapai) About 20 miles south of Prescott, this camp was established in 1863 when gold was discovered at nearby Placerita Gulch by Mexican miners. When the surface gold was gone the place was abandoned. Then during the 1880s a prospector named Callen found gold veins and the camp came to life once again, with a post office established in 1896. There were attempts at dredging in nearby French Gulch, but that project proved a failure. Placerita was short-lived again, for only small veins were unearthed, and by 1905 the place was deserted. About 30 years later placer mining was resumed, but the prospectors were only averaging about 50¢ a day per man.

POLAND (Yavapai) Davis R. Poland located the Poland Mine in 1865, but the camp wasn't established until 1900 along Big Bug Creek, about five miles from the property. It received a post office in 1901, and boasted a population of about 800 miners, with all the ordinary businesses and six saloons. After the ore ran its course, a tunnel was cut about two miles through the mountain to Walker, and the gold from those mines was hauled through the tunnel by mule train to be loaded onto the Prescott–Crown King Railroad a few miles from Poland. The Poland Mine contained both gold and silver, and between $122 and $310 per ton was recovered. Today the site is a summer community that goes by the name of Breezy Pines.

POLARIS (Yuma) On a slope of the Kofa Mountains Felix Mayhew discovered the Polaris Mine, so called because it was thought to have been located by its position to the North Star. It was also north of the Kofa Mine. Mayhew originally worked at the King of Arizona Mine, and during his off-time did his own prospecting. He discovered a ledge, staked his claim and worked it until 1907 when he sold it to the Golden Star Mining Company for about $350,000. The mine closed by 1911, and with its closing Polaris was soon deserted.

QUARTZITE (La Paz) The camp was originally a stage station on the road from Ehrenberg to Prescott, located in Tyson Wash. Mainly a placer mining district after the discovery in 1869, the claims were not extensively developed until 1880. Heikes wrote in 1912, "Surrounding the post office of Quartzite, in the Plomosa mining district, and extending in every direction, covering an area of about 7,500 acres, is found dry-placer ground.... The gold content per cubic yard is reported to average in coarse gold from ten cents to several dollars."[32] Total production from the district was approximately $500,000.

During 1855 more than 70 camels were imported for use in building a wagon route from Texas to California. Unfortunately, they camels didn't fare well in the rocky soil, and the mules panicked at the sight of the strange animals. Finally, one man named Hi Jolly took a few camels and began a freighting business. Some of the animals were later simply turned loose in the desert to fend for themselves, and the rest were sold to make jerky for the miners.

QUIJOTOA (Pima) This community developed with the discovery of silver and copper at nearby Ben Nevis Mountain. Four camps were established, and Quijotoa became a town with the merger of those camps. The post office was established on December 11, 1883 (Ransome Gibson, postmaster). James Flood and John Mackay (the "Comstock Kings") promoted the region and brought in investors. News of the supposedly great wealth brought in speculators and a few mining companies.

The *Prospector* wrote in 1884, "Where Quijotoa now stands there were, on the first of January, a couple of tents. All around was a dry and deserted land, no water within ten miles. There are now a mile of houses, and a population, in and around the camp, of at least 1200. The happy discovery of the wonderful mines on the summit of Quijotoa mountain has drawn the eyes of the whole Pacific Coast to this spot.... We are now waiting for the cutting of the ledge in the tunnels which are being run. Quijotoa will then take her real start.... The opening of 1885 will see a city here, compared with which Tombstone, Virginia City and Leadville are mere villages."[33]

A tunnel was bored through Nevis Mountain, and what ore was there played out within a few years. About all that's left is the hole in the mountain. Quijotoa is derived from the Papago words qui-ho and toa, for "carrying basket mountain," descriptive of the shape of the mountain. In 1889 most of the town was destroyed by fire.

SALERO (Santa Cruz) Located on the southwest slope of the Santa Rita Mountains, this site bears a Spanish name meaning "salt cellar," because, "The Bishop of Sonora is coming to visit our Tumacacori mission! This is an event to prepare for.... Be certain the best tableware is ready. But something special should mark the occasion. We will have an elegant salt-cellar, fashioned from the silver of our mine, ready near his plate."[34] Spanish Jesuits may have been the first to locate a silver lode on Salero Mountain during the 1690s. It was relocated before 1800 by Mexican miners. Then in 1857 a group of six prospectors purchased the mine, but one by one they were killed by hostile Apache Indians, save one, and he left. The mine produced only about $10,000.

SASCO (Pima) Sasco was a smelter town located along the Arizona Southern Railroad. Its name is an acronym for the Southern Arizona Smelting Company. The facility was built in the early 1900s and Sasco's post office opened in 1907. The company treated ore from the Silver Bell Mine and Picacho Mining Company properties. Sasco didn't last long because the smelters never showed a profit and closed down by 1910.

SEYMOUR (Maricopa) Named for an owner of the Vulture Mine, New Yorker James Seymour, this camp was established about 1878 when a 20-stamp mill was erected. It operated for only one year, then was dismantled and reconstructed at the Vulture Mine. When the mill went, so did Seymour.

SIGNAL (Mohave) While prospecting in 1874, Jackson McCrackin discovered the McCrackin

Present ruins of Sasco (courtesy Thomas Pritchard).

Mine. Signal was formed when a 20-stamp mill was built and a post office opened in 1877. The site was pretty isolated, as one visitor in 1878 noted: "freight ... at this time came by rail to the west side of the river at Yuma and thence by barge up the river to Aubrey Landing where it was loaded on wagons and hauled by long mule teams thirty-five miles upgrade to Signal. The merchants considered it necessary to send orders six months before the expected time of delivery."[35] The McCrackin produced over $6 million in silver.

SILENT (Yuma) The Silent Mine was discovered about 1880, named for mining man and attorney Judge Silent. A camp was established about the same time as was the post office. The mine was in operation about three years. Silent was very small with only a few buildings, or a facsimile thereof. Since there was little timber in the region most of the people lived in dugouts. They couldn't even build adobe dwellings because of lack of water. The ore was hauled to Norton's Landing along the Colorado River to a smelter, but it is not known how much ore was recovered. Today the Red Cloud Mine at Silent is the main supplier of wulfenite, named for Franz Xavier Wulfen, who wrote about the mineral's occurrence. It is a highly prized mineral among collectors.

SILVER KING (Pinal) Local farmers Isaac Copeland, William Long, Charles Mason and Ben Reagen discovered the Silver King Mine while they were on their way to Globe in 1875. The mining camp was formed at the base of a butte with a post office established in 1877. It was also a stage depot on the road to Globe. Long and Copeland sold their interests to Mason and Reagen for $65,000. The mine was producing up to $20,000 a week in silver bars hauled by mule teams to a mill at Pinal about five miles away. After the stages had been robbed a number of times, the bars were smelted and cast so they were too heavy for the thieves to steal them. The mine continued to produce well, yielding a good $25,000 a month in dividends, and ultimately recovered between $7 and $17 million.

STOCKTON (Mohave) During the 1860s silver ore was discovered in the region. As more and more miners moved in, they established the camp of Stockton at the foot of Stockton Hill. It may have been named for the town in California that honored Commodore Robert F. Stockton, who took possession of the territory for the U.S. Continual flow of ore ensured the camp's success, but production results are unknown.

SUPERIOR (Pinal) Superior was a seed born from an Army camp established in 1870 when a new road was being built to a mine. During 1873 a silver nugget was found at what would become the Silver Queen Mine. Silver and gold were produced for many years even with numerous Apache raids, and the site was formally established during the early 1900s. First named Hastings with a post office in 1902, it was changed when the Lake Superior Mining Company came in. The town went into a depression a number of years later when the ore was exhausted. Total gold production was about $8 million. It is not known how much silver was recovered. The site was revitalized when copper was discovered about 1912, and taken over by the Magma Copper Company.

TIGER (Yavapai) This camp lasted as long as the Tiger Mine. Located in the 1880s, the claim was touted as being one of the richest in silver. The property did produce, but not to the extent that some believed. The other smaller claims, along with the Tiger, recovered nearly $2 million.

TIP TOP (Yavapai) The Tip Top Mine was located in 1875, named because it was thought to be a tip-top prospect. It was discovered by Jack Moore and his partner, a Mr. Corning, and became one of the best producers in the region. The camp was established the same year, as was the post office. Located in the foothills of the Bradshaw Mountains, Tip Top had two general stores and the usual businesses. It also had a school. Ore from the mine was hauled to the town of Gillett for milling. The Tip Top yielded nearly $2 million in silver and gold before it closed in 1883. During World War I tungsten was mined at Tip Top. Total gold production from other properties came to about $200,000.

TOMBSTONE (Cochise) Ed Schieffelin came prospecting to a place called Goose Flats. With only $30 in his pocket and a prospector's outfit, he sought to find his wealth in silver. An Indian scout warned Schieffelin that if he kept prospecting in this area, it would become his tombstone because the Indians would surely get him. Schieffelin ignored the warning and discovered a silver lode he named the Tombstone, worth more than $75 million. Other mines in the region yielded about $40 million. Between 1881 and 1882, about $5 million had been recovered. By 1886 many of the large ore bodies were either depleted or mined to water level. Between 1879 and 1886, ore production was estimated at nearly $19 million.

Tombstone was once one of the largest towns in the West, boasting, some say, about 15,000 people, an enormous population for that era. The town endured many disasters: June 1881, a fire in a saloon from an exploding whisky barrel destroyed about four blocks; May 1882, someone dropped a cigarette and started a fire that almost wiped out the town; 1887, an underground river flooded the mine shafts, ending the glorious reign of silver mining; and in 1929, the county seat was moved to another city.

At the corner of 5th and Toughnut Streets is the "Million Dollar Stope." It was a huge hole where $1 million in silver was taken from the tunnels of a mine. The labyrinth led for miles to connect with other lodes. The tunnel caved in after a horse pulling an ice wagon crossed over it.

In October 1929, the residents celebrated the first Helldorado Days to honor the anniversary of the OK Corral gunfight. Bad blood between the Earp brothers and the Clanton and McLowery families caused Marshall Wyatt Earp to settle the feud. Three of the Clantons and the McLowery families were killed.

Residents later created the slogan, "Tombstone — The Town Too Tough to Die." It is now designated a Registered National Historic Landmark, a town that *is* a museum. The Rose Tree Inn boasts the largest rose bush in the world. Given as a gift of friendship, the Lady Banksia rose was planted in 1885. A trellis system was designed for the rose bush and as it grew, more material was added to the trellis (or arbor). Millions of white roses bloom in April, and the bush now covers more than 8,000 square feet.

Tombstone enjoys many events centered around the Old West, the most famous being the Gunfight at the OK Corral. Memorial Day Weekend brings Wyatt Earp Days, with gunfights and a chili cook-off. For more information, contact the Tombstone Chamber of Commerce at (520) 457-9317.

TOTAL WRECK (Pima) When John L. Dillon found a silver ledge with an odd conglomeration it looked to him like a total wreck, hence the camp's name. The ore was found during 1879 in the Empire mountains where the site was established with a post office in 1881. That same year a 70-ton mill was built. Development was hampered for a time when a band of Geronimo's Apache Indians attacked the miners, killing some of the men. By 1901

Present-day Tombstone (courtesy John R. McNally).

the silver had pinched out, but not before the Total Wreck brought out more than $500,000.

At one time a rare shooting occurred at Total Wreck. E.B. Salsig got into an argument with someone, which ended in a shootout. Salsig should have been killed since the other fellow was the first to draw. But he had a large packet of love letters from his lady love, and the bullet got only as far as the paper. Salsig later married the woman.

VEKOL (Pinal) This site was named for the Vekol Mine, discovered in 1880 by John D. Walker, P.R. Brady and Juan José Gradello. Walker started out as a trader with the Pima Indians. Because of his good relationship with them, one of the Indians showed Walker the location of some silver £ore he had found. When the trusting Indians found out Walker had told others of the claim they got angry. Walker was walking on eggs at this point and so he brought only his brother, Lucien. Gradello later sold his interest in the mine, and Lucien took over.

The camp was established in 1880 and had a population of about 150. Brady later sold his interest in the mine for $60,000. Vekol had the usual businesses, and also included one saloon and a public library. The saloon had so many problems with the workers getting drunk too often that they weren't showing up for work on time. As a result mine owners refused to allow their employees to frequent the establishment, forcing the owner to shut his doors. The mine realized about $2 million before litigation closed it down.

VULTURE (Maricopa) Prospectors searching the area for gold during 1862 had no luck, but a year later when gold-bearing quartz was discovered in the Vulture Mountains by Henry Wickenberg. The camp was formed shortly thereafter. Recovered ore had to be hauled about 15 miles to the mills along the Hassayampa River. In 1867 placer gold was found and more than 200 miners rushed in to stake their claims. A good number of the men were realizing between $25 to $50 a day. By 1880 most of the placers had been worked out, recovering a little over $6,000. James Seymour later purchased the Vulture Mine and began redevelopment. He spent about $380,000 building an 80-stamp mill and bringing water to the site. The Vulture became the highest yielding property in the area until 1888 when it closed. It reopened in 1910 and was in operation until 1917, during which time it recovered a little more than $1 million. Total production for the district was about $7.5 million, with the Vulture Mine accounting for approximately $3 million.

WALKER (Yavapai) A group of prospectors from California discovered placer gold in 1863. It was also thought that Army captain Joseph R. Walker (who built a small fort) and his men were the first discovers. Nearby Lynx Creek was full of the glittering stuff; the gold was so rich it could be dug out with a butcher knife. Word of mouth brought in about 200 prospectors. The richest placers were depleted in the early days, but small and intermittent mining continued for many years. This was followed by some hydraulic mining, but a flood destroyed the dam that held water for the operations. About 1900 the Speck Company came in and attempted dredging, but their equipment fell apart after recovering only $800 in gold. Placer output was about $1 million.

WHITE HILLS (Mohave) This town was founded in 1892 with a post office established the same year by postmaster William H. Taggart. Silver was found by a Hualapai Indian named Jeff, who was applying iron oxide as face paint when he noticed the shiny material. He showed the ore to Judge Henry Shaffer, who immediately went prospecting and located his mine. The town grew to more than 150 people, and about a dozen claims were staked. In 1893 a mining entrepreneur from Denver, David Moffat, came to White Hills to check out the prospects. He purchased most of the properties and consolidated them under the White Hills Mining Company, then ordered a 10-stamp mill brought in. The mines were good producers until a flash flood in 1899 inundated the mine shafts. Total yield was between $3 and $12 million.

WICKENBURG (Maricopa) This camp was named for Henry Wickenberg, discoverer of the Vulture Mine. It became a trading point for miners and cattlemen, in addition to serving as a station on the Atchison, Topeka and Santa Fe railroad. The post office was established in 1865. A mill was built using water from the Hassayampa River to process the nearby claims. Less than ten miles from Wickenburg was the Black Rock District, which had a gold production of $195,000.

YELLOW JACKET MINE (Santa Cruz) About 80 miles from Tucson, this mine was originally found in 1736 by the Spaniards. First known as the Mina Longorena, the property was worked until the Mexican–American War in 1854. The site was also used as an overnight stop by padres traveling between the Tumacacori and Sacric missions. The mine was relocated by Thomas G. Roddick in 1865 and renamed for the multitude of wasps in the region. It was abandoned until 1868 because of Apache Indian attacks. Production amounts came to a little over $500,000.

CALIFORNIA

ABRAMS (Trinity) In 1850 James Abrams moved to the Salmon-Trinity Alps and established a trading post, which served the miners working the diggings along the Salmon and Trinity rivers. The post office opened in 1895.

ACTON (Los Angeles) Acton was founded about 20 miles north of Los Angeles in 1875 when the Southern Pacific laid its tracks through Soledad Canyon. Its post office opened in 1887. The railroad may have named the town for a town in Massachusetts. Gold quartz discoveries brought in the miners about 1877 and Acton grew as a major supply center. The Red Rover and Governor mines opened at nearby Soledad Canyon and continued operations into the 1940s. California governor Henry T. Gage named the Red Rover for his dog. More than $1.5 million in gold was recovered.

ADAMS (Stanislaus) Adams was established as a way station between Tuolumne City and Empire City on the route to the Southern Mines. The location made it an ideal place as a supply and trading center. Although Adams served as county seat in 1854, that didn't stop the town from fading away.

AGUA FRIA (Mariposa) This camp was located near the town of Mariposa and took its name from the creek; *agua fria* is Spanish for "cold water." The camp began about 1851 with a post office established the same year. It was divided into two sections: Lower and Upper Agua Fria. Tradition says it was at this point that miner Alex Godey discovered gold on explorer General John Fremont's land grant. The camp lived its life as a trading center for the mining population. After a succession of fires, Agua Fria ceased to exist.

In 1862, Chinese who had been working the placers at Lower Agua Fria made the local headlines after an altercation over water rights to the claims: "Saturday last a fight took place between two companies of Chinese on the Lower Agua Fria, in which a Chinaman named Ty was seriously if not mortally wounded by a countryman named Tin See. The difficulty occurred about a water ditch and division of water. An examination of the affair took place before justice Bruce, who, in default of a $500 bond, committed the said Tin See to prison to await the action of the Grand Jury."[36]

ALABAMA FLAT (El Dorado) This site was situated on Johntown Creek and named for the state of Alabama, a Choctaw word for "thicket clearers." Not much placer mining occurred there, but a prospector named William F. Coe worked a small claim with moderate success. The area later became known for its fruit orchards.

ALABAMA HILLS (Inyo) The camp was named for the Confederate ship *Alabama* by southern sympathizers, and was settled between Angels Camp and Carson Hill. Much of the mining there consisted of lead, gold, and silver. Prospectors were accommodated by a trading post at nearby Lone Pine City.

ALBANY FLAT (Calaveras) This camp was also settled between Angels Camp and Carson Hill, and was a busy place during 1851 with over 1,500 miners working the diggings. By 1856 there were only a few miners and one store still operating. Albany is a Scottish word for "hill," or "highland."

ALDER CREEK (Sacramento) Gold was discov-

Agua Fria (Mariposa County Museum and History Center, Mariposa, CA).

ered in the area about 1867, bringing in a rush of miners who sunk more than 100 mine shafts in a week's time. The Sacramento Valley Railroad built its junction near the creek about 1880. When the gold played out five years later, the Chinese arrived to glean what remained. The old Coloma Road ran through Alder Creek, which was used by John Marshall to haul his gold from Sutters Mill. The site was named for the nearby alder trees.

ALGERINE CAMP (Tuolumne) Located south of the town of Sonora, this site was first called Providence Camp. It was later renamed Algiers (Arabic for "the islands"), and then corrupted to Algerine. French prospectors arrived in 1853 and extracted more than 200 ounces of gold the first few days they were there, which of course brought in other miners. By 1867 the placers were just about depleted, but not before realizing more than $2.5 million.

ALLEGHANY (Sierra) Founded on a fork of Kanaka Creek, Alleghany was named by miners from Pennsylvania. It is a Lenape expression defined as "people of the cave country," or "the fairest river." The town was settled in 1856 with a post office established the following year, and had a population of about 300. Communications included a telegraph office and wagon roads which connected Alleghany to Camptonville, Marysville, and Nevada City.

One of the richest mines in the district was the Sixteen-to-One, which brought out more than $25 million in gold and operated until the 1960s. It was one of the last active underground mines in the state. Operations ceased because, as the mining company's attorney said, "You cannot produce gold at a cost of $50 an ounce and sell it for $35 an ounce and stay in business forever."[37] The Oriental Mine just southwest of Alleghany produced more then $4 million.

Alleghany made the local headlines in 1910 when it got electricity: "This is to be a gala night at Alleghany.... Tonight the electric light will be turned on in the town for the first time permanently. The town will be brilliantly illuminated for the holidays. A number of extra lights have been strung along the street, which will resemble the 'great white way' of New York when the 'juice' is turned on."[38]

ALLISON DIGGINGS (Nevada) Located near Grass Valley, this site was named for a farmer. Gold was discovered about 1852 at a claim named the Phoenix Ledge, which six years later recovered more than $1.5 million. Rumor had it that in 1859 more than $42,000 had been extracted from an outcropping in a 12-day period. Total production from the surrounding mines was about $2.7 million.

ALPHA (Nevada) First discoveries of gold were made during 1850 by miners named Henderson and Rogers. A few years later Charles Phelps also located his claim, and the Alpha Mine began operating. Other prospectors who came seeking their wealth were averaging about $50–$70 per day, and one miner found a piece of gold worth about $750. The camp was settled a few miles southeast of Washington with a post office in 1855. During the 1860s hydraulic mining went into operation, and as a result the town was almost engulfed in water. The Alpha Mine produced more than $2 million.

ALTAVILLE (Calaveras) Established near Angels Camp, this site was given a Spanish name meaning "high." It had previously been called Forks of the Road, Winterton, and Cherokee Flat. The town prospered during the 1850s when a foundry was built to manufacture stamp mills and machinery for the outlying camps. The post office was established in 1904. Altaville was later absorbed by Angels Camp. At nearby Bald Hill some mining was conducted, but operations ceased by the 1940s.

This is the place where the infamous Calaveras Skull was found. In 1866 someone unearthed the skull about 130 feet down a mine shaft. Geologist Josiah D. Whitney maintained that the skull proved the existence of man in California during the Pliocene period, which has never been proven. During this time period an Angels Camp dentist discovered the skull from his office skeleton was missing.

ALVORD (San Bernardino) This site was named for Charles Alvord who hailed from New York and discovered "wire gold" in the 1860s. Wire gold consisted of thread-like forms that were curved or in a tangled mass of fine wire-like strands. Serious mining didn't really start until about 1885 when the Alvord Mine was located. Charles originally took a party of Mormons in search of a lost cliff of silver. When he was unable to find the spot, his party deserted him. Alvord's friend, Asahel Bennett, along with Lewis Manley, went back to look for Alvord and found him hungry and exhausted. Alvord later returned in his quest for the lost cliff, but as far as anyone knows he never returned. The site was located northeast of the town of Barstow.

AMADOR CITY (Amador) Settled just north of Sutter Creek, the town was named for Don Pedro Amador, a soldier of fortune with the Portola Expedition, or his Indian-fighter son, Jose Maria Amador. Development of the nearby mines gave rise to its rapid growth. During the 1850s three quartz mills were built and a post office opened in 1863. By 1881 there were three mills with 116 stamps. Mines in the region were the Bunker Hill that yielded $5 million, and the Amador which recovered more than $3 million in gold.

At a nearby gulch was the Minister's Claim, worked in the 1850s by the Reverend S.A. Davidson, a Baptist minister, and Lemeul Herbert and M.W. Glover, Methodist-Episcopal ministers. The Reverend Ashly was another preacher who tried his luck at mining because he wasn't having any success preaching to the miners. He failed miserably, as noted by one of his neighbors: "He did not have energy enough to dig a hole in a day big enough to bury a cat." So his friends raised enough money to send him back home "as a flower too frail for a new settlement."[39]

Amador is home to a number of historical buildings. One of the most notable is the Fleehart Building that dates back to the 1860s or earlier. It was the only structure to survive a fire in 1878 and now houses the Amador Whitney Museum. Also on display is an old arrastre that was dismantled and reassembled there in the 1960s. Fiesta Days is held on the second weekend in June, celebrating the gold rush history. For information, contact the Amador County Chamber of Commerce: (209) 223-0350. Sutter Creek nearby has an underground tour of the Sutter Gold Mine. Call (209) 736-2708, or (866) 762-2837 for times and directions.

AMALIE (Kern) This small mining camp was just north of Mojave, between Jawbone and Lone Tree Canyons. It had a post office in 1894 and served the Agua Cliente Mining District. Up until the early 1900s a stamp mill and arrastre were still operating. The principal mine was the Amalie, which produced $600,000 in gold.

AMERICAN BAR (Placer) This camp was established in the 1850s and located on the Middle Fork of the American River. In 1855 a 12-stamp mill was erected to serve the surrounding mines, which had an output of about $3 million. To transport the ore to the mill across the river, a wire rope was constructed in 1856. Andrew S. Hallidie, creator of this device, went on to became the builder of suspension bridges and developer of the Frisco cable car system. In 1858 the Bay State Mill was

Amador City (courtesy of the Amador County Archives, Amador, California).

erected and used new methods of amalgamating and saving gold.

AMERICAN RIVER (Placer/El Dorado/Sacramento) Beginning in the Sierra Nevada Mountains, the river contains three forks as it travels through Placer and El Dorado counties, then empties into the Sacramento River at Sacramento. Thousands of mining camps were settled along its banks. The river was named in 1837 by Governor Juan B. Alvarado. In 1855 the *Placer Herald* wrote, "The success of the river miners on the North Fork of the American River this season, has been beyond all expectation. From nearly every portion of the river, we hear of claims paying well.... Some companies are working their claims night and day, washing as much ground as they possibly can before the rainy season commences."[40]

ANDREWS CREEK (Shasta) This site was located on a branch of Clear Creek, just north of the camp of Igo. During the 1850s A.R. Andrews conducted mining there with little success. He was one of the signers of the California Constitution in 1879.

ANGELS CAMP (Calaveras) Henry Angel came from Rhode Island and started Angels Camp as a trading post in 1848. That same year prospectors arrived after hearing about the discovery of gold quartz. The post office was established in 1853. After the surface gold was exhausted, miners had to go to hard-rock mining. Tradition says Bennager Rasberry jammed a ramrod in his gun while hunting rabbits, and while attempting to remove the ramrod, the gun discharged, firing the rod into the ground. When he pulled the rod out, stuck on the end of it was a piece of quartz gold. He ultimately recovered more than $10,000 from the spot. The Gold Cliff mine located next to town yielded about $3 million. One of the largest mines in the area was the Utica, which took over some of the other claims and produced over $16.4 million in gold. Tents lined the streets that were full of potholes and ruts from wagons and stagecoaches. It wasn't until 1927 that the streets were paved and substantial buildings were constructed.

Bret Harte came to Angels Camp, and his experience in the surrounding gold camps prompted him to write the books *Luck of Roaring Camp* and *Outcasts of Poker Flat*. Samuel Clemens also visited from time to time. While there in 1865 he heard a story about a "jumpin' frog named Dan'l Webster." From this, he wrote his first successful short story under the name of Mark Twain, *The Celebrated Jumping Frog of Calaveras County*, which precipitated an annual frog-jumping contest.

Today Angels Camp continues to holds its annual Jumping Frog Jubilee. Information: Angels Camp at (209) 736-2561.

APPLEGATE (Placer) Discovery of quartz gold in the region gave rise to this small camp. It was named in the 1850s for Lisbon Applegate, with a post office in 1875. Quartz mining did not last long and the site was abandoned.

AQUEDUCT CITY (Amador) This town was established about 1850 with the discovery of placer gold; a post office opened in 1855. It was named for a 130-foot high aqueduct that carried water over a nearby ridge to the claims. By 1868 the mines were closed and the town was no more.

ARASTRAVILLE (Tuolumne) Named for the Mexican arrastre ("mining mill"), the camp developed about 1851 when a claim yielded between $85 and $100 a ton. A Mr. Williams came to Arastraville from Sweden after trying his luck at numerous mining sites with little success. He finally settled here and built an arrastre at his claim, but experienced only fair returns.

ARCATA (Humboldt) Founded as Union in 1850 by L.K. Wood, the name was changed to Arcata when the post office opened in 1860. It is a Wiyot word meaning "union," or "sunny spot." The Wiyot Indians were native inhabitants, but when Arcata became a supply station for packers and miners at the Klamath-Trinity mining camps, they were forced to move elsewhere. After the gold ran out, timber came into prominence. Arcata is located near Humboldt Bay, a deep-water harbor that encouraged a number of companies to start shipbuilding. Author Bret Harte got his start writing while living at Arcata. The town is also home to Humboldt State University.

ARGENTINE (Plumas) Argentine began in the late 1850s when a silver-bearing ledge was discovered. It was given the Latin name meaning "silver." Extensive gold hydraulic mining was conducted there in the 1860s, but its production history is not known.

ARGUS MOUNTAINS (Inyo) Gold was discovered in the mountains during the 1890s with the discovery of the Orondo and Ruth mines. The Ruth lode was the most productive, yielding about $750,000. A post office opened in 1897 to serve the mining companies whose operations continued into the 1930s. After 1939 there was some activity, with a total gold value of about $293,000.

ARKANSAS DAM AND BAR (Trinity) During the 1850s, the dam was built by men from Arkansas to divert the Trinity River in order to improve the nearby mining operations. The dam broke several times, and although costly, it was rebuilt because the mining claims along the bar were so rich.

ARRASTRE SPRING (San Bernardino) Located in the Avawatz Mountains, this was an old Mexican camp founded by miners in the 1860s. It was named for the arrastres (Spanish for "mining mill") used by the miners. There was a later attempt to revive activities, but not enough gold was found to make it worthwhile.

ARROWHEAD DISTRICT (San Bernardino) This region, located near the Providence Mountains, was worked in the 1880s mainly by Mexicans using arrastres. The Hidden Hill Mine was located in 1882, beginning a period of large-scale mining. But the boom busted because most of the claims were not profitable. In the early 1900s a pocket of gold was discovered valued at about $13,000.

ASHFORD (Inyo) In 1907 Harold, Henry and Louis Ashford discovered gold and named their claim the Golden Treasure Mine. However, it did not live up to its name, and after yielding a few tons of ore, the mine was shut down.

ASHLAND (Sacramento) First called Russville for a mining entrepreneur, Colonel Russ, the name was changed about 1860 to Ashland. Russ arrived about 1857 to search for gold, but the ore was so poor he abandoned the place. Ashland later became a station on the California Central Railroad between Folsom and Lincoln.

ATCHISON BAR (Yuba) This site was named for prospector J.H. Atchison. It was located on the North Fork of the Yuba River. There was not much ore taken out, although a few experienced some successful mining, but most left for better diggings.

AUBURN (Placer) Prior to the discovery of ore, the earliest non-native people to the region were hunters and fur trappers, followed by explorer John Fremont. In 1848 gold was discovered in a nearby ravine by a Frenchman named Claude Chana, along with his friend, Theodore Sigard. They recovered the ore using a *batea* (a large wooden bowl in which gold-bearing earth or crushed ore was washed, as in a pan, used by Spanish prospectors). Many of the claims were quite rich, and some of the miners were averaging between $800–$1,500 every day. One of the men was digging a ditch to carry water to his claim when the water flow stopped. He checked out the problem and noticed the water was running into a gopher hole. He ended up extracting about $40,000 from it. Other claims were yielding $2,000 a day. Just north of Auburn the Green Emigrant mine was discovered in 1864 and produced about $100,000. Total gold recovered from the district came to about $3 million.

Before becoming Auburn, the town was called many names: Rich Dry Diggings, North Fork Dry Diggings, and Woods Dry Diggings. The name was changed in 1849 for a town in New York. The post office opened in 1853. The town was centrally located and became a jumping off place for the remote gold-filled regions. Auburn experienced some fires, the most destructive occurring in 1855. It started in one of the Chinese houses causing the loss of the entire south side of town. After another fire in 1863, residents wised up and built their structures with brick and stone. The camp eventually became a center of trade and a major supply depot during construction of the Transcontinental Railroad. On December 23, 1867, heavy rains caused a nearby ravine to overflow, with the result that the water rushed through the town, lifting some of the buildings from their foundations, but Auburn was once again rebuilt. Its strength as a vigorous trading center assured Auburn's future.

Auburn has many historical buildings, one of which is the Hotel Auburn, built in the 1800s, and the Auburn Iron Works, built about 1865. The town celebrates an annual rodeo. For dates and directions, call the Auburn Chamber of Commerce: (530) 885-5616.

AURUM CITY (El Dorado) Located near the town of El Dorado, this camp was given the Latin name for "gold." Aurum's post office opened in 1852. There was some mining conducted and a 14-stamp quartz mill was built, which produced only fair results.

AVAWATZ MOUNTAINS (San Bernardino) These mountains are located in southeast Death Valley. Tradition says the Spanish were mining for gold there long before the arrival of other prospectors. Silver and gold claims were worked off and on during the 1850s. The mountains were named Ivawatch, which means "mountain sheep," by the Paiute Indians. The spelling was later changed to Avawatz. Indians who lived in the area resented white encroachment and had a saying to the effect that the Avawatz meant much gold for the white man, and much hell for the Indians.[41]

BACKBONE CREEK (Shasta) This creek takes its name from the mountain for its shape. The site was situated at the northeast corner of Shasta Lake. German miners found rich placers there, and soon after came the formation of the Backbone District. The largest producer was the Uncle Sam Mine, which realized more than $1 million.

BADGER HILL (Plumas) This site realized very little revenue, about $5 per day for each miner, although there were five drift companies in operation. Badger Hill was located about 15 miles northwest of Sawpit Flat, and named for the resident badgers.

BAGBY (Mariposa) Formerly known as Ridleys Ferry for ferry operator Thomas Ridley, the name was changed to Benton Mills to honor Missouri senator Thomas H. Benton, or perhaps the owner of a quartz mill that was established there. It was finally renamed Bagby for an owner of a local hotel, and the post office opened in 1897. The town served as a trading center for the surrounding mines and a number of stamp mills were erected. One of the top producers in the area was the Mountain King, which brought out more than $1 million in silver and gold. After the Yosemite Valley Railroad was built, Bagby became one of the stops for tours to Yosemite Valley. The site later disappeared under the backwaters of Lake McClure.

BAGDAD (San Bernardino) Bagdad was a station on the Santa Fe Railroad just east of Barstow. The camp was named for the Bagdad Mining and Milling Company, taken from the city in Iraq, Persian for "God's gift." It began its life as a trading and shipping point for the nearby gold mines with a post office in 1889. The settlement prospered quite well, and then the Bagdad-Chase Mines were developed at a place called Stedman in the early 1900s. The mills were handling more than 200 tons of ore a day from two of the producing mines, the War Eagle and Orange Blossom. Bagdad began its demise after a major fire burned down most of the buildings in 1918. The Bagdad-Chase Mine recovered more than $4 million in gold between 1903-04. Other lode mines were a little over $2 million.

BALAKLAVA HILL (Calaveras) This camp was located south of Vallecito. It was named after a seaport on the Black Sea, a Turkish word for "fish net." There was some drift and hydraulic mining conducted in the 1890s.

BALD HILL (Placer) Named for its lack of vegetation, Bald Hill was the scene of mining during 1856 when German miners discovered gold. They located a small claim and built an 8-stamp mill to crush the ore, which yielded about $28 per ton. Other prospectors who followed hit a rich lode and for a time were rewarded with about $5,000 a day.

BALLARAT (Inyo) This town was named for a mining district in Australia; its name is an Aboriginal expression for "camping (or resting) place." Located in the Panamint Valley, the camp was formed in about 1890 with a post office established in 1897. Ballarat served as a supply center for the nearby mines. During the early 1900s there were 60 stamp mills in operation. The Ratcliff Mine was the biggest producer, recovering between $300,000 and $1 million during the years 1897–1903. The town saw its demise sometime after 1904.

About three miles outside of Ballarat is a plaque erected in 1992 that reads "Now a ghost town, Ballarat served nearby mining camps from 1897 to 1917. They produced nearly a million in gold. The jail and a few adobe ruins remain.... On Sunday morning at 3 A.M., March 22, 1908, a car in the world's longest race, a Thomas Flyer, arrived in Ballarat. It won the New York to Paris race, covering 13,341 miles in 169 days. The car is now in Harrah's Museum in Reno."

BALLARDS HUMBUG (Amador) Although A.M. Ballard found a rich lead on a tributary to Sutter Creek in the 1850s, it did not produce much. Others followed Ballard, but only small-scale mining was conducted. The site was located south of Volcano.

BALLENA PLACERS (San Diego) Placers discovered near the town of Ramona were located in an old river bed about four miles long. Some mining was conducted in the 1890s, but the most active period was between 1906 and 1914. Because there was so little water to wash the gravel, not much gold was mined. The site was named Ballena, which is Spanish for "whale," taken from the mountain that resembled a humpbacked whale.

BALSAM FLAT (Sierra) Named for the balsam fir trees, the flat was located east of the town of Alleghany. Not much was taken out, although it was reported that one mining company was getting between 14–37 ounces a week. Most of the gold panned out in the 1860s.

BALTIC MINE (El Dorado) Situated near Grizzly Creek, the mine was named for the Baltic Sea in Northern Europe. A 10-stamp mill was erected to crush the ore and continued operating until about 1907, but there are no records of its production.

BALTIMORE RAVINE (Placer) Rich ore was discovered in the ravine during the 1850s. About 20 years later the gold began to pan out, until Austrian miners came upon a huge piece of quartz that contained gold and weighed 97 pounds. The site was located near the town of Ophir.

BANDERITA MINE (Mariposa) Development of the Banderita began about 1856, which was located at the confluence of the North Fork of the Merced River with Gentry Gulch. Banderita is Spanish for "little bandit." The mine yielded about $1.5 million in gold.

BANGOR (Butte) Located between North and South Honcut Creeks, Bangor was settled in 1855 by the Lumbert brothers, who established a general store. The post office opened two years later. Bangor is a transfer name from Maine, Irish for "place of pointed hills." It was 1857 when rich gravel deposits were found and the lucky miners were recovering about $100 a day.

BANNER (San Diego) Gold was discovered by a group of men while they were looking for wild grapes in 1870, and before long a number of mines were developed: Ready Relief, Hidden Treasure, and the most famous, the Golden Chariot. It was discovered in a canyon south of Banner and named in 1871 by George King who noticed a rock with a glint of yellow. Considered to be the richest gold strike in Southern California, the lode produced nearly $1 million within three years.

Banner City was established as a trading center and its post office opened in 1873. In 1872 a new road was built to the mines. But some of the claims were in such remote places that the only way supplies could be brought in was by sleds on ropes that were held back by heavy brakes to keep them from catapulting down the hillsides. By the 1880s the ore had played out and Banner ceased to exist. Total production in the district was about $4.5 million.

BANNERVILLE (Nevada) Named after the Star Spangled Banner Mine, the camp was located a few miles southeast of Nevada City. The mines realized more than $1 million.

BARNES BAR (Placer) Gold was discovered on the bar by G.A. Barnes in 1849 who later opened a general store to serve the miners and mining companies. The camp was located on the North Fork of the American River near Mineral Bar. When prospectors from Georgia brought their slaves with them to the diggings, the local prospectors objected, saying that no chattel could assume ownership to a mining claim. The majority ruled, and the men from Georgia folded their tents and left.

BARSTOW (San Bernardino) Barstow was named for Santa Fe Railroad president William Barstow Strong. Originally it was called Waterman Junction for state Governor Robert W. Waterman, who owned a nearby mine. It became a trading center when the mines at Calico opened in the 1880s, and the post office was established in 1886. One of the mines in the region was realizing anywhere from $20 to $200 a ton of gold. During the 1890s a stamp mill was established at Barstow. Today the town is a division point of the Santa Fe Railroad.

Barstow is located near a number of natural attractions: Rainbow Basin, Mojave National Preserve, Afton Canyon (called the Grand Canyon of the Mojave). It is also home to the Route 66 Mother Road Museum, which supports efforts to preserve the history of the road. Information: (760) 255-1890.

BARTONS BAR (Yuba) This camp was named in the 1850s for store owners John and Robert Barton, who served the miners in the region. It was located on the main Yuba River near Parks Bar. The diggings were fairly rich, but lasted just a few short years.

BASSETTS STATION (Sierra) Formerly known as Hancock House, the station was established near Sierra City during the 1860s. The name was changed to Howard Ranch for its owner, Howard C. Tegerman, then renamed Bassetts Station in the 1870s for owners Jacob and Mary Bassett. Never a gold mining camp, it was a place where prospectors were welcome and the owners often grubstaked them.

BATH (Placer) Established in the 1850s, the camp was situated near Forest Hill close to the Middle Fork of the American River. It was originally called Volcano, then Sarahsville. When the post office opened in 1858, the name was changed to Bath. It served the outlying mines, one of the most important being the Paragon. Its earnings were reported to be about $2.5 million. Bath began to see its demise in the 1870s when the town of Forest Hill increased in importance, and the post office closed its doors in 1899.

BEALS BAR (Placer) Located at the confluence of the North and South Forks of the American River, the bar was worked during 1849 by more than 25 men on shares during 1849, each share worth about $2,800. The claim was owned by Messrs. Beeroft, Baisley, Blinn, Kent and Small. When the military governor of the state, Colonel Richard B. Mason, came through touring the mines in 1848 he found a miner who had about 50 Indians working for him. They panned for gold using their Indian baskets and brought out about $16,000 in gold for the chap. It's not known if the prospector shared his wealth with any of the Indians.

BEAR MOUNTAINS (Calaveras) Supposedly the Bear Mountains were one of famous bandit Joaquin Murieta's hideouts. He was an elusive fellow, and tales told of his sightings at many gold camps have been suspect. The mountains are south of the Calaveras River. While a miner was hunting a grizzly bear in 1856, he accidentally came across a rich placer deposit by scratching out the gold with a hunting knife, immediately extracting about $10 in gold.

BEAR VALLEY (Mariposa) Bear Valley was originally part of explorer John C. Fremont's grant. Charles, David and Willard Hayden arrived in 1850 to what was initially called Haydensville, with establishment of a post office the following year. It was renamed Johnsonville in 1856 when the site was surveyed and named for John F. "Quartz" Johnson, then changed to Bear Valley in 1858. Mexican miners working their claims during the 1850s were driven out when American newcomers saw the amount of gold that was being taken out (about $250,000 worth). By 1852 most of the placers were depleted. But two years later a prospector came upon a piece of pure gold that weighed nearly three pounds. Fremont tried to have the boundaries changed on his land grant so it would include the region where all the gold had been discovered, and the Supreme Court upheld his claim. But problems continued to plague the decision, and finally Fremont sold more than 40,000 acres and was rid of it. Bear Valley later became a large trading center for local mining operations.

BEAR VALLEY (San Bernardino) Situated in the San Bernardino Mountains, the valley was discovered to have gold in the 1850s. Although a geologist noted there was a lot of gold there, most of the miners were receiving less than $5 a day. A 40-stamp mill that was built proved to be a failure, although there was some commercial mining. The building of a dam put a portion of the valley under water.

BELLEVILLE (San Bernardino) Most of the mining in this area was placer gold. The camp was named for Belle, the first child born there to a blacksmith named Jed Van Dusen. Some quartz mining was conducted in the 1870s, but the men were realizing only about 50 ounces of gold a week. By 1880 most of the people had left.

BEND CITY (Inyo) This short-lived camp was located along the Owens River near Independence and was one of the first settlements east of the Sierra Nevada Mountains. In 1862 a soldier from Camp Independence discovered gold, which brought in the San Carlos Mining & Exploration Company. When William Brewer with the Geological Survey came through in 1864 he wrote in his journal, "It is a miserable hole, of perhaps twenty or twenty-five adobe houses, built on the sand in the midst of the sagebrush, but there is a large city laid out — on paper."[42] By 1865 the site had disappeared. Had it not, the camp would have been demolished when an earthquake measuring 9.5 on the Richter scale hit the Owens Valley.

BENNETTVILLE (Mono) This hoped-for metropolis was located at the summit of Mono Pass along the old Mono Trail, and was established to serve the mines in Tioga Pass. The post office that opened in 1882 was named for Thomas Bennett, Jr., from Massachusetts, president of the Consolidated Silver Mining Company. The company dug a tunnel to get into solid granite where supposedly rich ore lay hidden, but financial woes forced operations to cease, probably because of the cost of transporting the machinery to the mines. Some of the equipment was hauled by pack mules. The rest that was too heavy for the animals had to be shipped via Reno through Lundy, then up the mountain by heavy sledges. More than $350,000 was spent building the town in 1883 and developing the Great Sierra Mine, but nothing was ever realized.

BENSONVILLE (Tuolumne) During the 1850s gold placers were doing well, with the miners getting about $10 a day in gold. Hydraulic mining also began during this time period, but production yields are unknown. Bensonville was settled just below the town of Columbia.

BENTON (Mono) First called Hot Springs for its mineral springs nearby, the camp lived its life as a supply and milling center for the mining districts. It was established in 1856 near the California-Nevada border and the post office opened in 1866. The name was changed to honor Missouri senator Thomas H. Benton. Although there was some gold, the majority of the ore was silver. Between 1862 and 1888 more than $4 million was taken from the nearby mines, most of it from the Blind Spring Hill. The Bodie & Benton Railroad was projected to be built through there in the 1880s which would have been a boon to the town, but track grading stopped more than 15 miles away. Benton was then

Benton Trading Post, ca. 1915 (photograph courtesy of the Mono County Historical Society, Bridgeport, CA).

relegated as a way station when the mining town of Aurora in Nevada began to grow.

BERRY CREEK BAR (Butte) Named for settler Henry Berry, this camp was established with a post office in 1875. Mining operations consisted of a 15-stamp mill and one arrastre during the early 1900s. The camp was located at the confluence of Berry Creek and the North Fork of the Feather River. Some mining continued into the 1940s.

BESTVILLE (Siskiyou) Bestville was named in 1851 for seaman and trader Captain John Best, and was located along the North Fork of the Salmon River. An Indian named Squirrel Jim showed the miners where he had found gold, and by 1855 the men were earning about $8 a bucket. Within a few years the camp was deserted when the gold declined and Sawyers Bar farther north was found to be richer.

BEVERIDGE (Inyo) In 1877 W.L. Hunter discovered gold in the region and the camp was settled in the Inyo Mountains. It was named for miner John Beveridge, who had also found gold in the Cerro Gordo area. The largest producer in the district was the Keynot Mine, which yielded about $500,000.

BIDWELL BAR (Butte) A friend of John Sutter, John Bidwell set off to the Feather River searching for gold, which he found on the Middle Fork of the river in March of 1848. Seven years earlier, Bidwell had been a schoolteacher in Iowa and Ohio. Upon his arrival in California he bought a ranch and became an agriculturist. While at the bar Bidwell made friends with the local Indians who helped him dig for gold, and in return were paid with beans. One of the Indian elders commented, that the white men were foolish to trade beans for a "handful of yellow rocks" as the rocks were plentiful and, unlike beans, weren't much good as a food source.[43] The population at the bar grew to about 3,000, warranting a post office that was established July 10, 1851. A suspension bridge of sorts was built across the Feather River at the bar. A boat was linked to each shore with a 12-inch plank so the miners could go back and forth to their claims.

The Camp brothers moved to the bar and set up a store loaded with more than 15,000 pounds of provisions for the prospectors. As with most of the camps, the cost of goods was astronomical. A jar of pickles and a few sweet potatoes cost up to $11, while a needle and spool of thread went for $7.50. A hard-to-get item was cranberry sauce, which went for $16 a can. One enterprising storekeeper overstepped his bounds. It was discovered that he was deceiving his customers by selling a salve for butter.

Tradition says that in 1853 a little boy named Elisha Brooks paid a dollar for an orange that had been brought in from Tahiti. He paid for the fruit with the gold he picked up from wagon ruts. Elisha planted the orange seeds in his parents' yard, and it may have been the first orange tree grown in California. Others believe a seed planted in Sacramento was transferred to the bar.

BIG BAR (Amador, Calaveras) Once known as Upper Ferry for a ferry that operated there in 1848, the site was located near Mokelumne Hill. The bar was first worked by James Martin and his party of eight men. In 1850 a miner found a piece of quartz that contained eight pounds of gold. By 1856 a 6-stamp mill and two arrastres were in operation. Gold production is not known.

BIG BAR (Trinity) A miner named Jones found gold here in 1849, and a year later more than 500 men were conducting placer mining, making between $25–$50 per day. Its post office opened in 1851. Named for its size, the bar was located along the Trinity River.

BIG BEND (Butte) This camp received its name because of its location on a bend of the North Fork of the Feather River, and was the scene of scores of mining claims. An attempt to divert water for mining operations by building a tunnel in the 1880s met with no success, so the prospectors left for greener pastures.

BIG CANYON (El Dorado) Located south of Shingle Springs, Big Canyon saw its first miners in 1851 with the arrival of Lucius Fairchild and Samuel F. Pond. Fairchild came out empty-handed, but Pond managed to get a little more than $5 a day in gold. The Big Canyon Mine was later discovered and during the 1880s realized more than $1 million in gold. When the mine was taken over by the Mountain Copper Company, more riches were recovered, amounting to more than $2.3 million by the 1940s.

BIG CANYON (Trinity) In 1850 prospector David Leeper and his friends settled at Big Canyon, which was located along the Trinity River. While getting water from the stream, Leeper took out $80 worth of gold. A few years later the Eagle Flume Company moved in, but it was averaging only about $30 a day for each miner. Big Canyon was not so big in gold, and was soon deserted.

BIG CANYON (Tuolumne) This canyon was also called Big Creek, and was tributary to the Tuolumne River near the town of Groveland. Miners experienced only fair success there. The only mine of any importance was the Hunter that reported gold-bearing quartz worth $300 a ton.

BIG CREVICE (Placer/El Dorado) This "crevice" crosses the Middle Fork of the American River and extends into El Dorado County. J.D. Galbraith was the first to find rich deposits, but they were difficult to work because the ore was on the bottom of an ancient riverbed, covered by a stratum of soapy sediment. In 1878 the claim was taken over by the American River Dredging Company.

BIG OAK FLAT (Tuolumne) In 1846 James D. Savage, who was one of Sutter's men, opened a trading post in the area, and was also the first to prospect the flat. This camp was located southwest of Groveland. Originally called Savage Diggings, it was renamed for a huge oak tree that had a base measuring about 11 feet in diameter. The post office was established in 1852. There were a number of mines in the region, and it was reported there was more than $85 a ton in free gold. The shiny ore was everywhere. A blacksmith named James Mecartea had no time for prospecting because he was too busy making implements for the miners. When his eleven children asked for spending money, he simply told them to pan for gold on the cellar floor.[44]

The old oak tree has its own history. Tradition says that miners tried to pan for gold around the tree. Attempts were made to protect it by town ordinance, but it eventually died. Another story says the tree was burned during an 1863 fire that just about destroyed the camp. Someone later put a monument near the grand old oak.

BIRCHVILLE (Nevada) First known as Johnson's Diggings for miner David Johnson, the name was changed in 1853 for settler L. Birch Adsit. The camp was located northwest of Nevada City on a gravel channel and settled in 1851 when gold was discovered in the vicinity. It wasn't until about 1857 that the camp saw any growth with the advent of hydraulic mining. Between the years 1865 and 1866 profits amounted to only about $325,000.

BIRD FLAT (Placer) Located high in the mountains near Iowa Hill, the site was named for store owner Mr. Bird. In 1850 news of rich discoveries brought in the miners, and the camp was filled with more than 2,000 hopefuls. They stayed near Bird's store when winter snows kept them from going to the mountains and canyons for gold. During 1854 a tunnel was built by the Bird Flat Company, but it did not reach the quartz ore until about 1866 at a cost of more than $50,000. Reports showed that in 1855 someone found a 35-ounce piece of gold worth a little over $700.

BIRDS VALLEY (Placer) Southwest of Michigan Bluffs, Birds Valley was found to contain gold about 1849. Although there were about 3,000 miners working their claims, it was a short-lived camp because the ore was not as plentiful as first believed.

BLACK BEAR (Siskiyou) Gold was discovered in the region between 1860 and 1865. Located on Black Bear Gulch, which is tributary to a fork of the Salmon River, the camp became a trading and supply center for the miners in the outlying claims. Its post office was established in 1869. The Black Bear Mine was one of the most productive lodes in the area. A local newspaper wrote a favorable report: "The stockholders in the 'Black bear' lead are erecting one of the most substantial and fine quarts [*sic*] mills in the State, and will have in full operation by the middle of September."[45] Machinery for a 16-stamp mill was brought in by pack trains from Trinidad. The property recovered more than $3 million, and continued operating until the 1940s.

BLOODY CANYON (Mono) The canyon takes its name from the red metamorphic rocks, and was situated near the old Mono Trail across the Sierra Nevada. Geological Survey member William Brewer suggested another origin for the name when he described the trail as "utterly inaccessible to horses, yet pack trains came down, but the bones of several horses or mules and the stench of another told that all had not passed safely. The trail comes down three thousand feet in less than four miles, over rocks and loose stones, in narrow canyons and along by precipices. It was a bold man who first took a horse up there. The horses were so cut by sharp rocks that they named it 'Bloody Canyon.'"[46] While the Army was chasing Indians from the re-

Loading ore cars at Black Bear, 1880s (photograph courtesy of the Siskiyou County Museum, Yreka, CA, PO4124-77).

gion in 1852 they found gold, but not much was done until 1879 when the Great Sierra Mining Company began operations.

BLOOMER BAR (Butte) While miners were building a water ditch for their claim in 1861, they accidentally dug into a vein, and much to their surprise got almost $20,000 out of it. The bar was located along the North Fork of the Feather River. In 1868 the Bloomer Mine began operating and was served by a 15-stamp mill that processed the ore until the early 1900s.

BLUE MOUNTAIN CITY (Calaveras) This town was established in the 1860s on the South Fork of the Mokelumne River, and served as a commerce center for the nearby gold and silver mines. The post office was established in 1865. More than ten stamp mills were operating during the 1870s and continued until the start of World War II. The name was descriptive of the mountain's color during certain atmospheric conditions.

BLUE TENT (Nevada) Established about 1850, Blue Tent was northeast of Nevada City with a post office in 1873. The Blue Tent Mining and Water Company operated on a large scale. In 1876 a 30-mile-long ditch was completed at a cost of $160,000. Hydraulic mining was conducted there, with more than $750,000 realized. Operations continued into the 1890s.

BODFISH (Kern) Originally called Vaughn, this camp began as a stage stop with a tavern in 1864 along the Kern River. It was renamed for store owner and miner George Bodfish who came from Massachusetts. The post office opened in 1892. Bodfish, who owned a mine in Keyesville, set up a stamp mill that was later destroyed. Near Bodfish Creek some mining was conducted in 1888. Total production in gold was estimated at about $700,000.

BODIE (Mono) Bodie was founded in 1859 near the town of Bridgeport after gold was found by a Dutchman named Waterman S. Body and his partner, E.S. "Black" Taylor. Body later lost his life in a snowstorm, and his name was given to the camp. The town received its post office in 1877. The town was filled with miners from the Mother Lode, which by that time had been saturated. J. Ross Browne, correspondent for *Harper's Monthly*, wrote of his visit to Bodie: "This was the famous Bodie Bluff. The entire hill as well as the surrounding country, is destitute of vegetation, with the exception of sage-brush and bunch grass — presenting even to the eye of a traveler ... a wonderfully refreshing picture of desolation."[47]

It was a wild town that boasted more than 65 saloons with countless shootings. In 1881 the *Bodie Daily Free Press* wrote, "Bodie is becoming a quiet summer resort — no one was killed here last week."[48] The great bonanza near Bodie did not occur until 1874. During the mining boom there were about 50 claims and ten stamp mills in operation. The most productive lodes were the Standard Consolidated Mines, which yielded an estimated $18 million, and the Fortuna that paid more than $5 million in dividends. By 1888 only three claims were being worked. Total gold production from the area was roughly $29 million.

Today Bodie is preserved in what is called arrested decay. Much of the town was destroyed by fire, but there are many good remains. The site is northeast of Yosemite, 13 miles east of Highway 395 on Bodie Road, a somewhat maintained dirt road. Bring your own food because there are no services or facilities except for restrooms. Bodie is a National Historic Site and State Historic Park.

BONDURANT (Mariposa) This camp was named for the first county judge, James A. Bondurant, and established near the North Fork of the Merced River. Gold was discovered about 1858 and the Bondurant Mine was located. Within a few years a 10-stamp mill was erected. The mine was yielding about $500 a ton in gold, operating off and on until the 1940s, and had a total production of about $390,000.

BONDVILLE (Mariposa) Bondville was established near Bagby along the Merced River. It was named for store owner Stephen Bond, who was also its postmaster when the office opened in 1855. Most of the mining conducted was placer, but it was very short-lived.

BOSTON RAVINE (Nevada) A group of men from Boston discovered gold in the ravine during 1849. The small camp had a post office established in 1889. More than $3 million was realized from the nearby mines. The site later became part of Grass Valley.

BOTTLE HILL (El Dorado) Settled north of Georgetown in the Hornblende Mountains during 1851, legend has it the camp received its name after miners found a full bottle of whiskey while search-

Preserved ghost town of Bodie (courtesy Kevin Leach).

ing for gold, and of course drank it. That's doubtful since no miner would have left a full bottle of liquor behind anywhere. It may have actually been for a nearby bottle-shaped hill. The post office opened in 1855 with Samuel M. Jamieson as postmaster. Prospectors had to conduct dry digging until water was brought in. Yields are unknown.

BOUQUET CANYON (Los Angeles) The canyon experienced some small-scale mining during the 1880s when claims were found near the town of Saugus. Then in 1933 the St. Francis Dam burst, eroding part of the canyon and revealing gold, and brought on another minor rush. A little more than $2 million in gold was recovered. The canyon was named for its millions of wildflowers.

BOX (Plumas) Originally named Letter Box (presumably for the post office established in 1896), the camp was between the North and Middle forks of the Feather River. It was established for the purpose of serving the local mines.

BRANDY CITY (Sierra) Gold discovered at nearby Cherokee Creek (tributary to the North Fork of the Yuba River) about 1850 brought about formation of this camp. Originally called Strychnine City, the name was changed when a post office opened in 1909. The mines at Canyon Creek were supplied by water from Brandy City via a flume. During the 1880s, hydraulic mining was conducted, but the anti-debris law put a stop to operations.

BRASS WIRE BAR (Nevada) Discovered in the 1850s, the bar was worked mainly by Chinese miners. It was located on a fork of the Yuba River near the town of Washington. Only about $50,000 worth of gold was taken out.

BREYFOGLE CANYON (Inyo) This canyon in Death Valley was named for Charles C. Breyfogle who claimed he found gold in the region. After being captured and held by the Indians for some time, he could not remember where the deposit was located.

BRISTOL MOUNTAINS (San Bernardino) It was in 1897 that the first gold was found by Hicorum, a Chemehuevi Indian. The place was located northeast of the community of Amboy and settled in the early 1900s. Some mining was conducted when the Orange Blossom and the Great Gold Belt mines were opened. Operations continued until the 1930s.

BROWN'S FLAT (Tuolumne) This camp was settled in the early 1850s along Woods Creek near the town of Columbia. Early on, one of the miners discovered a three-pound piece of gold. The Placerville *Courier* wrote, "And the little camp of Brown's Flat was in a positive fever of excitement." The *Union* also noted, "The old Italian led there, which has yielded so immensely during the past year, now bids fair to surpass anything heretofore found. On Saturday last they took out about $25,000 there and have taken out a larger amount since." [49] The mines in the area brought out a total of $4.5 million.

BROWNS HILL (Nevada) Giles S. Brown and his brother discovered gold on the hill during the 1850s, and within one year they recovered more than $30,000. By 1867 five mills with 42 stamps were operating. Later hydraulic operations pretty much decimated the hill.

BROWNS VALLEY (Yuba) About 1850 a miner named John E. Brown found gold by accident in a boulder situated near his campsite. After realizing more than $11,000, Brown figured he had enough and quit mining. Located northeast of Marysville, Browns Valley received its post office in 1864. It became a rich mining center, but by the 1890s most of the ore was gone. The district recovered roughly $18 million in gold.

Historian Hubert H. Bancroft's father came to the region and worked at a claim. Arriving at the mine for the first time, Bancroft wrote of his experience: "I found my father, in connection with other members of the Plymouth association, busily engaged in working this mine. He occupied a little cloth house in the vicinity of the ledge, and being the owner of a good mule team, he employed himself in hauling rock from the mine to the mill, about one mile apart.... The first night I spent with him in the hotel at Long Bar. Foremost among my recollections of the place are the fleas, which, together with the loud snorings and abominable smells proceeding from the great hairy unwashed strewed about on bunks, benches, tables, and floor, so disturbed my sleep that I rose and went out to select a soft place on the hill-side above the camp, where I rolled myself in a blanket and passed the night, my first in the open air of California."[50] After a few months of hard work with little reward, Bancroft's father returned to his home in Ohio.

BRUSH CREEK (Nevada) Brush Creek was settled on a tributary of the Yuba River, northwest of Nevada City. Mining began in 1850, but it wasn't until about four years later that a rich vein was found, which produced about $3 million.

BRUSH CREEK (Sierra) Taking its name from the creek, the camp was established south of Goodyears Bar. One lucky miner discovered a gold nugget valued at about $1,700. The Brush Creek Mine was discovered, followed in 1868 by the erection of a 10-stamp mill. Owners recovered about $1 million from the property during the 1870s. Total production of the region was approximately $4 million.

BRUSHVILLE (Calaveras) This little camp had a few small claims, but the miners were only earning about $7 a day in gold. The site lay south of the Calaveras River near Quail Hill.

BUCK (Plumas) Named for Horace "Buck" Bucklin, who owned Buck's Ranch, the camp served as a trading center for the outlying mines, and received a post office in 1868. It was located along Bucks Creek, a tributary of the Feather River. During the 1920s Buck was inundated by a reservoir built by the Feather River Power Company.

BUCKEYE (Shasta) Located just north of the town of Redding, Buckeye was the scene of rich quartz veins discovered during the 1850s by a Mr. Johnson. It received a post office in 1880. The diggings reported a yield of about $20 per day.

BUCKEYE HILL (Nevada) This camp was settled near Quaker Hill and Greenhorn Creek. Early prospectors were Eli Miller and Anthony Chabot. During 1859 a French gold miner named Eugene Caillaud decided he could make more money operating a store than busting his chops panning for gold. Chabot devised a method of washing loose gravel from the diggings that later became known as hydraulic mining. A man named Matteson later went to American Hill and improved on Chabot's method, no doubt taking credit for it. By 1879 most of the diggings had been exhausted, leaving the Chinese to glean what was left.

BUCKNER'S BAR (Placer) In 1847 Thomas M. Buckner moved from Oregon and settled along the bar where a camp was formed on a fork of the American River. Although the placers were shallow, they paid well. A group of men made plans to build a canal to bring water to the diggings. One of the rules was that if any shareholder got drunk while on duty he would have pay into the common treasury one ounce of gold dust and forfeit his dividends during that time. But a fellow named Porter Rockwell continued coming into the camp with his mule stocked up with whiskey. So many of the men became subject to those fines and forfeitures that the canal was never built.

BUELL'S FLAT (Siskiyou) William M. Buell arrived at the flat in the 1850s and opened a trading post to serve the miners. By the 1860s the gold had been depleted, and Buell's Flat was on its way out. It was located along a fork of the Salmon River.

BUENA VISTA (Amador) Just south of Ione, this camp bears a Spanish name for "good view." Serving as a supply center for the miners between 1848 and 1868, the place deteriorated after the gold ran out. A post office established in 1868 lasted another ten years.

BULL DIGGINGS (El Dorado) Named for a pair of bulls that were used to haul the gravel, the site was located on a ridge north of Mameluke Hill. Prospectors Edward and Greenfield Jones found ore in 1851, and brought out about $20,000.

BULLARDS BAR (Yuba) A Dr. Bullard ended up at the bar after a ship he was on headed to Hawaii was wrecked off the California coast. He and three other men dammed up part of the North Fork of the Yuba River and began placer mining. They acquired about $15,000 in gold in less than two months. The post office opened in 1866 and closed in 1914. The bar is now covered by Bullard's Reservoir.

BUNKER HILL (Sierra) During the 1850s a tunnel was dug through Bunker Hill where miners were recovering about $10 in their pans a day, which brought in scores of other prospectors in their quest. The camp was located near the town of Portwine. One of the deposits was yielding about $100,000 a year.

BURNS (Mariposa) In 1847 John and Robert Burns traveled from the East to seek their fortunes. They purchased a ranch and later discovered gold on the land. Until the 1940s, mainly placer mining was conducted. In 1949 dredging operations began and by 1951 more than $650,000 in gold had been processed. The site was located southwest of the town of Hornitos.

BUSTERS GULCH (Calaveras) Folklore says a black man named Buster found more than $40,000 in gold and buried it somewhere nearby. The site was along San Antonio Creek southeast of San Andreas.

CABBAGE PATCH (Yuba) Located on Dry Creek (a tributary of the Bear River) near the Nevada stateline, most of the mining operations were quite small and were active during the 1870s. The site was so named because two men attempted to cultivate a large cabbage patch. When the post office was established in 1898, the name was changed to Waldo.

CALAVERAS RIVER (Calaveras) The river flows into the San Joaquin River near Stockton. It was discovered by the Spaniards and initially called Rio de la Pasion ("river of the Passion") by Alferez Gabriel Moraga and Father Pedro Muñoz during their exploring expedition to the San Joaquin Valley in 1806. The name was changed to Calaveras (Spanish for "skulls") after numerous skulls were found, relics of a battle between the Indians. Gold placers along the river were worked using the hydraulic method until legislation made it illegal, then drift mining began. Total gold production was more than $2 million.

CALDWELLS GARDEN (Tuolumne) A Mr. Caldwell discovered gold there, taking out more than $150,000 before he sold his claim. During 1856, the new owners recovered another $15,000 within a few weeks. The site was located near Shaw's Flat.

CALICO (San Bernardino) The town took its name from the nearby calico-colored bluffs in shades of green and rose. Both silver and gold were found in the region during the 1880s. John McBride, Charlie Meacham, and Larry Silvia were thought to be the first to find ore deposits. By 1881 Calico was pretty well established, and its post office opened the next year. Many of the miners lived in the nearby dugouts because it was a lot cooler inside, since it could get into the triple digits during the summer. The mining district owned about 60 lodes, mainly silver. Between 1881 and 1907 more than $83 million in silver had been realized.

Miners' dugouts at Calico (courtesy Francis Nestor).

One of the old-time miners, Howard Kegley, wrote: "Walter Knott may not know it, but I was in on the earliest diggin's at Calico. In fact I and Slug Dawson mined the first yard of Calico out of the West Expansion, then known as Dontgiveacuss Mine."[51]

Dorsey the dog became famous during the early 1900s as a mail carrier. Dorsey was a collie that just wandered into town one day. The postmaster, Mr. Stacy, was the first to spot him. He fed the dog and then left the animal to his own devices, but Dorsey wouldn't leave. Over a period of months Dorsey followed Stacy as he delivered the mail to East Calico seven miles away. When Mr. Stacy got sick and was unable to take the mail, someone suggested that Dorsey do it. So he rubbed the mail across the dog's nose and pointed to East Calico. Off he went and back in record time. From then on, Dorsey delivered that mail and even had a special harness to carry the letters.[52]

Calico is still a living ghost town that was restored by Knott's Berry Farm and is a mecca for tourists. The cemetery has some most interesting epitaphs, e.g., "Here lies Zackera Peas in the shade under the trees. Peas isn't here, only his pod. Peas shelled out and went home to God." Calico has a number of events during the year: The Spring Festival (Mother's Day weekend); Annual Calico Days with burro races (Columbus Day weekend); and Heritage Fest (Thanksgiving weekend). Information: 1-800-TO-CALICO.

CALIENTE (Kern) This camp took its name, Spanish for "hot," from nearby Agua Caliente Creek. It was established as a trading center for the mines to the east. Goods were transferred from wagons, loaded onto pack horses and carried up the trails to the Kern River gold camps. A post office opened about 1875. The site later became a station on the Southern Pacific Railroad.

CALLAHAN (Siskiyou) This site was named for the owner of a hotel and store, M.B. Callahan, who built his cabin there and fed travelers for a dime. After gold was found nearby in 1851, Callahan made much of his earnings furnishing meat to the miners. The post office was established in 1858. Callahan later became a trading center and supply point. Placer mining was confined to an old gravel channel and bars of the South Fork of Scott River. Two of the productive mines were the Montezuma River claim, which brought out more than $50,000 a year, and the Beaudry Mine, yielding between $25,000 and $40,000 annually. Two other nearby mines, the Black Bear and the McKeen, recovered $3.1 million and $250,000, respectively. The camp was located at the junction of the East and South forks of the Scott River. The hotel continues to serve modern motorists.

CAMANCHE (Calaveras) This is a transfer name from Iowa, so called for the Comanche Indians, whose name means "distinctive," or "different." It was also interpreted as a Ute word for "enemy." The site was originally called Limerick, Gaelic for "bare land." Its post office began operations in 1864. The camp was about a mile and a half south of the Mokelumne River. Not much mining occurred until water was brought in during 1868. Later, dredging operations began. Estimated production was between $5 million and $20 million. The site was later inundated by the Camanche Reservoir.

CAMPO SECO (Calaveras) Bearing a Spanish name for "dry field (or camp)" because of the dry diggings, Campo Seco was settled in Oregon Gulch near Mokelumne Hill. It had its beginnings in 1850, with a post office established four years later. During 1853 someone discovered lumps of pure gold worth more than $900, and the following year a nugget weighing 93 ounces was found. Small-scale mining continued until about 1856, when water was provided by the Mokelumne Water Company. By 1899 more than $4 million had been realized. Between 1899 and 1919 another $1 million was recovered.

CAMPTONVILLE (Yuba) Camptonville was named for blacksmith Robert Campton, and got its start in 1850, acquiring a post office in 1854. The town was located between the North and Middle forks of the Yuba River. Between 1850 and 1851 placer gold was discovered on a ridge near Downieville and Nevada City. When a rich lode was located in 1852, it brought in a rush of miners. The Sierra Mountain Water Company furnished water for hydraulics about 1858, and the first shipment of gold came to more than $5,000. In 1862 Brewer called the place "a miserable, dilapidated town" in reference to the immense hydraulic diggings. "Bluffs sixty to a hundred feet thick have been washed away for hundreds of acres together."[53]

CANTON (El Dorado) Canton was established near Coloma and named for the Chinese who lived there. Its name was interpreted as "large, broad territory." A prospector named Stephen Wing arrived at Canton about 1853, accompanied by a tax collector. The Chinese refused to pay, so the tax man seized much of their belongings and sold them. Wing later conducted some mining, but with little success.

CANYON (El Dorado) Taking its name from Big Canyon Creek (tributary to the Cosumnes River), this site was just south of Shingle Springs. The post office was established in 1897, with William A. Green, postmaster. A miner named Samuel F. Pond tried his luck in 1850, but left empty-handed. Canyon mainly served as a trading and supply center for the outlying mines.

CANYON CREEK (El Dorado) This camp was named for the creek, which flows into the Middle Fork of the American River. There was some successful mining in 1848. The Gold Bug Mine opened in 1888 and yielded more than $1.5 million, continuing operations until 1984.

CARGO MUCHACHO MOUNTAINS (Imperial) These mountains are located about ten miles from the Colorado River where gold was found by the Spanish during the 1600s. They were followed by Father Francisco Garcis who reported rich-looking ore in the 1770s. Later, Mexican miners worked the diggings before the Americans arrived. The name is Spanish for "loaded boy," or "errand boy," which described gold seekers with their pockets filled with nuggets. About 1862 a railroad worker named Peter Walters discovered gold, and when word got out the mountain was besieged by prospectors. One of Walters' mines, the Black Butte, was estimated to be worth $130 a ton. His other claims were valued between $9,000 and $12,000 a ton. During the 1870s mining companies came in and built numerous stamp mills, which crushed ore yielding more than $4 million in gold.

CARIBOU (Plumas) Caribou was settled along the North Fork of the Feather River in the 1870s, but its post office didn't open until 1922. Locals called the place Cap Orsburne's Ravine after a sea captain who found his riches there. Some successful mining was conducted. Caribou is a transfer name from the Yukon, derived from the Micmac word for "scratchers," descriptive of the animals' habit of scratching the snow with their hooves to reach the grass underneath.

CAROLINE DIGGINGS (Placer) When gold was discovered in the 1850s, this little camp was formed as Last Chance, then renamed Caroline for a local young lady. The diggings were quite rich, with the miners taking out between $1,500 and $2,000 per week.

CARRVILLE (Trinity) Located along the Trinity River, Carrville served as a trading and supply center for the nearby mines during the 1850s. Its post office was established in 1882. The camp was named for landowner James E. Carr. One of the biggest producers in the area was the Bonanza King Mine, which realized more than $1.2 million in gold.

CARSON (Mariposa) Gold was discovered a few miles from the town of Mariposa in 1849 by Alex Goday, who formed the camp and named it for frontiersman Kit Carson. Between 1850 and 1854, prospectors were getting out only about $15 to $20 a day.

CARSON HILL (Calaveras) Just south of Angels Camp, this site was established on the slope of a hill. It was thought Mexican miners discovered gold there early in 1848. That same year a U.S. artillery sergeant on furlough, James H. Carson, found gold after an Indian guided him to the site. In less than ten days Carson and his partners took out about 180 ounces of gold. When William Morgan and John Hance founded the Union Company in 1850, they hit a bonanza of quartz ore. The first piece taken out weighed 112 pounds, loaded with gold worth about $10,000. The riches were astounding. In less than two years more than $2.8 million had been realized. Four years later a 195-pound gold rock was found, valued at more than $43,000. Total net profits from the mines around Carson were estimated at $26 million.

At nearby Carson Creek gold was discovered in a most unusual way. One of the miners had died and his friends were taking him to a spot to be buried. The minister began the service, but tended to be overlong in his monologue. Out of boredom, the kneeling miners began to scratch the dirt that had been dug for the grave. The minister looked down at the ground and hollered, "Gold! Gold!—and the richest kind of diggings! The congregation is dismissed!" The dead miner was removed from his grave to be buried somewhere else. The preacher and the rest of the men didn't hesitate long before staking their claims.[54]

CAT CAMP (Calaveras) This place was near the camp of Camanche and named for trading post owner Samuel Catts. During the 1860s miners claimed their diggings with hopes that a water ditch would be built, but it never occurred, so they left. When the Sierra Railroad built its tracks through in 1883, the name was changed to Wallace.

CAVE CITY (Calaveras) Cave City was named for the limestone caves that were purportedly used by Indians as a jail. The camp came into existence about 1850 east of the town of San Andreas. Gold was found in the region about 1852, but the ore ran out about 1867, and Cave City ceased to exist.

CECILVILLE (Siskiyou) Named for settler John B. Sissel, the camp was established 30 miles from Callahan along a fork of the Salmon River in the 1840s, but the post office didn't open until 1879. It served mainly as a trading and supply center for a number of quartz and hydraulic mines in the area.

CERRO GORDO (Inyo) Bearing a Spanish name for "big (or fat) hill," Cerro Gordo was founded near the town of Keeler in the Inyo Mountains at an elevation of 8,500 feet. Pablo Flores and two of his prospector friends made the first discovery about 1865. It was also thought that a packer named Savariano, who worked for a Comstock Mine owner, founded Cerro Gordo after absconding with his boss's bullion. Although gold was found there, the main production was silver.

In 1868 mining engineers Mortimer Belshaw and Victory Beaudry created the Union Mining Company and built a smelter. Because of the site's elevation, machinery had to be brought up with a block-and-tackle-type contraption. A freighter named Remi Nadeau hauled bullion down a rough grade to Los Angeles. In order for the wagons to make it down, the wheels were chained together so the wagons could slide down slowly held back by a span of mules tied to the back. There was so much bullion that it had to be piled up like cordwood to await shipment. By 1880 most of the ore had been depleted, but not before recovering more than $17 million.

CHAMISAL (Mariposa) Named for the dense chapparal, this small camp was formed after placer deposits were found during the 1850s along Temperance Creek. Later, hard rock mining was conducted. The site was inundated by Lake McClure.

CHEROKEE (Butte) Just north of Oroville, gold

was discovered about 1849 near the Feather River. Water brought in via the Grizzly Ditch in 1852 gave rise to a settlement with a post office established in 1854. Prosperity came to the region when more water was brought in from the Concow Valley during 1870, and by 1886 more than $15 million had been recovered from the mines.

This camp was the site of one of the largest hydraulic mines covering more than 25,000 acres. About 1,000 miners lived at Cherokee, named for the Indians who were led there by a New England schoolmaster. In 1863 about 300 diamonds were discovered, most of which were of industrial quality.

CHICKEN RANCH (Tuolumne) The camp was established on a slope of Table Mountain on the Chicken Ranch Indian Reservation. A mining company came in about 1855 and struck gold, then dug two tunnels at a cost of $20,000. Unfortunately, no more gold was located and the site was abandoned.

CHILI CAMP (Tuolumne) This camp was started about 1851 near Campo Seco by a prospector named Daniel B. Woods. The placers were quite rich, and nearby Poverty Hill brought out about $4 million in gold during the 1890s.

CHILI GULCH (Calaveras) The gulch is near San Andreas and extends to the southern end of Mokelumne. Originally called Chilean Gulch, at some point in time it was changed to just Chili, named for prospectors from Chile, South America. The post office opened in 1857. When a Dr. Concha brought in Chilean miners during 1850 who were treated as slaves, it precipitated what was called the "Chilean War." The Americans protested because a law had been passed that no claims could be owned in the name of slaves. The Chileans were expelled from the diggings, which resulted in a war between them and the American miners. Most of the foreigners were put on trial, some of them were hanged, others had their ears cut off, while the rest were flogged and run out of town.

CHINESE CAMP (Tuolumne) This was the oldest camp established by Chinese miners. It is located between the communities of Montezuma and Jacksonville. Originally the Chinese were at a place called Camp Salvado until the whites forced them out and they came to this site. At first the Americans had little interest in the camp because there was no water available for sluicing, leaving the Chinese to make a success of the rich diggings. But eventually the Americans moved in. Total production of gold in 1899 was estimated at $2.5 million. The Eagle Shawmut Mine discovered in Blue Gulch near camp was active between 1890 and 1920. A 100-stamp mill was brought in to crush the ore, which yielded more than $7 million.

This is where the famous Tong war began on September 26, 1826. A few days before, one of the factions had the Columbia *Gazette* print their threat: "Challenge from the Sam-Yap Tong, at Rock River Ranch, to the Yan-Wo Tong at Chinese Camp. There are a great many now existing in the world who ought to be exterminated. We, by this, give you challenge, and inform you beforehand that we are the strongest and you are too weak to oppose us. We can therefore wrest your claim ... it is our intention to drive you away before us.... You are nothing compared to us.... You won't stand like men; you are perfect worms; or, like the dog that sits in the door and barks but will go no further."[55] The Chinese purchased guns, but since they didn't know how to use them properly, only four men were killed. Most of them attacked each other with crude spikes, tridents and daggers. After they had their one-day fill of battle, a journalist wrote in the San Francisco *Bulletin*, "It was a very bad battle, as so few were killed."[56] There were 900 members of the Yan-Wo and about 1,200 of the Sam Yap. More than 200 of them were arrested and punished.

At one time John Studebaker worked at a forge in Chinese Camp making nails, hammers, and other implements for the miners. He later went on to become the founder of the famous Studebaker car company.

CHIPS FLAT (Sierra) Named for an English miner called Chips who found ore in 1852, this site was located along a fork of the Yuba River. Its post office began operating in 1857. Chips made quite a fortune, but unfortunately most of it went to drink, from which he succumbed. The nearby Plumbago Mine reaped more than $3 million.

CHLORIDE (Inyo) One of Death Valley's earliest towns, Chloride began its life in the early 1900s although a ledge of silver was discovered in 1871 by August Franklin. Two of the biggest producers were the Keane Wonder and Chloride Cliff mines located in the Funeral Range. The mills were crushing about 1,800 tons of ore a month. The Keane was named for its discoverer, Jack Keane, who later sold it for about $250,000. More than $1.2 million

Clearinghouse (Mariposa County Museum and History Center, Mariposa, CA).

in silver was produced, most of it from the Keane Wonder.

CHRYSOPOLIS (Inyo) Both gold and silver were discovered during the 1860s. Bearing a Greek name meaning "golden city," the camp was located below the Inyo Range north of the town of San Carlos. Its post office opened in 1866, and the same year a 20-stamp mill was built. Because of the camp's isolation and problems with the Indians, Chrysopolis began to decline, until not much was left by 1910.

CHURNTOWN (Shasta) Churntown was located north of the town of Redding, and took its name from the creek, because the stream's waterfall resembled the movements of an old-fashioned churn. Settled during the 1860s with a post office in 1863, the site was never really considered a town, although there was a store and school. Not much gold was taken out except from one or two small mining claims.

CLARAVILLE (Kern) Originally called Kelso, the camp was renamed for Clara Munckton, the first white child born there. During 1861 it became a trading center for the outlying mines. The town was located near the Paiute Mountains, between Johannesburg and Bakersfield. During this time period the first placers were discovered by Roger Palmer and Wade H. Williams. Other diggings were bringing out up to $300 a ton of ore. By 1869 not much was left of the camp.

CLAWHAMMER BAR (Siskiyou) Located near Sawyers Bar on the Salmon River, the bar was one of the richest along the river, but production rates are not known. It was named in 1854 for a prospector who wore a clawhammer (swallow-tailed) coat on the Fourth of July.

CLEARINGHOUSE (Mariposa) Initially called Ferguson, this camp was established along the Merced River after gold was discovered during the 1860s by the Ferguson brothers. It was later renamed for a mine owned by Frank Egenhoff because the place was a clearing house for gold bullion during the Panic of 1907. Its post office was established in 1913. The mine yielded more than $3 million and continued to produce into the 1950s.

COARSE GOLD (Madera) The town took its name from the river, for the coarse gold found in the streambeds during the 1850s. Harassment by Indians kept the miners at bay for a time, but the gold drew them back. One prospector found about 40 ounces of pure gold in one pan, while another found a nugget worth about $160. A stamp mill was later built to treat the ore. One of the mines recovered about $185,000. When the claims played out, Coarse Gold continued to thrive as a stagecoach station. After Highway 41 was built the town prospered as a farming and ranching community.

COEUR (Trinity) French miners arrived in the region during the 1880s, and named the camp Coeur, French for "heart." The post office was established in 1885, and the camp lived its life mainly as a trading post for the local miners.

COFFEE CREEK (Trinity) Taking its name from the stream, Coffee Creek came into existence about 1881, and was a trading center for the outlying mining camps. Its post office opened in 1882. The creek was so named because someone spilled a barrel of coffee into the stream.

COLFAX (Placer) Northeast of Auburn, this town was established as a trading and supply center for the area's mines. Originally known as Alder Grove, it was renamed for U.S. vice president Schuyler Colfax in 1865. The Central Pacific Railroad completed its tracks to Colfax the same year, and a post office opened in 1866. During this time the Rising Sun mine was located about a mile west of Colfax. It produced continuously until 1884, estimated at $2 million in gold. Another nearby mine also yielded more than $2 million. A fire all but destroyed the main street of town in 1874, but it was rebuilt.

COLOMA (El Dorado) Coloma has the distinction of being the site that sparked the California gold rush of 1848 (although it may have actually been a strike near Mokelumne Hill that led to the influx of fortune seekers). Before that time, it was home to the peaceful Maidu Indians, who called it Cullomah, meaning "beautiful valley." The post office opened in 1849 with Jacob T. Little as postmaster. John Sutter hired James Marshall to build and operate a sawmill at Coloma. During construction Marshall noticed something shiny in the water, and the rest is history. Marshall wrote in his diary, "Towards the end of August, 1847, Captain Sutter and I formed a copartnership to build and run a sawmill.... January, discovered the gold near the lower end of the race, about two hundred yards below the mill."[57] Little attention was given to the discovery until the *Californian* wrote in 1849, "Gold Mine Found — In a newly-made raceway of the saw-mill recently erected by Captain Sutter, on the American Fork, gold has been found in considerable quantities ... California, no doubt, is rich in mineral wealth; great chances here for scientific capitalists. Gold has been found in almost every part of the country."[58] It was at Coloma where the first gold rocker was used at the diggings, so miners no longer had to hand-pick the gold grain by grain.

From Sutter's letters, he was not exactly happy about the riches since he was in the process of establishing an agricultural empire: "What a great misfortune was this sudden gold discovery to me! It has just broken up and ruined my hard, industrious, and restless labors, connected with many dangers of life, as I had many narrow escapes before I became properly established. From my mill buildings I reaped no benefit whatever; the millstones, even, have been stolen from me. My tannery, which was then in a flourishing condition, and was carried on very profitably, was deserted."[59] John Sutter and James Marshall may have been the first to discover the gold, but neither prospered from it. Sutter died a broken man in a Washington, D.C., hotel room on June 18, 1880, and Marshall died a poor man five years later at Kelsey.

Coloma hosts its annual Gold Discovery Days in January, and the Coloma Fest in June, in addition to their Christmas in Coloma. For times and directions, call the Placerville Chamber of Commerce at 1-800-457-6279.

COLUMBIA (Tuolumne) Known as the "Gem of the Southern Mines," Columbia was settled northwest of Sonora. The diggings were first worked in 1849 by Mexicans, then taken over a year later by Dr. Thaddeus Hildreth and his party of men. First known as Hildreth's Diggings, the name was changed to Columbia and a post office opened in 1852. A year later, news of even richer ore brought more than 6,000 miners seeking their fortunes, which grew to almost 15,000 people. "It was one of the most rapid developments of a great and prosperous mining-camp ever known in California. Within a fortnight, the need for some system of government was manifest. A public meeting was called to talk up the subject; but nothing in particular was done except to give the camp a name,— Columbia."[60]

One of the men at the diggings found a 75-pound lump of pure gold worth about $14,000. Another discovered a nugget weighing close to 33 pounds, which brought him more than $7,000. In one 300-acre area, property owners brought out more than 100 tons which yielded an astounding $55 million. It was estimated that between 1853 and 1857 more than $100,000 a week in gold was being shipped out, but by 1868 it had dropped to less than half that. The Cascade Mine was reported to have taken out more than 60 ounces of gold in less than a week. In all, it was estimated that more than $87 million had been extracted. Eventually, lack of water contributed to the demise of Columbia and the nearby mines. The Tuolumne Water Company built a system of ditches to bring in water but the venture failed. In 1945 the town became part of the California State Park system.

The first American woman to live at Columbia was a Mrs. Denouille. "The miners were so eager to welcome her that they decorated the streets with arches and marched four miles down the road to escort her into town, preceded by a brass band."[61]

Although most of the miners were not churchgoers, they gave freely whenever there was need to build a new church, and Columbia was no exception. After the building was finished, the Reverend Long came to preach his sermons, which turned out to be quite tedious. As a result, fewer and fewer people came to services on Sunday. One day he rang the church bell that sounded different, a signal known by everyone that designated a fire. When the people ran to the church they asked where the fire was. The Reverend Long replied, "When I rang the call to church this morning, there wasn't a soul in this town of Sodom that answered. But you've all come to find out about the fire, haven't you? And I'll tell you where it is. That fire's in hell, and if you don't come to church and repent your sins, you'll all go there and burn in it!"[62]

One of the best preserved gold-rush towns, Columbia celebrates its Gold Rush Days each month. For details and directions, call the Columbia Chamber of Commerce at (209) 536-1672.

CONDEMNED BAR (Yuba) Prospector Joseph Stuart conducted some mining for a short time, then opened a trading post in 1851. The site was near the junction of Dobbins Creek and the main Yuba Creek. Some placer mining was conducted with fair success, but by the 1870s very few were working the diggings.

CONFIDENCE (Tuolumne) Named for the Confidence Mine, this community was established in the 1850s and received its post office in 1899. The name reflected the anticipation of finding great riches, which came to fruition. The site was located east of the town of Columbia. In 1869 a 30-stamp mill was built, followed two years later by more than 35 stamps. The Confidence Mine brought out more than $4 million in gold.

CONN VALLEY (Napa) Only moderate profits of silver and gold were taken at this site, which was located near the town of Saint Helena. It was named for settler John Conn who arrived about 1843.

CONTRERAS (Amador) The first person to find gold on the North Fork of the Mokelumne River in 1851 was a Baptist minister. This camp was a few miles southeast of Volcano and named for prospector Pablo Contreras. Although most of the veins were small, the miners were extracting a good $1,000 from them, and what was called the Belding vein yielded about $20,000. Most of the gold was crushed by arrastres. By 1880 the town had disappeared.

Arrastre, Amador County (courtesy of the Amador County Archives, Amador, California).

COOKS BAR (Sacramento) This was a trading post established in 1849 and named for its owner, Dennis Cook. It served the mining population along the Cosumnes River near Michigan Bar. The gold ran out in the 1860s.

COOL (El Dorado) First called Cave Valley because of its limestone caves, Cool was an early tourist attraction. Located just north of Pilot Hill, it was another of the small placer mining camps and supply centers in the 1850s and lasted until about 1900. When the town of Cave Valley applied for a post office in 1885, postal officials would not allow the name. So the residents changed it to Cool, after Aaron Cool who was a traveling minister and the son-in-law of Senator William M. Gwin, one of California's early senators. Henrietta Lewis was the postmaster.

In 1979 Russ Stowers and his business associate, Ken Haskins, who had purchased the townsite a few years earlier, joked with a news reporter and told him that Cool was on sale for $1 million. Believing this to be true, the reporter sent out on the national news wire that Cool was on sale for a cool million. Hoping for a bit of publicity, a local real estate agent played along with the hoax, causing pandemonium with the residents. Stowers and Haskins were out of town at the time. When they returned, they found the people in an uproar, and hurriedly left until the noise died down. Today, Cool has a population of more than 2,500.

COOLGARDIE PLACER (San Bernardino) North of Barstow, mining operations started in the early 1900s. The site was named from a location in Australia. Since there was no water, dry mining was conducted. About $100,000 was realized.

COON HOLLOW (El Dorado) South of Placerville, the camp was formed after rich gravel deposits were discovered about 1850 and named for the common raccoon. In 1851 a canal was constructed to bring water to the placers. Although not verified, it was thought the diggings brought out about $25 million in gold. Much of it was accomplished using hydraulic methods. One of the best producing mines was the Coon Hollow, which yielded about $10 million.

COTTONWOOD (Siskiyou) Initially called Henley, the camp later took its name from the creek for its growth of cottonwood trees. After discovery of gold near Yreka, John Thomas and another miner went prospecting on an ancient volcano which was said to contain gold, but there was nothing there. They ended up at Cottonwood with better luck in the shallow diggings for a short period of time. The camp was established in 1852 along the creek near the Klamath River with a post office four years later. During the winter of 1852-53, heavy snow caused food prices to soar, with flour selling for $2.00 a pound, and one potato costing 50¢. One very important commodity was salt, packed over the mountains from Oregon which cost $1.00 an ounce.

At a claim owned by a miner named Stiles, a little more than 12 pounds of gold was extracted. Then in 1864 at nearby Printer's Gulch, someone found a piece of gold that was worth more than $1,900. The gulches around the town became known for their rich, shallow diggings. Large-scale dredging operations commenced in the 1850s, bringing out more than $4 million in gold. The Golden Eagle Mine produced $1 million before 1931. Cottonwood thrived for about five years and then began to decline. It later became a station on a branch of the Southern Pacific Railroad.

COULTERVILLE (Mariposa) This settlement was founded near Mariposa about 1850 when George W. Coulter established a trading post, followed by Stephen Davis who opened a store. Coulterville served as a trading and supply center. It was first named Maxwell for George Maxwell, who was postmaster when the office opened in 1852. Major mining was conducted between 1894 and 1900. One of the important mines was the Potosi (Spanish for "great wealth"), which realized more than $1 million. The Mary Harrison Mine was located by a French mining company and began operating in 1853. It was later purchased by a Boston interest that further developed the mine and made a profit of over $15 million in gold. When the smelter could no longer get fuel, the mine closed in 1904. The Argo Mine, discovered in 1850, was the scene of a devastating fire in 1922. One of the tunnels caught fire, killing more than 45 miners, but it continued to operate until the 1940s, yielding close to $25 million.

COYOTEVILLE (Nevada) Gold was discovered nearby in 1850, and within a year more than $7 million was realized. Coyote is Aztec for "prairie wolf." The name was used by miners for a method of mining called "coyoteing," descriptive of a burrowing coyote. After shafts were sunk and bedrock was encountered, the miners had to get on their hands and knees and literally burrow in and dig out

Magnolia Hotel, Coulterville (Mariposa County Museum and History Center, Mariposa, CA).

the ore, bringing the dirt up with a windlass. One of the claims paid out $40,000 in a few days, and a nearby gulch yielded more than $8 million. Coyoteville was close to Nevada City and eventually became part of it.

CRACKERJACK (San Bernardino) Established in the early 1900s, this camp was located in the Avawatz Mountains with a post office in 1907. News of gold discoveries was published in a local paper: "In the Avawatz mountains, San Bernardino county, adjoining the westerly edge of Death Valley, a new mining region is being opened up; and yet it is not so new, for, as widely scattered arastras [*sic*] show Spanish miners worked the surface nearly 100 years ago.... The district is rapidly filling up.... Joining the Crackerjack claims on the southeast Messrs. John Anderson, L.A.W. Johnson and A.H. Schliebetz, Los Angeles mining capitalists, have a property called the Tom Boy. Only a little development work has been done on this property, but assays of the ore taken out run up to $5000 per ton, and the continuity of the ore body is shown by a canyon outcrop 200 feet below the surface. The new town will also be connected shortly by telephone with the railroad station, a company for the constructing of such a line having been practically organized."[63] The glowing reports were short-lived, however, as the mines closed and Crackerjack disappeared by 1918.

CRESCENT CITY (Del Norte) Crescent City experienced some small-scale mining along its beaches during the 1850s. The gold had been washed down from the Smith and Klamath rivers. Tradition says a miner reaped great riches along the beach, then hid the gold near his cabin. Indians came upon him one day, beat the poor man, and set fire to his cabin. Before he died the miner told his friends of his great wealth, but to this day it's never been found. After the gold was depleted in the surrounding mines, Crescent City became a lumber town. Its post office opened in 1853 and was named by J.F. Wendell because of the town's location on a crescent-shaped bay.

CRESCENT MILLS (Plumas) Crescent Mills was settled in the 1860s and the post office was established in 1870. There was fair success at gold mining until larger deposits were found at Indian Valley. Two stamp mills were established, and the area ultimately yielded more than $7 million. Crescent Mills was located along Indian Creek near Lake Almanor, and served as a supply and trade center for the outlying mines.

CROMBERG (Plumas) A corruption of Krumberg, the name is German for "curved mountain." The camp served as a trading center for the mines, and its post office was established in 1880.

The old Twenty-Mile House was an inn that welcomed travelers and prospectors alike. It is still open for rentals, including a couple of cabins for those interested in wedding parties or weekend fishing (530-836-0375 for more information).

CRUMBECKER RAVINE (Nevada) Located south of Alpha, this ravine contains a small tributary of Scotchman Creek. The only person known to have mined here was a Mr. Crumbecker who hailed from Kentucky. He reaped about $100,000 of gold, then returned home.

CURTISVILLE (Tuolumne) Taking its name from Curtis Creek, which honored a prospector, Mr. Curtis, the site was situated south of Sonora and had a post office in 1853. Prospectors experienced moderate success. Then in 1894 someone found a vein that was reported to yield almost $10,000 a day. As the gold petered out, so did Curtisville.

CUT-EYE FOSTERS (Sierra) Named for an unscrupulous prospector with a scar across his face, the camp was established along the North Fork of the Yuba River near Goodyears Bar. Old Cut-Eye was thought to be a horse thief. There was some mining conducted between 1851-1852, but the recovery amounts are not known.

DAGGETT (San Bernardino) Founded in the 1860s, the place was first called Calico Junction. It was changed in 1883 when the post office was established to honor John Daggett, state lieutenant governor. The town was mainly a supply point on the Santa Fe Railroad for the silver mines located in the Calico Hills, Death Valley, and the Mojave Desert. Between 1855 and 1882 two stamp mills were built to process the ore. Wagons hauled water and supplies to the mining town of Calico and brought ore back. After the gold was depleted, Daggett served the borax mines that began operating about 1883. It later became the site of a U.S. Marine Corps installation.

DALE (San Bernardino) Gold discovered during the 1880s near a dry lake brought in the placer miners. The site was located near the Pinto Mountains, east of Twentynine Palms, and its post office opened in 1896. The only mine of consequence was the Virginia Dale, which was discovered about 1886 and produced more than $1 million.

DAMASCUS (Placer) Founded by Dr. D.W. Strong in 1852, the camp was initially named Strong's Diggings. It was changed in 1856 when the post office was established, and bears a Hebrew word for "activity," an appropriate name since the place was very busy. Damascus was built on a steep hillside overlooking the Blue Canyon and North Fork of the American River. A 10-stamp mill and two arrastres were built to crush the ore. The mines in the area were good producers, with more than $15 million realized by the early 1900s. The Pioneer Mine recovered about $1 million, and the Mountain Gate Mine yielded more than $1.5 million by 1860. There was occasional mining until the 1930s.

DANBY (San Bernardino) Ore found in the Old Woman Mountains caused this site to be established as a trading and supply center. Danby's post office opened in 1898. The camp was located along old Route 66. The only remnant left is an old building that still has a mural painted on its side, depicting lofty mountains and mountain men. Nearby is the Old Woman Statue at the 5,090 foot level composed of three giant boulders, one on top of another, resembling an old woman with a shawl. The formation is about 100 feet tall.

DARDANELLES DIGGINGS (Placer) Named for the narrows between Europe and Asia Minor, this camp was formed about 1854. It was about four years later that some good strikes were reported, bringing in a few prospectors. Some of the men were getting more than $19,000 in one day. The Dardanelles Mine later recovered more than $2 million. After the placers were depleted, hydraulic mining took over.

DARWIN (Inyo) Located along Owens Lake west of Death Valley, this town was named for Dr. Darwin French who led a group of miners there in 1860 with hopes of finding the lost Gunsight Lode. This mine was a phantom gold or silver claim allegedly discovered in 1894 by a group of men traveling through the Death Valley region. The men didn't find the mine, but they did discover rich silver outcrops that assayed out at about $700 a ton. Darwin was settled about 1875 with a post office, and served as a supply and trading point for the local mines.

DEAD MAN'S BAR (Placer) Tradition says the bar received its name when miners from Oregon had a fight with men from Iowa over who owned a mining claim, and one of the men was killed during the melee. The rest ended up working together and called it Dead Man's Bar. It had previously been called Neptune's Bar for a group of sailors who had earlier tried their luck there. The site was located on a fork of the American River near the town of Auburn.

Two-stamp mill, Amador County (courtesy of the Amador County Archives, Amador, California).

DEADWOOD (Placer) Deadwood was settled near Michigan Bluff in 1852 and received its name because the miners were so happy to find ore, they were sure they had the "deadwood" upon securing a fortune. When word got out, more than 500 men arrived to work the placers. The road to Deadwood was nearly 4,000 feet high up a circuitous route that traversed to Secret Spring, around the head of El Dorado Canyon, and then down a narrow ridge to the camp. During 1860 David Davis and John Williams, who owned a water ditch, planned to bring water down for the miners. They got caught in a fearsome snowstorm in December and never came out. It was near the end of January before their bodies were found. Some hydraulic mining was conducted at Deadwood, but the principal mines were worked through tunnels by drifting.

DEADWOOD (Sierra) This minor camp was established along a fork of the Yuba River. Only one mine was recorded as recovering about $10,000 with an initial investment of $115,000.

DEER CREEK CROSSING (Nevada) Located near the confluence of Deer Creek and the Yuba River, some gold was found at the crossing in 1849. The prospectors were taking out less than $10 a day, and by 1856 the site was abandoned.

DEER FLAT (Tuolumne) Successful mining was conducted at this site which was located near the town of Groveland. During the 1850s mining operations slowed down because of lack of water. But during 1865 a 5-stamp mill was established and the men took out about $10,000. By the time the gold had run out, the region had yielded about $25 million.

DEIDESHEIMER (Placer) A German mining engineer named Philip Diedesheimer discovered a gold lode at the site, and by 1865 had realized more than $650,000 of ore. It was situated near the camp of Forest Hill.

DEPOT HILL (Yuba) Settled near the town of Camptonville by the Yuba River, this camp was the scene of hydraulic mining during the 1850s. It was estimated that between 1855 and 1927 more than $1 million in gold had been recovered.

DIAMOND MOUNTAIN (Lassen) Placer mining was first conducted during the 1850s, followed by lode mining about ten years later. The lodes recovered nearly $1 million in gold. Diamond Mountain was located south of Susanville.

DIAMOND SPRING (El Dorado) This name was thought to be for quartz crystals (called diamonds) that were found in the 1850s, or for the clarity of the nearby natural springs. It acquired a post office in 1853 with Chauncy N. Noteware as postmaster. Situated just south of Placerville, Diamond Spring was also used as a Pony Express station during its short life. One of the most profitable mines was the Griffith Consolidated located in the 1850s and composed of eight claims, which yielded anywhere from $8 to $65 per ton of ore.

DICKSBURG RAVINE (Yuba) A miner named Dick found gold in the area near Honcut Creek about 1850. The following year a mill was built, but not enough gold was found to make continued work worthwhile, and it closed down less than a year later.

DILLON CREEK (Siskiyou) Dillon Creek is a tributary of the lower Klamath River where placer gold was found during the 1850s. The Siskon Mine was located in 1851 and brought out more than $1 million in gold. Mining operations continued into the 1950s. Years later it was discovered that tailings from the mine entered Copper Creek, a stream in which salmon spawn, greatly endangering the fish.

DOBBINS (Yuba) Mark and William Dobbins found gold in the vicinity about 1850, and the camp was formed with establishment of a post office in 1851. William served as the first postmaster. The site was located near Bullards Bar. During the 1860s miners were realizing about $30 a ton in gold.

DOBLE (San Bernardino) Doble was located in the San Bernardino Mountains, and served mainly as a supply and trading center for the nearby mines. It was originally called Bairdstown for Samuel H. Baird who staked a nearby claim. It received its post office in 1875. The name was later changed to Doble for Budd Doble, son-in-law of the Gold Mountain Mine owner, E.J. Baldwin. The camp was gone by 1888 when the ore ran out.

DON PEDRO'S BAR (Tuolumne) Pierre Sainsvain arrived about 1839, followed by a number of prospectors. It acquired a post office in 1853 and was given Sainsvain's nickname. The bar is located along the Tuolumne River, northeast of La Grange. Sainsvain came here from Sutters Mill and later became a member of the California Constitutional Convention. A number of prospectors worked the claims amounting to $1,000–$1,500 each day. The site went under water when the San Francisco Power and Light Company built its reservoir.

DOUGLAS FLAT (Calaveras) Before the advent of gold seekers, Miwok chief Walker and his tribe lived near the flat. When gold was found in 1849 on Coyote Creek, the chief and his people left. Douglas Flat had its post office established in 1879. The site was located between the camps of Vallecito and Murphys. During 1856 one of the mining companies was recovering only about 30 ounces a day, but another mine realized about $130,000. The area was later destroyed by hydraulic mining. After the ore ran out, residents prospered by cultivating fruit orchards and vineyards since the soil was exceptionally rich.

DOWNIEVILLE (Sierra) Downieville was established near the junction of the North Fork of the Yuba River, and named for William Downie who had brought a group of gold seekers into the region. The place experienced great prosperity after the discovery of gold in 1849. Within two years more than 5,000 people were living there, which warranted a post office that was established in 1852. In less than 18 months more than $15 million in gold had been recovered. The ore was so common that a woman, while cleaning her earthen floor, swept up nearly $500 of gold dust in one morning. In 1859 the Reis brothers found gold in Hog Canyon, which yielded them about $300,000. At nearby Gold Canyon, the mines produced between $750,000 and $1 million.

Downieville also had its bad moments. During the Fourth of July in 1851, the partying got out of hand. A hot-tempered dance-hall girl named Juanita was being harassed by a miner, so she stabbed him. Juanita was immediately apprehended and sentenced to hang. Showing no remorse, Juanita went to the gallows, put the rope around her neck, said "Adios, Señors," and flung herself off the scaffold. A slightly different account of the event was published in the *Times and Transcript*: "The act for which the victim suffered was one entirely justifiable under the provocation. She had stabbed a man who had persisted in making a disturbance in her house and had greatly outraged her rights. The violent proceedings of an indignant and excited mob, led on by the enemies of the unfortunate woman, were a blot upon the history of the state."[64] A placard in town reads: "The Spanish woman also known as Josepa, was hung off the Jersey Bridge July 5, 1851, a short distance down stream from this spot, for the murder of Frederick Alexander Augustus Cannon. Cannon and his friends were celebrating Independence day and after closing most of the saloons they passed Jose and Josefa's cabin ... her last words were, 'I would do the same again if I was so provoked.'"

Under more pleasant circumstances, when the first Yankee woman arrived at Downieville, the miners were ecstatic. Her name was Signora Elise Biscaccianti, a well-known pianist. The men picked her up and hoisted her on their shoulders, carrying her down Main Street. They did the same with her piano.[65]

DRAGOON FLAT (Tuolumne) The site was named in 1849 for a number of Dragoons who had deserted their posts. Dragoons were part of the U.S. Cavalry who carried standards upon which dragons were imprinted. One of the soldiers discovered a huge lump of gold, bringing in the inevitable swarm of gold-hungry men. The flat was located near the confluence of Sonora and Woods creeks.

DRY DIGGINGS (El Dorado) This camp was located along Weber Creek south of Placerville, where placer gold was found in 1848 by Perry McCoon, J. Sheldon and William Daylor (or Daly), who once worked with Sutter. They extracted about $15,000 in gold, and were followed by Indians who took out more than 130 ounces of gold in a day. A Colonel Mason reported the discovery to Washington officials: "A small gutter not more than 100 yards long, but four feet wide and two or three feet deep, William Daly and Perry McCoon, had a short time before, obtained in seven days $17,000 worth of gold. Captain Weber informed me, that he knew that these two men had employed four white men and one hundred Indians, and that at the end of one week's work, they paid off their party and left with $10,000 worth of this gold. Another small ravine was shown to me from which had been taken $12,000 worth of gold. Hundreds of similar ravines, to all appearances, are as yet untouched."[66]

DRYTOWN (Amador) Founded along Dry Creek about ten miles from Jackson, Drytown became a placer mining camp in 1848 with more than 50 prospectors, and was the oldest gold camp in the county. The post office was established in 1852. The name Drytown belies its true nature, since in had more than 25 saloons. Nearly 250 miners arrived during that winter, some who realized a little more than $100 of ore in a few days, but the average was a few ounces a day. During 1851 the men left because there was no water, but the following year a flume was built to bring water to the mines. The mining camp at Drytown lasted until 1857 when a fire wiped out the town.

DULZURA (San Diego) Dulzura was settled in the 1860s and bears a Spanish name meaning "sweetness." It is an appropriate name because John S. Harbison established a honey business there in 1869. The camp was located near the Mexican border in the Otay mountains, and received its post office in 1886. A Mexican boy first discovered gold in 1877, but it would be more than ten years before any significant mining took place. Operations continued into the 1930s, and included some placer mining.

DURGAN'S FLAT (Sierra) First named Washington, the name was changed in 1850 when James Durgan moved there to establish a sawmill. A year later gold was found, bringing in more than 500 miners seeking their fortunes. The camp was located south of the town of Downieville on a fork of the Yuba River. More than $8 million in gold was produced from the nearby lode mines.

DUTCH FLAT (Placer) Dutch Flat was named for Dutch settlers Charles and Joseph Dornbach, and was located high up in the mountains. It became a major trading center for the mines that were very active in the 1850s. About ten years later a company was organized and built a wagon road from Dutch Flat to the eastern slope of the mountains to accommodate travelers headed into Nevada and its silver mines. The road provided additional profits for the town. While working at the Polar Star Mine in 1876, someone found a boulder that held $5,760 in gold. A year later a Chinese miner found a nugget worth $12,000. With a population of more than 2,000, Dutch Flat warranted a post office which was established in 1859 with Charles Steffens as postmaster. An unlimited supply of water enabled extensive hydraulic mining in the region beginning about 1857. It continued until 1883 when the companies were slapped with the anti-debris injunction. Total production from the mines up until 1883 was more than $5 million. Still visible today are the scars where the mountainsides were washed away by the hydraulics.

Hotel at Drytown (courtesy of the Amador County Archives, Amador, California).

Dutch Flat is home to the Golden Drift Historical Society, which holds exhibits from the early mining days, and of the history of the Chinese who mined there. Information: (530) 889-6500, or (530) 889-6500.

EDDY'S GULCH (Siskiyou) Rich diggings were discovered in the 1870s, and as a result a number of stamp mills were constructed to crush the ore. The materials for the mill had to be hauled by oxen because of the rough terrain. The gulch was tributary to a fork of the Salmon River, south of Sawyers Bar. Lode mines near the gulch brought out about $4 million.

EL DORADO (El Dorado) This camp was established in the 1850s as a crossroads for freight and stage lines. When the post office was established in 1855, it was initially named Mudville. Darwin Chase was the postmaster. A few years later the name was changed to El Dorado, Spanish for "the gilded one." Legend has it the king in Mexico had his body covered with gold dust each day, leading white miners to believe the region was rich in gold. El Dorado was a few miles southwest of Placerville. Mexican miners were the first to work the diggings in 1852. The Church Mine, discovered in 1850, was just southeast of town. It later consolidated with the Union Mine, and between the two more than $600,000 in gold was recovered. Another claim yielded between $800,000–$900,000. Total production from the local lodes was about $6 million.

ELIZA (Yuba) This site was named for Anna Eliza, John Sutter's daughter. It was situated along the Feather River just south of Marysville. The camp was created about 1850 because a mining company from the east asked Sutter to found a new place where the river was more navigable. Their plan was to rival Marysville as the new site at the head of navigation and a successful supply and trading center for the nearby mines, but their venture came to naught.

ELIZABETHTOWN (Plumas) Elizabethtown was named to honor the daughter of settler Lewis Stark. It was first called Tate's Ravine, for miners Alex and Frank Tate. The place was settled just northwest of Quincy. The discovery of gold in 1852 brought in the fevered prospectors. A 3-stamp mill was built to crush the ore, and the post office was opened in 1855. But most of the men had waited too late because the nearby ravines had been depleted of their wealth, and Elizabethtown began its downward spiral.

ENGLISH BAR (Yuba) Named for two Englishmen who prospected there in 1851, the bar was situated along the North Fork of the Yuba River. Unfortunately they didn't even find a prospect hole and left. Other miners who arrived there had better luck and found gold that yielded them about $90,000.

ETNA (Siskiyou) This camp was settled about 1852 at the bottom of the Etna Mountains in Scott Valley. The site was a supply and trading spot for the Salmon River mines and did a brisk business. Its post office opened in 1861. The camp consisted of five stores, two blacksmiths, a brewery and a number of saloons. Etna is Greek for "to burn," and the freighters really burned during 1878, hauling more than 200,000 pounds of produce and 84,000 pounds of general merchandise to the mines. The 200 mules packed about 600,000 pounds of goods to the diggings annually.

EXPERIMENTAL GULCH (Tuolumne) During the 1850s placer gold was found along the gulch, which is a tributary of a fork of the Stanislaus River. When word got out about the strike, hungry prospectors flocked to the region, and an 8-stamp mill was built. Some of the claims were yielding $100–$200 per man. The larger mines brought out more than $3 million in gold.

FAIR PLAY (El Dorado) Located near the Middle Fork of the Cosumnes River, this camp came into existence during the 1850s. Fairplay became a trading center for the local miners and received its post office in 1860, with George Merkindollar serving as postmaster. The camp received its name because the miners made a pact that there would be fair play for all. The 1880s brought the decline of gold, and residents turned to agriculture for their living.

FEATHER RIVER (Plumas/Butte) Rising in Plumas County, the river forms the boundary between Sutter and Yuba counties. Camps dotted the river and a post office was established in 1919. During the 1820s a Spanish expedition came through, led by Captain Luis Arguello who named the river El Rio de las Plumas ("the river of the feathers"). Sutter later Americanized the name: "I named Feather River ... from the fact that the Indians went profusely decorated with feathers, and there were piles of feathers lying around everywhere."[67] Gravels of the Feather River yielded a total of $44 million in gold.

FENNER (San Bernardino) Fenner was established in the 1890s mainly as a supply point and trading center for mines in the nearby Clark and Providence mountains. Its post office opened in 1892. After a branch of the Santa Fe Railroad was built, Fenner's importance as a place of commerce ceased and it faded away.

FIDDLETOWN (Amador) Fiddletown may have received its name when prospectors from Missouri were trying to decide what to call the town. One of the residents complained that the younger generation was "always fiddling around, it should be called Fiddletown." Others thought it was named by a German fiddler. The post office opened in 1853 and Dennis Townsend served as postmaster. It became the main trading center for miners after gold was discovered in 1852. Most of the deposits were recovered by drift mining and dredging. Records are not clear on the total amount, but it was estimated between $206,000 and $2 million.[68]

In 1870 a local judge thought Fiddletown wasn't proper and renamed it Olete. It reverted to its original name in 1932.

FINE GOLD GULCH (Madera) An Indian fighter named Jim Savage came to the gulch in 1849 and discovered gold deposits. He hired Chinese miners to work his diggings along the San Joaquin River. The place was called Fine Gold because the ore was mainly in the form of extremely fine dust. For a period of time the prospectors experienced problems with the Indians, but as the area became more populated, they moved away. A 10-stamp mill was built in 1868, and a post office began operating in 1881.

FLORIDA HOUSE (Sierra) This camp housed the Florida House, a hotel operated by a woman from Florida, Mrs. Romargi. It became a supply point for the pack trains headed to Goodyears Bar, and also served as a stagecoach transfer point.

Feather River mining, Butte County (Butte County Historical Society, Oroville, CA).

FOLSOM (Sacramento) In 1849 gravels along the American River were found to contain riches. Theodore D. Judah, engineer with the Central Pacific Railroad, established the town in 1855 and its post office opened the following year. In 1899 the first bucketline dredge used in the county was installed at Folsom. Prior to that time, the gravels were mined by hydraulic methods. Rising costs gradually forced curtailment of those activities.

Folsom once had a Chinese community of more than 2,500 people. Those who worked on the Transcontinental Railroad came here and set up their own little town. During 1856 the Sacramento Valley Railroad laid its tracks which linked Folsom to Sacramento, bringing more prosperity to the camp as a commerce center for the foothill mining camps. Total production was approximately $62 million. Folsom was named for a member of Stevenson's Volunteers and landowner, Captain Joseph L. Folsom.

FORBESTOWN (Butte) First settled by a Mr. Toll about 1849 and known as Toll's Old Diggings, the camp's name was changed in 1854 when the post office was established. It was named for Ben F. Forbes, who moved there in 1850 and opened the first store. A short time later Ben's brother, James, moved to Forbestown and built the United States Hotel, and later a stage line. The site was located east of the town of Oroville. Rich deposits of placers and quartz mines were discovered in the nearby ravines. One of the best producing mines was the Gold Bar, which yielded more than $2 million.

FOREST CITY (Sierra) Forest City was established south of Downieville about 1853 after miner Michael Savage found gold nearby. It was initially called Brownsville and then Elizaville before it was changed to Forest City, a descriptive name. Its post office opened in 1858. Rich ore was to be found in the nearby mountains, as experienced by some of the men who were extracting almost $100 a week. At one of the claims a 42-ounce lump of pure gold was discovered. The nearby town of Alleghany began to grow rapidly and drew many of the residents from Forest City, which became all but deserted. But when new ore was discovered in the Bald Mountains just north of town, the camp experienced a slight revival. Most of the mining operations consisted of tunneling. The Bald Mountain Mine had over 8,000 feet of tunnels and yielded more than $3 million. Because of the property's location, only locomotive engines could be used to haul out the loaded ore cars. Other claims were recovering about $100,000 a year in gold.

Forbestown (Butte County Historical Society, Oroville, CA).

FOREST HILL (Placer) This site was established between a fork of the American River and Shirttail Canyon when gold was discovered in 1849, followed by a post office ten years later. A trading post was established for the miners by R.S. Johnson, and James and M. Fannan. During the winter of 1853 violent storms broke off a large outcropping and the boom was on. After the Jenny Lind Mine began operations, it was yielding about $2,000 a day. The most productive lodes were the Mayflower ($1 million) and Paragon (more than $2 million). Other combined lodes recovered nearly $13 million in gold. Forest Hill was named by traders for its location in a timberland. When the mines played out, lumber became its main industry.

FOREST HOME (Amador) This camp was named for the Forest Home Tavern, which was established in the 1850s. It grew up with two stores, saloon and hotel. Placers were discovered nearby but they drew little interest because the diggings were dry, and brought in less than a dozen miners. After water was brought in from the Cosumnes River, others came in to work their claims. The camp was located north of Ione. By 1868 the diggings had been depleted and the place was abandoned.

FOREST SPRINGS (Nevada) Mining operations began here about 1850 when the Norumbagua Mine was opened. It operated until the 1920s and brought out about $1 million in gold. James Whitesides, P.H. Lee and others located a gold ledge in 1850, which was sold to A.C. Peachy for $100,000 in 1866.

FORKS OF SALMON (Siskiyou) This camp was named because of its location at the confluence of the South and North forks of the Salmon River, and the early discoveries nearby proved quite rich. Forks of Salmon spent its life as a trading and supply center, and acquired a post office in 1858. During the early mining days there were more than 500 acres of placer diggings loaded with gold. Water from a fork of the Salmon River was brought in via ditches. The hydraulic operations were among the largest in the region and produced thousands of ounces in placers. Gold that was recovered between the Forks of Salmon and Sawyers Bar was estimated at about $25 million.

FORMANS (Calaveras) This site was located north of Double Springs and named for David Foreman who, along with Alexander Beritzhoff, established a hotel and operated a ferry to serve the miners moving back and forth between the gold fields. Some small placers were discovered in the vicinity during 1854, but they amounted to only a few ounces in a pan.

FORT JONES (Siskiyou) Located in the Scott River Valley, Fort Jones took its name from an old fort built to protect settlers against the Indians. It was named for an Army officer, Colonel Roger Jones. The camp began as a stop for the stages running between Yreka and Shasta. It was initially called Scottsville, then Wheelock for a Mr. Wheelock, and finally changed. Fort Jones was founded about 1851 after the nearby camps of Deadwood and Hooperville began to decline. It served as a supply point on the route from Yreka to Shasta, in addition to the mines on the Scott and Salmon rivers. The post office was established in 1860.

FOSTERS BAR (Yuba) This site, which is now submerged under Bullard Bar Reservoir, was named for a member of the famous Donner party, William Foster. Foster moved to the area in 1847 and established a furniture store. He arrived at Fosters Bar a year later and was one of the first miners in the region. Its post office opened in 1852. The bar was somewhat prosperous but mining lasted only about ten years.

FOURTH CROSSING (Calaveras) Rich placers were discovered in the region about 1848, bringing in the miners, which precipitated the establishment of a trading post owned by David Foreman. First known as Foremans Ranch, the name was changed to Fourth Crossing, since the camp was situated where the Calaveras River had to be crossed four times from the town of Stockton to the mines. The post office was established in 1855. When the placers ran out, lode mining began. Fourth Crossing was a busy trading center for the Southern Mines until the late 1800s. After the gold was depleted, the camp continued for a time serving as a stage and freighting stop.

FOX GULCH (Tuolumne) Gold was mined there during the 1850s, and was located not far from Experimental Gulch. The few miners who worked the diggings reported only about $18 and $20 for their claims. There simply wasn't enough gold and by 1861 the place was abandoned.

FRAZIER MOUNTAIN DISTRICT (Ventura) Placer mining began in the 1860s, followed by lode mining up until 1895. While hunter Warren Frazier was tracking a wounded deer, he discovered a rock full of gold, forgot about the deer, and began digging. More than $1 million was realized from the Frazier Mine.

FRENCH CAMP (Amador) A few miles northeast of Lancha Plana, this camp was established during the 1850s and named for French miners who came to claim their fortunes. During 1854 a band of Yaqui Indians also worked for the gold and, contrary to other Indians, lived peaceably with the prospectors. In 1856 Joe Septen found a rich deposit, recovering a few thousand dollars a week. Other than his claim, not much gold was taken out, and by 1870 the place ceased to exist.

FRENCH CORRAL (Nevada) Established southwest of Birchville near a fork of the Yuba River, this site was named because a Frenchman built a corral in 1849 to contain his mules. When gold was discovered in 1851, French Corral became an important trading center. Some of the miners hired Chinese to work their claims, which incensed the others. The angry miners went to their cabins and burned the homes of the Chinese, driving them away. In 1878 the first long-distance telephone line in the West was begun here, stretching through 60 miles of mining camps, enabling communication between mining companies. Some hydraulic mining was conducted.

FRENCH GULCH (Shasta) Just northwest of Whiskeytown along Clear Creek, French Gulch was a successful mining camp during the 1850s. Its post office opened in 1856. A 10-stamp mill was built in 1863, and within five years two more mills with 32 stamps had been constructed. The *Shasta Courier* wrote in 1852 that "such rich diggings have been struck that miners are tearing down their homes to pursue the leads which extend under them."[69] Gravels of Clear Creek and its gold-bearing veins produced about $1.6 million.

FRIANT (Fresno) Friant was a late bloomer. It was settled by Charles Converse in 1852 who established a ferry, and it later became a station on a branch of the Southern Pacific Railroad. The site was located along the San Joaquin River. When the White-Friant Lumber Company located there, the camp was named for one of its owners, Thomas Friant. Between 1940 and 1942 more than $195,000 in gold was recovered from the sand and gravel that was processed during construction of Friant Dam on the San Joaquin River. Up until about 1969, the annual yield was $25,000 a year.

GARDEN VALLEY (El Dorado) This camp was settled about 1850 between the towns of Coloma and Georgetown, and acquired a post office in 1852. The Black Oak started out as just a pocket mine and went on to produce more than $1 million. A miner also found a gold nugget in 1857 worth more than $500. In 1865 the Taylor Mine was located and the owners recovered about $1 million before it closed. When the gold began to pinch out, residents turned to farming, which turned out to be more profitable than mining, thus the name Garden Valley.

GARLOCK (Kern) Eugene Garlock built the first ore mill, followed by founding of the camp that took his name. It began to decline during the late 1800s when a 28-mile railroad spur was completed to Johannesburg, connecting the Randsburg Mining Company's Yellow Aster Mine with the Atchison, Topeka & Santa Fe Railroad. After a 100-stamp mill was established at the mine, it enabled the ore to be shipped to Barstow for more efficient milling. As a result Garlock's little mill was no longer needed. What was left of the site disappeared when the 1992 Landers earthquake occurred.

GENTRY GULCH (Mariposa) This gulch was located near the Merced River, east of Coulterville. One lucky miner found a piece of quartz that was filled with more than 130 ounces of gold, worth about $2,600. The first mine that was developed averaged about $100 per ton of ore. Another property that was located by a Cherokee Indian brought out about the same amount. By 1859 there was an 8-stamp mill and two arrastres, but a year later the mill was shut down and then sold. The Hasloe Mine, discovered by John Funk in 1854, yielded more than $3 million in gold.

GEORGETOWN (El Dorado) A miner named Hudson was the first to find gold and was lucky enough to acquire in one month more than $20,000. This site was first called Growlersburg because the gold nuggets were so big they made a growling sound in the miners' gold pans. A post office was established in 1851 with William T. Gibbs as postmaster, and the site was renamed for either George Phipps or George Ehrenhaft. Georgetown became a major trading center for more than 100 camps and about 10,000 miners. In 1852 the town was almost destroyed by a fire started by a photographer's flash powder. Residents rebuilt the place, but made the streets extra wide to serve as a firebreak just in case.

The nearby mines were very rich. In one instance a solid mass of gold was struck which yielded more than $40,000. During 1853 one of the mines produced about $2 million, and another a little more than $2 million. At nearby Hudson Gulch, placers realized more than $1 million. Just north of Georgetown, the Beebe Mine produced more than $1 million. By 1887 four mills with 35 stamps were in operation, with a few continuing into the early 1900s. The Alpine Mine began its operations in the 1860s. By 1884 it was recovering more than $72 per ton of ore of gold. Using a 10-stamp mill and a dredge, the mine continued working into the 1930s, until the Beebe Gold Mining Company took over. Under new management, it produced more than $450,000 worth of gold.

Georgetown celebrates its Annual Founder's Day each August. For times and dates, call the El Dorado County Chamber of Commerce at (530) 621-5885.

GEORGIA HILL (Placer) The site was named for miners from Georgia who accidentally found gold in 1851. The claims proved to be very rich and a majority of the ore was extracted using hydraulics. Georgia Hill was established between forks of the American River just south of Yankee Jim's. By 1868 more than $5 million in gold had been recovered.

GEORGIA SLIDE (El Dorado) Just north of Georgetown, the discovery of rich gold seams brought in about 100 miners during 1850. Most of the gold was extracted by hydraulicking between 1860 and 1870. The Georgia Slide Mine was named for the miners who hailed from that state, and recovered almost $6 million.

GIBSONVILLE (Sierra) This camp was situated on an ancient gold-bearing river channel at Little Slate Creek. A miner named Gibson discovered a deposit along a ridge during the 1850s, and the prospectors came en masse to stake their claims. The population warranted a post office which was established in 1855. Gibsonville was a busy trading center, but by the 1860s the gold had played out and the town died.

GILTA (Siskiyou) Placer and quartz mining in the southwest corner of the county was very successful. The camp itself was mainly a supply and trading point for the miners, and the post office was opened in 1892. Nearby was the Gilta Mine (owned by August and Henry Dannenbrink), which yielded about $1 million before 1900.

GLENCOE (Calaveras) First called Mosquito Gulch, the camp later took its name from a place in Scotland, a site of the Massacre of Glencoe of 1692, in which members of Clan McDonald were killed by Clan Campbell. It was named by settler L.P. Terwilliger and a post office opened in 1878. The site was located about ten miles northeast of Mokelumne Hill. Glencoe served as a trading center for the nearby mines.

GOLD BLUFF (Humboldt) After gold was discovered there in 1850 by Herman Ehrenberg, the camp was established near the town of Orick. Miners who followed went on to work the gold placers in the Klamath-Trinity area. Gold Bluff's deposits came to a little over $1 million.

GOLD FLAT (Nevada) Just south of Nevada City, gold placers were found in 1851, and Gold Flat became a settlement of more than 300 people. For some unknown reason, it was deserted two years later. During 1880 people moved back when new quartz mines were opened. Some of the claims were extracting ore valued at $60–$180 per ton. The largest producer was the Gold Flat-Posoti Mine, which realized about $90,000.

GOLD HILL (Placer) Situated below Ophir in the Auburn Range, Gold Hill attracted miners who found placer deposits in 1851. Most of the placers were superficial and not very rich; it was the lodes later discovered that contained high values. One of the mines located in 1860 by Sandy Bowers recovered close to $10,000 in gold. The Plato Mine was also a good producer, yielding between $150,000 and $200,000. When the ore ran out the camp was gone.

GOLD RUN (Placer) Gold Run was first called Mountain Springs. Its post office opened in 1863 with O.W. Hollenbeck as postmaster (he later became county treasurer). The Central Pacific Railroad's line reached the camp in July of 1866. Gold Run received its name because of the existence of a "run" of auriferous gravel. So rich were the small claims that some of them were bringing out about $18,000. Hydraulic mining later came into play, but a suit was brought against a mining company by the attorney general for the vast amount of debris that was being washed into the American River. A temporary halt to hydraulicking was put in place. Between 1865 and 1878 more than $7 million had been extracted from the nearby mines.

GOLDSTONE (San Bernardino) Goldstone didn't get started until the early 1900s. It was settled about 30 miles north of the town of Barstow. Most of the placer mining was conducted in the gulches by dry washing since there was no water. The Red Bridge Mine was opened during this time, recovering ore valued at about $200 a ton in gold.

This is the place where NASA has some of its installations, most notably a station that performs radio astronomy on the synchrotron emission from Jupiter's radiation belt, searching for disturbances caused by dust from the impact of Comet Shoemaker-Levy.

GOODYEARS BAR (Sierra) Miles Goodyear and his brother arrived about 1847 and discovered gold on a sand bar along the Yuba River. News spread fast and hungry gold seekers flocked to the area, causing a camp to be formed. It was located at the foothills of Grizzly Peak, which the Indians once used as a lookout. During the height of the gold rush, more than 10,000 people lived at Goodyears Bar. The ground was so rich that on one section of the bar some miners found over $2,000 worth of gold in one wheelbarrow full of dirt. Some of the camps on the bar had unique names: Cut-Throat Bar and Hoodoo Bar are two examples.

By 1851 Goodyears Bar had grown considerably, and the simple miners' digs were replaced by trade buildings, saloons, and hotels. To meet the needs of the growing population, a post office was established in 1851 and named for Miles Goodyear. The Langton brothers also had a very prosperous business there, hauling their pack trains full of supplies to the outlying mining camps.

GOTTVILLE (Siskiyou) Established in 1887, the camp was named for mine owner William N. Gott, who was also postmaster when the office opened in 1887. Located along the Klamath River northwest of Yreka, it served as a supply and trading point for the small, nearby camps.

GOUGE EYE (Nevada) This interesting name came about after a French miner jumped a prospector's claim and lost his eye while fighting with the owner. The site was located north of Dutch Flat, and named by Thomas Concord. Mining was conducted about 1855, but with little results, and by 1892 Chinese miners were there gleaning what was left of the placers.

GRAHAMS RANCH (Santa Cruz) This was a short-lived camp. Prospectors were drawn to the area in 1854 after a quartz boulder was found that assayed out at more than $27,000. A mining district was established, but aside from the boulder, not much ore was found and the place was soon deserted.

GRANITEVILLE (Nevada) Located near the Middle Fork of Yuba River, this camp was first called Eureka, but when the post office was established in 1867, it was changed to Graniteville. Total production of the nearby mines was estimated at about $5 million.

Some of the gold seekers didn't fare well. Michael Brennan and his wife moved to Grass Valley, where he began his search for gold. Upon discovering a vein where the ore could be extracted only by drift mining, he wrote to a New York firm for money. Brennan also invested all the money he had in the project, which failed miserably. Destitute and disgraced, Brennan shot himself, but not before he killed the rest of his family.[70]

GRASS VALLEY (Nevada) The first man to discover gold there was D. Stump from Oregon and two of his companions in 1848. Then in 1850, through dumb luck a German was carrying a bucket of water and stumbled over a piece of gold-bearing quartz. Another version of the story is that George Knight stubbed his toe and when he looked down there was a piece of gold-laced quartz. At any rate, news spread about the excitement over the discovery and Grass Valley was established just south of Nevada City. Within six years more than $4 million in gold had been realized. In 1850 George Roberts accidentally found some gold while surveying trees. A year later he sold the mine for about $350, not knowing it would bring more than $1 million a number of years later. About 1855 miners named Moulder and Bunch located the Lone Jack Mine, which yielded more than $500,000. A year later the Heuston brothers discovered a lode which was sold a number of times, and realized more than $500,000 before operations ceased. Just southeast of Grass Valley on Ophir Hill, the Empire Mine was discovered in 1850. It operated continuously until 1956, yielding more than $130 million.

Famous dancer Lola Montez came to Grass Valley during its heyday, but according to residents, she wasn't really that popular. One of the settlers said, "She had heavy black hair and the most brilliant flashing eyes I ever beheld. But ordinarily she was such a slattern that to me she was frankly disgusting. When attired in a low-necked gown as was her usual custom, even her liberal use of powder failed to conceal the fact that she stood much in need of a good application of soap and water."[71] She spent her last days in New York, doing rescue work among the local women there.

One of Grass Valley's miners had a most unique mule named Jasper who was addicted to chewing tobacco. His owner gave Jasper the tobacco so he would quit mooching food scraps from other residents. When Jasper wanted more tobacco he would quit his work in the mine, lift up his rear hooves and kick the side of the drift.[72] Another mule that was well-loved by her owner made headlines in the local paper: "Fifteen-year-old Mabel is dead. Dr. E.M. Roessner made her as comfortable as possible during her last few hours on earth, but the grim angel of death yesterday morning summoned Mabel to Mules' Paradise. For years Mabel has been a faithful worker at the North Star mine.... She was pulling an ore train Tuesday evening when she accidentally slipped on a board and plunged fifteen feet to the bottom of an ore bin. She landed on her haunches somewhat 'feet up' in the apex of the chute."[73]

There were so few women at Grass Valley that those who were of marriageable age were scarcer than hens' teeth. A gentleman who took up his residence in 1852 said, "When I came here there were only two girls in the town, and one of them was engaged, so I had to take the other. You see that it was a clear case of Hobson's choice."[74]

Grass Valley hosts a St. Piran's Day Festival in March. Other activities include the Bluegrass Festival, Music in the Mountains Summer Festival, Draft Horse Classic and Harvest Fair, and Celtic Festival. The Empire Mine also has a small museum. For times and dates, call the Grass Valley–Nevada County Chamber of Commerce at (530) 273-4667. Toll free in California, 1-800-655-4667.

GREENWOOD (El Dorado) This place was named for either mountaineer Caleb Greenwood, or his son John Greenwood, who discovered gold along Greenwood Creek in 1848. When the Greenwood Mine began operations, a 10-stamp mill was built. The post office opened in 1852 with George C. Blodgett as postmaster. Greenwood later found that hunting venison for the miners was more profitable than mining.

GRIZZLY FLATS (El Dorado) This site was discovered in 1850 and named for a grizzly bear that miner Buck Ramsey shot while he and his friends were having an evening meal at their camp. The place was located along Steely Fork, tributary to the Cosumnes River. About 1852 miner Victor Steely found gold on an outcrop. The miners who followed worked the diggings which averaged about $20 a day. That same year three stamp mills were built when the mines opened. The post office was established in 1855. A good producer was the Mount Pleasant Mine discovered in 1851 that yielded more than $1 million in gold. Just north of Grizzly Flats was the Cosumnes Mine (Miwok for "salmon people"), where a 4-stamp mill was built that processed ore valued at about $25 a ton in gold. Some mining continued into the 1930s. A lode that was located in 1948 recovered more than $1 million.

GROVELAND (Tuolumne) Mexican miners discovered gold here in 1849, and a year later more than 2,000 of them were working the rich placers. Groveland was first named Second Garrote, referring to someone who was hanged there. Near Groveland the Longfellow Mine was discovered which recovered about $500,000 in gold. When the Indians began harassing the miners, most of them left. By the 1860s not much was left of the place.

GRUB GULCH (Madera) This name was descriptive of the miners having to grub out their meager living. Settled about 1850 a few miles north of Coarsegold, the camp acquired a post office in 1883. Some mining was conducted during the 1880s and when the Gambetta Mine opened, Grub Gulch began to prosper. Although the place touted hotels, saloons, general stores, and a rooming house, the gold lasted only until the early 1900s. The Gambetta was worked until 1904 when it was abandoned, but not before yielding about $490,000. Another lode that produced well was the Josephine which recovered nearly $360,000 until heavy rains caused a cave-in and the mine closed down in 1914. The combined production of surrounding mines came to a little over $1 million.

HALLORAN SPRINGS (San Bernardino) Gold was first discovered in 1900. During 1930 the Telegraph mine was located, but produced only about $100,000 within an eight-year period. Gold seekers who came later to check out the prospects did not fare well and left. The camp was named for J.F. Halloran, publisher of *Mining & Scientific Press.*

HAMBURG BAR (Siskiyou) Founder Sigmund Simon named the place for a town in Germany in 1850. Simon opened the first store and a post office was established in 1878. The site was located along the Klamath River by its confluence with the Scott River. Hamburg spent its life as a trading and supply center for the placer and hydraulic mines. Two nearby lodes yielded moderately well: The Grider, with a monthly production of about $2,000, and the Portuguese, which was recovering between $10,000 and $15,000 a year. By 1863 the gold was depleted and the town was abandoned. The last traces of the place were washed away by a flood.

Mule hauling ore from the Empire Mine, Grass Valley, ca. 1910 (courtesy California Geological Survey).

HAPPY CAMP (Siskiyou) This community lies along the wild and scenic Klamath River about 20 miles south of the Oregon border. A group of forty-niners traveling up the river camped on Indian Creek, and were confident their gold-mining efforts were going to be very prosperous and were so exuberant they named the campground Happy Camp. Prior to that, it had been called Murderer's Bar for claim jumpers who killed a number of the owners. Its post office was established in 1858. Miners were making between $9 and $15 per day at the diggings. Happy Camp became very prosperous by serving as a trading center for all the local mining camps. It was estimated that more than $4 million was recovered from lode and placer mines. By 1913 most of the diggings had panned out and were left for the Chinese to clean up. This region is the ancestral home of the Karuk Indians, who are recognized throughout the world for their basketry.

HARDSCRABBLE GULCH (Siskiyou) This short-lived camp bore a transfer name from Wisconsin from where two of the miners hailed. The Hi You Gulch Mining Company, led by superintendent Noah Williams, began operations in 1854 along McAdams Creek, but the gold began to pinch out the following year.

HARRISBURG (Inyo) Located in Death Valley National Monument, this tent town was named for Frank "Shorty" Harris who found gold deposits during the early 1900s, along with his prospecting friend, Peter Aguerreberry. Harris was a colorful character of Death Valley lore, who quite enjoyed the drink. He wrote his own epitaph that says it all: "Here lies Shorty Harris, a single-blanket-jackass prospector."[75] When the camp of Skidoo came into prominence, Harrisburg faded into the sunset.

HARRISON GULCH (Shasta) Mining began about 1894 at the gulch which is a tributary of Cottonwood Creek, and was named for settler W. Harrison. The most important producer was the Midas Mine discovered in 1894, which yielded about $3.5 million in gold, but in 1914 a fire forced it to close. It was reopened the following year under new ownership, and operations continued until 1920 when another $500,000 was recovered.

HATCHET CREEK (Trinity) This small camp was settled during the 1850s, receiving its name after an Indian stole a hatchet from a party of immigrants passing through. Placer mining was conducted by prospectors who reported making between $5–$20 each. Hydraulic mining later began, but brought only about 33¢ a cubic yard of gravel. The site was located along Hatchet Creek north of Trinity Center.

HAVILAH (Kern) Bearing an Arabic name meaning "sand region," Havilah was established by Asbury Harpeding near the Kern River after gold was discovered about 1864. The camp warranted a post office in 1866 after it acquired a population of about 3,000. In addition to becoming a trading center for the miners, it also crushed ore with its three stamp mills. Havilah became county seat in 1866 but the designation was taken away and given to Bakersfield in 1874. The town then quickly declined and the region later became farming land.

HAWKINSVILLE (Siskiyou) This short-lived camp was settled along Yreka Creek just north of the town of Yreka in 1854. First called Frog Town, it was renamed for miner Jacob Hawkins, and the post office opened in 1888. It served as a supply center for the mines as did Yreka. The Hawkinsville diggings were described as "the richest and most extensive that have been discovered north of the trinity range of mountains. They have yielded an immense amount since their discovery, but this is a small amount in comparison to that which these placers will yet yield."[76] A 99-mile ditch was built from Parks Creek to serve the placer mines. One of the men found a nugget worth a little more than $2,000. Along Yreka Creek 19 buckets of gravel yielded $25,000. But because Hawkinsville was so close to the fast-growing town of Yreka, it couldn't compete and quietly passed away.

HAYDEN HILL (Lassen) Discovery of gold in the 1860s by prospectors searching for the Lost Cabin Mine caused the camp to be founded on the Modoc Plateau with a post office in 1878. It was named for its postmaster, J.W. Hayden. During the 1870s lode mines were discovered that contained both gold and silver. The Eagle extracted more than $1 million. Other properties realized more than $2.5 million through the early 1900s. Most of the town was destroyed by fire in 1910.

HAYWARD MINE (Amador) This mine was discovered by Alvinza Hayward, who developed it into the best paying lode in the region. It was also the deepest, reaching a depth of more than 2,000 feet. Charity H. Hayward married Alvinza and lived at the old Sutter Diggings that would become the famous Hayward Mine. "Alvinza felt very keenly about asking me to share such a primitive cabin

Chinese placer miners at Hawkinsville, ca. 1880s (photograph courtesy of the Siskiyou County Museum, Yreka, CA, #P11917–4).

with him, but I laughed and told him I could be just as happy there as I could be in a millionaire's house." They struggled in their quest for gold until they were heavily in debt. Charity later wrote, "I did everything I could to cheer him. Finally only two Mexicans were left, and they were leaving, starved out, and Alvinza said he would dig alone. Then one of them drove his pick into the big vein and the world burst open." [77] Alvinza made his fortune, recovering between $28,000 and $65,000 per month from his property.

HELLS HOLLOW (Mariposa) This is an appropriate name because the ore found during the 1860s was located in a nasty, steep gulch near the Merced River. The hollow contained mainly placers, but it is not known how successful they were. In order to get the gold out of the precipitous ravine, a small railroad was built by John Fremont. The cars were controlled by multiple brakes and descended into the ravine by gravity, then the cars were pulled back up by mules.

HERBERTVILLE (Amador) Only moderate profits were made from the small mines near town. One of the first was discovered by two men named Jones and Davis who reported ore yielding about $50 a ton. But overall the claims met with little success, although there were 20 stamp mills in operation. The camp was located south of Amador City. Herbertville was named for a miner and minister, the Reverend Lemuel Herbert. Only a few ruins remain.

HESSE'S CROSSING (Yuba) This site was named for Thomas Hesse, who built a bridge across the Yuba River in 1851. Two years later gold was found by two miners who extracted about $500 in a week's time. In 1854 this short-lived camp made news in the local paper: "New and rich diggings have been discovered at Hesse's Crossing. In one claim — two young men have been at work for nine weeks. The first weeks' work netted 91 dollars and the production has steadily increased up to last week, when they took out 500 dollars. There are but few miners in that region at present."[78]

HINCKLEY MINE (Amador) Named for a Mr. Hinckley, this pocket mine was accidentally found in 1863 while he was digging a post hole for a fence. Costing about $6,000 to get the mine into operation, Hinckley realized more than $18,000 in gold.

HITES COVE (Mariposa) Located on the South Fork of the Merced River, this site was named for John R. Hite who found gold in 1862. A post office was established six years later. When Hite's mine began operations the ore was crushed by arrastres, and as it got larger the arrastres were replaced with a 20-stamp mill. Hite realized more than $3 million from his claim. In 1924 a fire destroyed all that was left of the camp.

HOBOKEN (Trinity) When gold was discovered along the New River during the 1850s, Hoboken sprang up to became a supply and trading point for the miners, with a post office established in 1892. Hoboken is a transfer name from New Jersey, a Lenape name for "tobacco pipe."

HODSON (Calaveras) The Hodson area contained a number of well-producing mines, and acquired a post office in 1898. The district encompassed a good ten miles of land from Copperopolis to Salt Springs Valley. In 1904 Clayburn and Willis Womble with the Mountain King Gold Mining Company built a 120-stamp mill, which was one of the largest mills in the state. The rich-bearing mines recovered more than $6 million in gold. The mill was later refitted to treat copper ore that was discovered nearby.

HOGGS DIGGINGS (El Dorado) Miner John B. Hogg discovered gold nearby in 1849, and a small camp was formed just southeast of Auburn. Prospectors experienced only fair success working the placers, although one lucky man found a lump of pure gold worth $400, and another discovered a nugget that weighed more than 80 ounces. In addition to spending time mining, Hogg had earlier served as deputy secretary of state in his home state of Tennessee.

HOLCOMB VALLEY (San Bernardino) This site is located just north of Big Bear Lake. The gold in the valley was discovered by William F. Holcomb who came to try his luck at the gold fields in Northern California. Having no success, he headed back to the valley where he discovered gold about 1859. It didn't take long for word to get out and soon the place was thick with miners. The valley was remote, and a pack trail up Santa Ana Canyon was its only thread of communication with the outside world. When the miners got ready to haul in their equipment, Jed Van Dusen, who was the town blacksmith, built a road for them. The men paid him $2 in gold dust for his efforts. During the 1860s the placers yielded about $7 million in gold.

HORNITOS (Mariposa) This town bears a Spanish name that means "little ovens." It was first settled by Mexicans about 1848, and a post office was established in 1856. The Mexicans may have named the site because of the heat, or for the little baking ovens of rock and mud made by Germans who cooked their food in them. The camp was well known for its fandango halls and fiestas.

The mines that were discovered about 1851 gave high yields. One claim was bringing out about $1,000 a day, and another vein more than $75,000 from surface diggings. Another produced more than $600,000. Hornitos had three stamp mills and two arrastres to process the ore and Wells Fargo was shipping out more than $40,000 of ore a day. Total production was estimated at about $30 million.

In 1873 a fire destroyed a number of buildings, as described in a local paper: "Fire in Hornitos. A fire broke out in the lard factory in rear of Geo. Reeb's butcher-shop, in Hornitos, on Tuesday last about 3 o'clock P.M., which destroyed the butcher-shop, George Reeb's residence in rear of the shop.... Considerable damage was also done to a number of fire-proof buildings, besides burning fences and outhouses.... The loss is a serious one to the sufferers, some of whom are totally unable to rebuild and resume business. Loss $14,000, insurance about $7,500."[79]

This is where the famous Ghirardelli Chocolate got its start. Domingo Ghirardelli came from Italy with 600 pounds of his chocolate. He first tried his luck at mining, but was a miserable failure, so he opened a store. In 1852 he moved to San Francisco and there he made his fortune.

Legend has it that famous bandit Joaquin Murieta spent time at Hornitos. He supposedly had a tunnel built from one side of town to the other so he could elude his pursuers. But a store owner scoffed at that, saying, "The tunnel was built to roll barrels from the brewery to the stable below. I used to play in it when I was kid. Murieta! Huh!"[80]

HORSESHOE BEND (Mariposa) So named for its location on a horseshoe-shaped bend along the Merced River, this camp was established in 1850.

Reeb's store, Hornitos (Mariposa County Museum and History Center, Mariposa, CA).

Most of the mining was conducted using hydraulics, but the workers were receiving only about $4 a day. The site is now under Lake McClure.

HOWLAND FLAT (Sierra) Settled during 1850 in Gold Canyon, the camp had good mining prospects. By 1867 more than $3.5 million had been realized from the nearby mines. It experienced growth when the Yuba River Canal was built in 1856.

HUESTON HILL (Nevada) These diggings were named for the Hueston brothers who discovered gold about 1853. Located south of Grass Valley, about $500,000 was recovered in a three-year period.

HUMBUG CANYON (Placer) This canyon was tributary to the North Fork of the American River, and was first explored in 1850. But the gold was insufficient so prospectors didn't bother to try and locate other claims. First named Mississippi Canyon, it was changed to Humbug because the early prospectors found minimal gold. Humbug was a word used by miners to express disappointment. Later prospectors found the banks and bed of the canyon to have enough gold to continue digging.

HUMBUG CITY/CREEK (Siskiyou) The creek has its origins in the Humbug Mountains and is a tributary to the Klamath River. The bars on the river were found to be rich in gold; between $500,000 and $1 million of gold was realized from the placer and lode mines. Humbug was located along the creek about ten miles from the town of Yreka. During 1851 prospectors heading to this area were met by others leaving saying there was nothing to be found, that it was just a humbug. But the new men did indeed find gold, and before long there was a migration from Yreka of more than 600 men to work the diggings.

This camp served as a trading and supply point for the miners. "Humbug City was full of miners drinking and gambling and carousing. There was a man named Nobles keeping a drinking saloon and he played on the banjo and gambled, and he had a Fancy Woman, very good looking.... She was a stout robust splendid looking German girl from Strausburg, France, and many a miner paid 50 cents for a drink to get to see her."[81] The town was abandoned after gold was discovered in Idaho. Then a fire burned most of the place down in 1862.

HUNTER CANYON (Inyo) Gold discovered in this remote section of the Inyo Mountains generated very little interest. It was named for miner W.L. Hunter who worked three arrastres until the 1890s. There simply was not enough gold and the place was deserted.

HUNTER VALLEY (Mariposa) During the 1850s, mainly placer mining was conducted here. The site was named for William W. Hunter and the post office opened in 1907. The camp was located south of Horseshoe Bend near Bear Valley. Two of the largest lode mines produced nearly $600,000 between 1863 and 1870.

HURLETON (Butte) This little place served as a trading center for the local mines. It received a post office in 1880. Located east of Oroville, the camp was named for rancher Smith H. Hurles.

IGO (Shasta) This interesting name comes from tradition that says a miner named McPherson had a son who followed him everywhere. Whenever the boy wanted to go to the mines he would say "I go," but his father would not allow him there because it was too dangerous. The origin of the name has been disputed, and some believe it is a Maidu word for "head," or named for a miner named Igo. The camp was located southwest of Reading and received its post office in 1873. Placer gold was discovered in the 1850s, followed by hydraulic mining. "Gravels along Clear Creek near Igo were worked in the early days, but there is no published account of their production. The district was revived in 1933 and through 1942 produced more than 113,000 ounces of placer gold, and based on $20 an ounce, amounted to $2.2 million."[82]

ILLINOIS CANYON (El Dorado) Also known as Thousand Dollar Canyon because someone found a nugget worth $1,000 in 1849, the name honored the Illinois Indians, whose name means "human beings." Most of the claims were seam diggings, and little gold was found in the canyon.

INDIANA BAR (Plumas) Gold was found by a man named Windeler and his friends in 1851. The camp was settled near Rich Bar on a fork of the Feather River. Windeler and company made less than $5 a day at the diggings and Indian Bar was soon abandoned.

INSKIP (Butte) Inskip received its name from a Dutchman named Eenskip who discovered gold in the 1850s. Its post office was opened in 1862. The camp was situated near Butte Creek and served as a trading center and stage station. The Inskip Hotel is listed on the National Register of Historic Places.

IONE (Amador) Located on Sutter Creek just west of Jackson, gold was discovered about 1848. Prospectors were recovering about $1,000 a week from their placers. When the gold began to pinch out, hydraulics were brought in to recover what remained. William Hicks and Moses Childers were the first to settle at Ione, and its post office opened in 1857. It became a trading center, in addition to serving as a stop for the stagecoaches transporting ore. Before 1876 Ione was accessible only by the worst kind of wagon roads. When the Central Pacific Railroad built its line near town, Ione had easy access to the outside world. After the gold ran out, coal and copper came into prominence. First called Bedbug and Freezeout, Ione was named for the heroine in Edward Bulwer-Lyttons' novel *The Last Days of Pompeii.*

During 1861-62 an immense freshet caused the mine tailings to overflow. Accumulations from thousands of mining claims were swept away from Dry and Sutter creeks. Farmers and wine growers were flooded with debris, and Sutter Creek enlarged to more than 100 feet wide with the result that many of Ione's buildings were swept away.

Famous jazz pianist Dave Brubeck was born in Ione. Daniel Stewart established the general store in 1855, which is still owned and operated by his descendants. The Ione Hotel that was built during the 1850s is still open for business. Ione celebrates its annual Homecoming every May. It is a continuation of the centennial celebration of 1876, and has a different theme each year depicting the gold rush. For dates and directions, call Ione at (209) 274-2412.

IOWA HILL (Placer) Although gold was discovered nearby in 1849, rich placers were not found until about 1854. Iowa Hill was east of the town of Colfax, and received a post office in 1854. The first discovery to make news was the Jamison Claim, which was bringing out more than 200 ounces of gold a day. During 1856 about $100,000 a week was being realized from the various claims. A miner found a 58-ounce lump of gold that brought more than $1,000. Two of the better producing mines were the Big Dipper and the Gleason, each yielding more than $1 million. The Morning Star, largest of the drift mines, recovered about $1.7 million. Total production in the district came to about $10 million.

Iowa Hill began its downward slide after it experienced a fire in 1857, in addition to the hydraulic mining that almost washed out the town. The *Iowa Hill News* wrote, "This morning at 3 o'clock the alarm of fire was given. In a few moments the central portion of the town was in flames. The fire was

Stage at Ione (courtesy of the Amador County Archives, Amador, California).

first discovered in the back part of the City Bakery, and is supposed to have been the work of an incendiary.... It is almost impossible to approximate anything near the loss sustained. So rapidly did the flames extend that the fire-proof cellars under the most of the large stores were of little use, there being no time to remove merchandise into them."[83]

IRISH HILL (Amador) Just north of Ione, the Irish Hill diggings were reported to be very rich. Three miners named Hawley, Nelson, and Millner extracted about $9,000 in seven months. Water for the mines was brought in via the Plymouth ditch. The ground was owned by Alvinza Hayward, who also owned the Hayward Mine. Hawley worked his claim until he made enough money, then headed back East. Millner wasn't so lucky. He was killed when a bank of his claim caved in. Hydraulic mining eventually destroyed the site.

ISABELLA (Kern) Named for the Spanish queen and patron of Christopher Columbus, Isabella served as a supply and trading center for the area's mines. It was located northeast of Bakersfield, and acquired a post office in 1896, with George King as postmaster. When the mines played out, Isabella was temporarily saved from extinction because it was situated along a tourist route.

ITALIAN BAR (Tuolumne) This rich bar northeast of Columbia was discovered in the 1850s. Two stamp mills were erected to treat the ore. Total value of the gold taken from nearby mines amounted to about $3.5 million by 1899.

IVANPAH (San Bernardino) Ivanpah is Southern Paiute for "good (or clear) water." Local legend says the name means "dove spring," though there's little evidence to support this origin. The mountains were exploited for silver in the 1800s, and the camp was established at the foot of the Clark Mountains. A 10-stamp mill was erected to crush the ore from mines that recovered about $4 million. The silver was transported to the San Francisco smelters for reduction. Financial problems later plagued the mining companies, and by 1880 Ivanpah was totally deserted.

The Ivanpah Mountains contain the only known dinosaur tracks in California, and also the world's largest Joshua tree forest.

JACKASS GULCH (Tuolumne) Tradition says this site was named when a miner chasing his obstinate jackass found a rich deposit of gold. The gulch was near the Stanislauw River. Although the miners experienced some success, this was a short-lived camp.

JACKASS HILL (Tuolumne) This camp was settled west of Tuttletown and experienced a successful but short life. A five-stamp mill was built to treat ore from some claims that brought out as much as $10,000. One of the miners had a spot that garnered him between $100 and $300 a day for a number of years. When the gold was depleted, prospectors took their searches elsewhere. The place was named for the pack trains that stopped there to graze the more than 2,000 jackasses going to and from the mines. Samuel Clemens came to Jackass Hill in 1846 and lived there for a short time.

JACKSON (Amador) Jackson is situated between Sutter Creek and the Mokelumne River. It was originally called Bottileas, Spanish for "bottles," because teamsters hauling freight and miners on their way to the gold fields rested here and "So convivial were these overnight stops that piles of bottles collected and gave the town its first name."[84] The post office was established in 1851 with Henry Mann as postmaster. It was changed to Jackson for miner Colonel Aldan Apollo Jackson, who was there in 1848. The mines in the region were very rich, one of the biggest being the Kennedy, which recovered $34.2 million. This property had four remarkable wheels built to lift the tailings over a mountain ridge for disposal, quite an engineering feat.

A number of lodes were extremely good producers: The Argonaut yielded more than $25 million. The Consolidated Keystone was discovered in 1853 and yielded $24 million although it was plagued with ownership problems. Other lodes yielded nearly $8 million in gold. The Zeile Mine was discovered by Leonard Coney. He later sold it to a Dr. Deile for about $75,000. Operating until the early 1900s, it recovered about $5 million in gold. Near Jackson the Oneida Mine was discovered while a hunter was chasing a rabbit and came across some gold quartz outcrops. It was located between Sutter Creek and the town of Jackson and there were high expectations of high yields. The mine was leased to Dr. E.B. Harris, who later sold it to Fullen, Flemming, Bergon & Company, which met with moderate success. In 1865 William Stewart purchased the mine for about $100,000 but he had no luck. Finally, in 1881 a Boston company took over with more success, recovering about $2.5 million.

During 1861 Jackson experienced a flood when it rained beginning about the first of December and continued for several weeks. The ground was so saturated that more than 20 buildings were washed off their foundations.

In 1851 Jackson became the county seat. Early county elections took too much time and surreptitious methods were often used to move the seats from one place to another. Two of the town's residents, Charles Boynton and Theodore Mudge, traveled to Double Springs with their cohorts, went to the bar and invited everyone to have a drink on them. "While one detachment of the enemy artfully engaged the attention of Colonel Collyer, who was county clerk, another detachment at the other end of the room gathered the archives under his arm, tumbled them into a buggy and ran away with them to Jackson. The archives were deposited with the proper ceremonies, the liquors being remarkably fine, and Jackson became the center of government."[85] The county seat was later moved to Mokelumne Hill.

The Kennedy Gold Mine is open for self-guided tours between mid–March and October on weekends. For directions and tour reservations, call (209) 223-9542.

JACKSONVILLE (Tuolumne) This camp was settled in 1849 and was also named for Colonel Aldan Apollo Jackson. It was founded by Julian

Kennedy Mine, Jackson (courtesy Kennedy Mine Foundation, Jackson, CA).

Smart who planted the first orchard in the county. Its post office was established in 1851. Located along the Tuolumne River, gold in the region produced about $9 million. The site was inundated by the Don Pedro Reservoir.

JAMESTOWN (Tuolumne) Jamestown was established about 1849 near the town of Sonora with a post office in 1853. It was named for Colonel George F. James, an attorney from San Francisco. After he moved here, James established a store and hotel. He was also the *alcalde* (judge) for the area. About 1849, James was run out of town after it was discovered he used investors' money to set up mining operations and paid the workers with scrip. When he discovered there wasn't enough money to repay his creditors, James skedaddled out of town in the middle of the night, and the investors and miners were left holding the bag.[86] It's not really clear if he left at night, was ordered out of town, or was just ahead of a lynch mob.

By 1850 most of the placers had run their course. A few years later the New York Tunnel drift mine began operations, bringing renewed life to the town. The App Mine was located in the Quartz Mountains near Jamestown and was named for John App, who found the gold with prospector friends. App moved to California from Illinois via wagon train and married Leanna Donner, of the famed Donner Party. During the 1860s miners were pulling out ore at about $15 per ton. App later sold his interest for about $40,000. Other nearby mines recovered about $30 million in gold.

Jamestown Railtown has steam-powered train excursions every weekend between April and October. For information, call the Railtown Historic Park at (209) 984-3953, or the Tuolumne County Chamber of Commerce at (209) 532-4212.

JAWBONE CANYON (Kern) Some gold mining was conducted in the canyon through the 1930s, but it is not known what the miners realized from their claims. Situated about fifteen miles north of Mojave, the canyon received its name because its shape resembled a jawbone.

JELLY (Tehama) Most of the gold placers along the Sacramento River here were worked by Chinese miners, but they didn't stay long because little gold was recovered. The site was named for Andrew Jelly who became postmaster when the office opened in 1901. He also operated a ferry across the river.

JENNY LIND FLAT (Placer) Settled in the 1850s, the site received a post office in 1857. It was located near the town of Forest Hill. Placer mining was at first conducted on the nearby hillsides, then the mines opened. It was reported that about $1 million was realized. There were also rumors that some claims were paying from $6,000 to $10,000 a day. The camp was named for the Swedish singer Jenny Lind, and although it was rumored she was in some of the camps, Lind never set foot in California.

JILLSONVILLE (Shasta) Gold was found during the 1860s north of French Gulch. After the Goldstone Mine had been located, it was purchased by I.O. Gillson in 1900. The town was named for him but spelled with a "J." The property yielded more than $4 million.

JOHANNESBURG (Kern) This town was founded about three miles from Randsburg by a man named Depew who, along with his associates, had a railroad built to connect with the Santa Fe. The post office was established in 1897. Johannesburg is a transfer name from South Africa, which honored government surveyor Johannes Rissik, or mining commissioner Johannes Meyer. The town was a supply point for the mines in the Randsburg and Kern River mining camps.

JOUBERTS DIGGINGS (Sierra) These placers were found by miner J. Joubert in 1852 north of the town of Camptonville. By 1927 more than $1 million had been taken out using hydraulics.

JULIAN (San Diego) During 1869 Fred Coleman discovered gold near the future camp. A year later another claim was found by Mike Julian and Drury Baily. When word got out about the discovery of a 10-pound lump of gold-bearing quartz, the money-hungry miners flocked to the site and a camp was born northeast of San Diego with a post office established in 1870. It was named for Mike Julian, who later became a mine recorder. A number of mines were opened, but by 1876 the boom ended until the 1880s when more significant gold was discovered. Total yield of the major mines (Stonewall, Gold Queen, and Gold King) was about $5 million. When the gold played out Julian became a farming community.

Julian has a number of historical buildings. The Grosskopf House was owned by Chris Grosskopf, who was a miner and blacksmith. The Julian Jail that was built in 1914 at one time contained the only indoor toilet in town. Housed in an old brewery built in the 1880s is the Julian Pioneer Museum. Julian has an interesting History Hunt. It is an ongoing event, where visitors sleuth through town in search of answers to questions about historic sites. The prize is $50. For more information, directions and dates, call the museum at (760) 765-0227.

KAWEAH RIVER (Tulare) First called Four Creeks, the name was changed in 1853 to Kaweeyah, a Yokut word meaning "I sit down," "I squat here," or "here I rest." It was also interpreted as "crow water," indicative of the thousands of buzzards and crows that inhabited the site. A post office was established in 1879 to serve the mining community. The river rises in the Sierra Nevada and empties into the Tulare Lake Basin. There was some gold mining along the river, but not much was realized and the men left for more productive gold fields. The most gold recovered was about $70,000 during 1884. The river was discovered by the Gabriel Moraga Expedition in 1806.

KEARSARGE (Inyo) This camp was established in the Sierra Nevada and received its name because during the time the Union ship *Kearsarge* had destroyed the Confederate ship *Alabama*. Gold discoveries were fairly rich with some of the ore assaying at $900 a ton, but Kearsarge was a short-lived camp and by 1888 not much was left.

KEELER (Inyo) Named for steamboat operator Captain J.M. Keeler, the camp was mainly a shipping center for ore. Steamer ships on Owens Lake picked up the gold at Swansea and carried it to Keeler, where it was shipped to the Carson and Colorado Railroad. When the mines were depleted, the terminal at Keeler was abandoned.

KELLYS DIGGINGS (El Dorado) This site was named for a miner named Kelly, who may have been the only person to conduct placer mining there, acquiring about $15–$20 a day. The diggings were located near Volcano.

KELSEY (El Dorado) Settled about 1848 in Kelsey Canyon, the camp was named for Benjamin and Samuel Kelsey. Its post office was established in 1882. The mines along a fork of the American River were good producers and a number of stamp mills were brought in. A few miles from town was the Alhambra, discovered in 1883, which recovered more than $1 million. Another claim yielded between $1 million and $2 million. When the Kelsey Mine began operations it was producing only about $1.80 to $6.40 a ton and in later years didn't fare much better. This is the place where James Marshall of Coloma fame died in 1885.

KENNEBEC BAR (Yuba) In 1849 the Kennebec Company from Massachusetts founded the site, and gave it an the Abnaki name for "long, quiet water." Between 1849 and 1850 some gold was discovered, but none of it was worth the effort, and within a year the bar was deserted.

KENWORTHY (Riverside) This camp was established in the San Jacinto Mountains to supply provisions to the miners at Hemet. It was named for a British subject, Eugene Kenworthy, and the post office was established in 1897. By 1899 the gold had been exhausted, and the camp of Kenworthy passed away.

KERNVILLE (Kern) When gold was found in the Kern River during 1851, it brought a deluge of prospectors. The town was first called Whiskey Flat after a settler opened the first saloon. It was renamed for topographer Edward M. Kern, and the post office was established in 1868. Placer mining was initially conducted, but was later replaced by lode mines. The Big Blue Mine was discovered a few miles from town and a 16-stamp mill was erected to treat the ore. It became the major producer and was credited with $1.7 million in gold by 1933. Most of the workings of the mill were destroyed by fire in 1883. The mines were depleted by 1879, but Kernville continued to survive a while longer until it was inundated by the Isabella Reservoir. The town was then moved to a new site.

KINCADES FLAT (Tuolumne) Mining operations began about 1850 at the flat, which was located east of Jamestown. The rich ore rewarded the owners with more than $2 million by 1867. Total production came to about $3 million.

KLAMATH CITY (Del Norte) This camp was established at the mouth of the Klamath River in 1850 by Herman Ehrenberg after gold was found in the nearby hills. It served as a trading center for the miners, but the pickings were poor and by 1851 most of the men had left. The name is a Klamath expression for "encamped," or "yellow water lily," a major food source for the Indians.

KNAPPS RANCH (Tuolumne) Most of the gold

deposits nearby were surface because they were located in volcanic material. The site was situated southeast of Columbia. While placer mining was being conducted in the 1860s, one of the claim owners was realizing about $1,000 a month. Although the diggings were shallow, one piece of ground of about 5 acres brought out nearly $200,000. There were also some gold nuggets found that were valued at about $8,000 apiece.

KNIGHTS FERRY (Stanislaus) William Knight came from Indiana, worked as a fur trader and scout, then established a ferry on the Stanislaus River in 1848. He also built a trading post, and from there a camp developed with the establishment of a post office in 1851. The ferry transported the miners and their pack animals across the river to the Southern mines. The ferry could hold only 5–6 people, and wagons had to be taken apart and carried over piece-meal. After Knight died, the ferry was taken over by John and Lewis Dent, brothers-in-law of Ulysses S. Grant. About 1858 a solid bridge was built and the ferry was no longer needed.

Then in 1862 a flood came, the Stanislaus River roared down the channel and its mass hit the Knight's Ferry bridge, hurling it downstream. A local newspaper wrote: "We are informed by a gentleman who arrived last evening from Stockton that the whole town of Knights Ferry has been swept away — mill and everything clean swept off!"[87] The region received almost 50 inches of rain and the rivers flooded everything, but the town managed to survive as a trading center for ranchers and cattlemen after the placers dried up.

LA COMMODEDAD (Tuolumne) Settled in the 1850s, this mining camp was populated mainly by Mexican and French miners. Not much is known of the place except the miners were earning low wages from the site.

LA GRANGE (Stanislaus) La Grange was located along the Tuolumne River where French miners discovered placer gold in 1849. Originally called French Bar, the name was changed in 1854 when the post office was established. Louis M. Booth served as postmaster. It is a French expression for "the barn," although it may have been named for a mine owner named Baron La Grange. The camp became a central trading center for the miners, and included stage lines that served Chinese Camp, Knights Ferry, Mariposa and other outlying camps. The placers finally ran out, but not before realizing about $13 million. Settlers later made their living in agriculture.

LA PANZA (San Luis Obispo) The camp took its name from the nearby La Panza Range, Spanish meaning "paunch," for a piece of beef the hunters used to catch bear. The post office was established in 1879. During the 1880s yields from dredging operations in the area amounted to about $100,000.

LA PORTE (Plumas) Established in the 1850s as Rabbit Creek, the name was changed to La Porte, after banker Frank Everts' hometown in Indiana. The site acquired a post office in 1875. La Porte was located in the southwest part of the county, and served as a supply and trading center for the nearby mines. Gold production during the 1850s and 1860s yielded nearly $60 million. After the enactment of the anti-debris control laws, the gravels were worked by drift mining on a much reduced scale.

Lotta Crabtree's mother brought her daughter to Rabbit Creek in the summer of 1854, where Mrs. Crabtree established a boarding house. Lotta began her career when she heard music at a little log theater owned by Martin Taylor, who noticed her dancing to the music. From there she began her tours to the many outlying camps, singing and dancing for the miners who were starved for entertainment. Others believe Lotta began her career at Rescue.

LAFAYETTE HILL (Nevada) Located south of Grass Valley, the Lafayette Mine was discovered by French miners in 1851. They named it for the Marquis de Lafayette. During the 1860s the mine was recovering about $12,000 a month.

LANCHA PLANA (Amador) Mexicans settled here in 1848 and called their camp Sonora Bar. The name was changed to Lancha Plana when the post office opened in 1859. The name is Spanish for "flat boat," descriptive of the makeshift raft of casks that served as a ferry for the miners. The ferry was built by Messrs. Kaiser and Winter, who charged 50¢ a trip. The first store was established by a French-Canadian named Frank. Built with tents and shanties, in 1853 Lancha Plana experienced its only fire, which burned the entire town, but it was immediately rebuilt. Some of the claims near town paid well; one in particular, owned by William Cook, yielded about $100 a week for each man who worked the diggings.

White miners tried to expel the Mexicans in

Lancha Plana (courtesy of the Amador County Archives, Amador, California).

1850, and held a meeting where a resolution was drawn up for the Mexicans' immediate expulsion. But a General Stedman opposed the measure because he said it was unfair, and his influence caused the resolution to be dropped. Shortly thereafter, Stedman's trusted Mexican servant robbed his camp of money and other valuables.

LAST CHANCE (Placer) Located northeast of Michigan Bluff on a fork of the American River, Last Chance was founded two years after gold was discovered nearby in 1850. The post office was established in 1856. During 1871 production from hydraulic mining came to more than $2 million, the majority of the profits from the El Dorado Mine. Last Chance received its name because prospectors were ready to quit but decided to take one more turn at mining, which paid off when they struck gold.

LEWISTON (Trinity) Named for William Lewis, who settled there in 1854 and became the first postmaster, Lewiston was situated along the Trinity River. Not a lot of gold was found, although a few men were getting about $1,000 from their claims. This short-lived camp was gone by the 1870s, and residents turned to farming for their living.

LIFE PRESERVER (Placer) A wedding ring made a miner named Pike Bell rich. After going broke, he pawned his wife's wedding ring, continued hunting for gold and struck it rich. As a result, he called his claim Life Preserver. During the 1870s Bell made about $30,000 in a few days. His claim was located just north of Auburn.

LITTLE YORK (Nevada) This once vital camp was settled along an old emigrant trail a few miles from Dutch Flat. Mining was conducted nearby at Scott's Ravine. During 1850 a party of miners worked the gravel beds, but since they were newcomers and very inexperienced, they left empty-handed and their claims were not relocated until 1852 by William Starr and John Robinson. Although they were getting only about $1 to the pan, news of their discovery brought some attention and a number of other miners came in to check out the prospects. Little York was established a few years later, followed by a post office in 1855. The diggings prospered after water was brought in from the Bear River. During the winter of 1852 provisions became very scarce, and most of the men abandoned their claims because they couldn't support themselves. After the placers declined, hydraulic mining took over until it became illegal in 1884. The town was later moved to another site because the original location was right over a gravel channel that began to wash away.

LIVE YANKEE CLAIM (Sierra) Gold was discovered just north of Forest City during the 1850s. Some of the men were picking up nuggets weighing up to 60 ounces. Others found rocks close to 100 ounces. During 1856 one mining company acquired more than $115,000 in two years. Total recovery came to almost $1 million.

LONG'S BAR (Butte) This camp was established along the Feather River, west of Bidwell Bar. It was a prosperous little place during 1852 and had a population of about 200 miners who were making good wages. One of the early prospectors was Peter H. Burnett, who worked his claim for one month, making about $20 a day. When he had enough capital he went to San Francisco, practiced law, and ended up becoming California's first governor from 1849–1851.

LOS BURROS (Monterey) Named for the Mexican burros, this site was located near the camp of Jolon in the Santa Lucia Range and settled about 1887. By the early 1900s only about $150,000 had been realized.

LOTUS (El Dorado) Lotus was settled near the town of Coloma on Highway 49. Although there were rich diggings Lotus was a short-lived mining camp. Until its demise there were more than 2,000 people living there. Originally named Marshall, then Uniontown, the name was changed to Lotus when the post office opened on January 6, 1881. Someone named it because he thought the miners were as easygoing as the lotus eaters of the *Odyssey*. At the nearby Stuckslager Mine, a rare vanadium mica called roscoelite was found. It was named for Henry Enfield Roscoe of Manchester, England, who was the first to prepare pure vanadium.

LUDLOW (San Bernardino) Ludlow spent its life as a supply point and trading center for the nearby mines. It was named for a Central Pacific Railroad repair man, William B. Ludlow, and acquired a post office in 1926. The town also served as a terminus of the Tonopah and Tidewater shortline railroad. When the ore was depleted, Ludlow began to decline, and after railroad traffic stopped in the 1940s, it was gone.

LUNDY (Mono) Lundy was the center for a rich gold and silver mining district near Mono Lake. It was named for May Lundy, daughter of timber man William O. Lundy. The Lundy Mine that was discovered in 1877 continued operating until 1911, and recovered more than $3 million in gold. William Lundy became a mining recorder for the Homer Mining District when it was established in 1879. Proponents of the Homer District said, "In many respects Homer District is a remarkable region. It abounds in 'big things.' Big mountains, big trees, big trout, and a sprinkling of big talkers."[88]

LYTLE CREEK (San Bernardino) This site was discovered by Captain Andrew Lytle while leading a group of Mormons through the area in 1851. When the post office was established two years later, Lytle was appointed postmaster. Gold discovered nearby proved to be fairly rich, with some of the miners extracting roughly $30 a day. Hydraulic mining began later, but a flood caused operations to cease.

MAGALIA (Butte) Magalia is located on a fork of the Feather River just north of the town of Paradise, and was founded in 1850 by E.B. Vinson and Charles Chamberlin. It was first called Mountain View, then changed to Dogtown because Mrs. Bassett raised dogs, and then to Butte Mills. A newspaper article eloquently tells the story: "Our ancient village of Dogtown is as unfortunate as a yearly widow — eternally finding it necessary or convenient to change her name. When the old Dog died, Butte Mills slid in through the post office requisition, and now, through the same channel, Butte Mills is crowded out of the kennel, and the euphonious Magalia usurps the hour. However, the thing is fixed now, and the Dog's dead."[89]

When the post office was established in 1861, the town was renamed Magalia, Latin for "cottages," or from an Indian word, mahalas, which described a plant called the dwarf ceanothus, better known as squaw's carpet. By 1852 there were more than 1,500 prospectors looking for their fortunes. A gold nugget weighing in at 54 pounds was discovered in 1859, valued at $10,690. Called the Willard Nugget, it was named for Phineas Willard, who owned a hydraulic mine. A major mine was the Perschbaker, which produced more than $1 million by 1910. Today, Magalia is a small community composed of nice homes and a golf course.

MAHONEY MINE (Amador) The mine was located by Alvinza Hayward, who also owned the

Mill at Lundy, ca. 1910 (photograph courtesy of the Mono County Historical Society, Bridgeport, CA).

Hayward Mine. He later sold it to the Mahoney brothers for $1,000, who realized good dividends with their prospects. After working the vein down to about 800 feet, activities were suspended after one of the brothers died from consumption. The property was later sold to Nevada senator William Stewart, but by that time most of the ore had been taken out, and he received only moderate rewards.

MALAKOFF DIGGINGS (Nevada) Discovered in 1851 by three prospectors, the Malakoff mine was one of the richest north of West Bloomfield. A tunnel was built in the 1850s by hydraulic engineer Hamilton Smith. This mine was extremely large, measuring more than 7,000 feet long, 3,000 feet wide and 600 feet deep. Since the ore was low grade, it could be recovered only with hydraulics, and seven huge hoses were used to wash the hillsides. Lovett wrote, "It was so destructive, however, that it eroded millions of tons in debris down the Yuba and Feather rivers, threatening to drown the valley below in mud, and raising the bed of the Yuba River high enough that at one time it was above street level in Marysville."[90] More than $5 million was taken out before the anti-debris law was enacted.

MAMELUKE HILL (El Dorado) Gold was discovered near Oregon Canyon by prospectors named Keiser and Klepstein in 1851. Shortly thereafter, the camp was established and named for the corps of Mamluks used by Napoleon. Although the veins were small, they were so rich that more than $2 million in gold was recovered.

MAMMOTH (Mono) Mammoth is one of the highest towns in the area, at an elevation of 9,000 feet. Miners who were searching for the Lost Cement Mine in 1877 accidentally found rich gold outcroppings, and a year later a mining company came in and began operations at the Mammoth Mine. The community was established with a post office in 1879. The Mammoth Mine yielded $200,000, but investors lost money because the expenses were more than the earnings and operations shut down. Ironically, a miner wrote in 1879, "I believe the Mammoth Mine here to be the best buy in the lot.... They don't try to take out bullion and grand show but sack up all the rich ore and work the front ore and work the poorest ... the ledge is immense and rich. They are running a tunnel at the base of the mountain and if they find the same ledge of gold ore as in the upper tunnel there is no telling the value of this mine."[91]

MANZANITA HILL (Nevada) Gold was discovered on the hill in 1852, and within five years more than $3 million had been recovered. During 1868 McKeeley & Company began hydraulic operations. It continued until the anti-debris act went into effect, but not before realizing another $350,000.

MARBLE SPRINGS (Mariposa) Marble Springs was established about 1851 after coarse-gold placers were found along the Merced River. The place was east of the town of Coulterville. Some of the gold deposits also contained some silver. Miners realized about $200,000 in gold.

MARIPOSA (Mariposa) Mariposa is situated between Merced and Yosemite Valley off Highway 49. In 1849 Alex Goday found gold on what was known as Fremont's Mariposa Grant. Before the miners arrived, squatters on Fremont's land managed to extract about $200,000. Lode mining in the county was inhibited by the controversial Las Mariposas Land Grant which gave title to 14 miles of the Mother Lode to General John C. Fremont. Long before Fremont attempted to establish his right, the grant was overrun by miners who were reluctant to give up what they considered just claims. After years of conflict, Fremont's claim to the grant was formally recognized by the courts. But by then the property was plagued by mismanagement and the mines never fulfilled expectations.[92] The town received its post office in 1851.

The Princeton Mine, discovered during the 1890s, recovered more than $3 million in gold before it shut down in the early 1900s. Mariposa has the distinction of having the first recorded hard-rock gold mine (the Mariposa), which yielded more than $2 million. Operations began in 1851 with a vein that at places had gold three feet thick. Since it was so rich, some high-grading occurred. Miners working in secret took away as much gold as they could stuff in their pockets, but no one ever found out how much they got away with.

Hauling material for the Princeton Mine, Mariposa (Mariposa County Museum and History Center, Mariposa, CA).

The first stamp mill in the state was built at Mariposa (July 1849). John S. Diltz located the Diltz Mine north of town and proved to be quite successful. Active from the 1850s, the claim was consolidated with the Oro Grande Mine. About 1895 the property was sold to the Sierra Buttes Gold Mining Company of London, which built a 20-stamp mill. Total production was estimated between $500,000 and $1 million. Mariposa is a Spanish word for "butterfly," for the multitude of butterflies that fluttered around the golden poppy fields.

MARYSVILLE (Yuba) Long before the town was founded, it was leased in 1842 by John Sutter of Coloma fame. Marysville was named for Mary Murphy Covillaud, a survivor of the famous Donner Party. The town is located at the junction of the Feather and Yuba rivers, and was once the head of steam navigation north of Sacramento. It was the debarkation point for riverboats loaded with miners headed to the gold fields. By 1854 more than 20 freighting companies were in operation, with 400 wagons and 4,000 mules that hauled supplies to the Yuba and Feather river diggings. More than $10 million in gold was shipped out of Marysville to the San Francisco mint in 1887 alone.

The river town was changed with the advent of hydraulic mining near the Malakoff Diggings because the new technique altered the topography. Villages and farmlands were buried under the debris and river bottoms were filled with silt. The tremendous amount of silt caused the Yuba River to rise about 75 feet, forcing earthen dikes to be constructed.

Women were considerably outnumbered by men and were pretty much able to pick and choose their mates. One of the local ladies was looking for a husband and put an advertisement in the Marysville newspaper: "A Husband Wanted by a lady who can wash, cook, scour, sew, milk, spin, weave, hoe, (can't plow).... Now for her terms. Her age is none of your business. She is neither handsome nor a fright, yet an old man need not apply, nor any who have not a little more education that she has, and a great deal more gold, for there must be $20,000 settled on her before she will bind herself to perform all the above."[93]

MASONIC (Mono) This site was name for a group of Masons who came searching for gold in 1862. It was located near the Nevada-California border. Not much mining was conducted until the early 1900s when the Pittsburg-Liberty Mine was discovered. Its post office was established in 1906 and a 10-stamp mill was brought in to crush the ore. Mining operations continued until about 1910, with the Pittsburg yielding about $470,000. Another nearby lode, the Serita, produced $500,000.

MASSACHUSETTS HILL (Nevada) Gold was discovered just southwest of Grass Valley in 1850. A few of the lucky miners were extracting more than $5000 in a week on the hill. One of the mines located about ten years later began recovering gold valued at $90 a ton. Litigations over ownership of the property brought a halt to mining operations, but it ultimately yielded more than $3 million. A mill was built by Dr. J.C. Delavan, an agent from a New York company, but it failed. The company sent a Mr. Seyton to take Delavan's place in hopes he would have better luck. Unfortunately, Seyton

Upper Masonic, ca. 1904 (photograph courtesy of the Mono County Historical Society, Bridgeport, CA).

was an extravagant man and pocketed the proceeds for himself. After he disappeared, more honest managers arrived but they didn't fare any better.

MATELOT GULCH (Tuolumne) The gulch runs from Columbia to the Matelot Reservoir; its name is a French word meaning "sailor," an appropriate name since French sailors were early prospectors. The French were soon followed by the Irish. The mines were worked during the 1850s, but it wasn't until about 1861 when operations picked up after the Columbia Gulch flume was built. One of the miners pulled out $100 worth of gold in a few hours, while others were gleaning only ounces a day. Another was yielding $10–$20 a day for each claim.

There were constant clashes between French and Irish miners. After returning to their claims, the Irish would discover that the French miners had stolen from their sluices. One day, an Irishman caught a Frenchman digging through his claim. The Irishman hollered, "There's the son of a bitch that robbed us." He was nabbed and immediately put on trial. Of the course the outcome was preordained. With a rope around his neck ready to swing, a Frenchwoman got upset, grabbed a pick and went for the Irish vigilante. Since it was virtually a mortal sin for a miner to attack a woman under any circumstances, the men stood there while she took the noose from the Frenchman's neck and took him to his shack. Because of the damage done to the poor guy's neck, he lived for only two more years.[94]

MELONES (Calaveras) Located along the Stanislaus River, Melones was established in 1850 when gold was discovered on Carson Hill, but it would be 1902 before a post office opened. It grew to be one of the largest camps in California. The Melones Mine was located in 1895 and continued operating until the 1940s. The name is derived from the Spanish word for melons, descriptive of the melon seed-shaped gold flakes in the region. The camp was first known as Robinson's Ferry. Established by John W. Robinson in 1848, the ferry took the miners across the Stanislaus River to and from the nearby mines. He later sold the ferry to Harvey Wood. In less than six months, Wood made more than $10,000 operating his business. His earnings enabled him to build toll roads to the ferry.

MICHIGAN BLUFF (Placer) This place was established in 1850 and named by miners who came from Michigan. The site was located near the Middle Fork of the American River north of Forest Hill, and had a post office established in 1854. Although gold had been discovered in 1850, it would be two years before major operations began when companies were organized to build ditches to bring water for the claims. The years between 1853 and 1858 were times of most activity when considerable hydraulic and drift mining was conducted in the gravels. At one spot, a 40-acre area recovered $5 million in gold. Some of the mines were bringing out anywhere from $25,000 to $100,000 a month. In 1864 a pure gold nugget that weighed 226 ounces was found and assayed out at more than $4,000. George Cameron discovered the Hidden Treasure Mine in 1875, which produced about $4 million in gold.

Originally Michigan Bluff was located on an extensive flat. But when hydraulics were put in place, it began washing away a rim of the bluff and undermining the town. By 1858 it started to slide down the honeycombed hill, with some of the cabins cracking in half. A resident who lived at Forest Hill expressed his contempt for the rival camp: "Michigan Bluff perches upon a steep mountainside. When a house falls down here [Forest Hill] we burn it up, haul it away, or get rid of it somehow. But over in Michigan Bluff, they just let 'em lay where they are."[95] The town was later moved to another steep mountainside, with a yawning view down a gorge of the American River.

MINERSVILLE (Trinity) Formerly called Diggerville, this camp was located on the East Fork of the Trinity River, and took its name for the miners who lived there. Numerous placer and quartz mines were worked as early as the 1850s. Its post office opened in 1856 and closed in 1954. Most of the later operations entailed hydraulic mining.

MINISTERS GULCH (Amador) The gulch was named for a Baptist preacher, the Reverend Davidson, who, along with his friends, discovered quartz gold there about 1851. It was located near Amador City.

MINNESOTA (Sierra) During 1850 a prospector hailing from Minnesota discovered gold. Not long after his arrival, other miners had staked their claims and were making between $12 and $15 a day. One of the men found a lump of pure gold weighing 37 ounces, and another at 225 ounces that was valued over $5,000. Minnesota was established between a fork of the Yuba River and Kanaka Creek. There was only moderate success, and by the 1860s Minnesota was on its way out.

MISSISSIPPI BAR (Sacramento) Gold was discovered along the American River near Folsom about 1850. In order to get water to work the diggings, a ditch was built to the bar three years later. One of the mines nearby produced about $500,000 worth of gold.

MISSOURI HILL (Nevada) Located between the Bear River and Steep Hollow Creek, Missouri Hill was established after a lode mine was discovered in 1853. The owners sold their claim to Thomas Hartery who built a mill, plus hoisting and pumping machinery, but he failed to produce anything. Not long after that the mine flooded. William Loutzenheiser and Edward McLaughlin later purchased the property, then leased it out. A tunnel was built to drain the water, but that too failed. Then the mill burned down in 1860. Even with all those problems, the owners managed to recover about $200,000 in gold.

MOJAVE (Kern) Mojave got its start in 1876 when the Central Pacific Railroad (now Southern Pacific) made the site one of its stations and construction camps. Before the railroad was established, Mojave was an old stagecoach stop called Cactus Castle. The town's economy was based on serving as a mining center after the Exposed Treasure Mine on Standard Hill was found in 1894 by George Bowers. Total production of gold (and some silver) amounted to about $23 million. After the mines declined, Mojave continued its prosperity with the borax industry. Borax brought from Death Valley was hauled by 20-mule team to Mojave. The town later served as headquarters for Los Angeles during construction of the Los Angeles Aqueduct.

The Yuman tribe once made their home in this region. *Hamok avi*, from which Mojave is derived, means "three mountains." Others ascribe the meaning to "people who live along the river."

MOKELUMNE HILL (Calaveras) When gold was discovered in 1848, prospectors flooded to town, creating a population of about 15,000. Its post office opened in 1851 and the hill became one of the important camps of the Southern mines. A man who was with Stevenson's regiment found a nugget along the Mokelumne River that weighed over 24 pounds and assayed out at over $5,000. It was sent to Colonel E.F. Beale at Washington by a Colonel Mason who wrote there was gold in the West. Although ore had previously been found at Coloma, news of that discovery didn't generate a lot of interest at first. The nugget was shown at a public display, and may have been what really sparked the gold rush to California.

A group of Frenchmen arrived in 1848 and found almost 140 ounces of gold in one gulch near the hill, while others recovered in just a few days ore that was worth more than $175,000. The Boston and the Quaker City mines produced more than $2 million. After the gold played out, Mokelumne was consigned as a bedroom community.

The camp's name came from the Miwok Indian tribe living along the Mokelumne River. It may have been distorted from wakalumitoh, meaning "people of the village of muk-kel," but its origin is unknown. The town is home to the oldest Congregational Church building in the state, which was constructed in 1856.

MONO (Mono) This camp came to life near Mono Lake about 1859 when gold was discovered at a place called Dogtown, with a post office established in 1882. Placer mining was conducted when ditches were built to bring water to the site. Mormon miners from Nevada came to the region when word got out that gold was just lying on the ground. But when the town of Aurora in Nevada was found to have greater riches, all the prospectors headed there and Mono was no more. The name is derived from a Yokut word which means "flies," descriptive of the brine fly grubs the local Indians harvested from the lake and traded with the Yokuts in exchange for acorns.

MONTE CRISTO (Sierra) Just northwest of Downieville, this camp was established and took its name from the novel *The Count of Monte Cristo*, written by Alexander Dumas. During 1852 gold was found high up on a very steep cliff where only mules could go. To circumvent the problem, tunnels were dug through a cliff so the ore could be hauled out, which averaged more than $15,000 a week. As the gold pinched out, hydraulic mining was attempted, but because water couldn't reach the diggings it was abandoned. By 1868 the place was gone.

MONTEZUMA (Tuolumne) Gold discovered in 1859 caused Montezuma to be established just south of Jamestown. It was named for a trading post opened by Solomon Miller and Peter K. Aurand, which they called the Montezuma Tent. The placers produced a little over $1 million. In 1866 Montezuma experienced a serious fire that destroyed everything.

MONTEZUMA RIVER CLAIM (Siskiyou) During the 1850s gold was discovered on a fork of the Scott River near the Callahan Ranch. Total values are unknown, but during the 1890s Chinese miners were thought to have gleaned between $500 and $4000 a week till the gold ran out.

MONUMENTAL CLAIM (Sierra) This lode was discovered in 1855 near the Sierra Buttes. The mining conducted there was called "breasting," or side tunneling. Numerous lumps of pure gold were found there in 1860, one of them weighing out at 4,596 troy ounces. R.B. Woodward, who lived in San Francisco, purchased the piece for more than $21,000 and had it on exhibit for a time. Eventually the nugget was melted and brought a little over $17,000 in cash.

MOORES FLAT (Nevada) The flat was established southwest of Graniteville in 1851 and became a trading center for the nearby mines. It was named for store owner Henry Moore, who was also the first postmaster in 1857. Miners achieved moderate success at the diggings, but by 1895 the gold had pretty much been depleted.

Isaac Zellerbach was born at Moores Flat. He later went on to become the founder of the famous Crown-Zellerbach Corporation. Another notable was Patrick Manogue, who arrived in 1853 to search for gold. He was appointed alcalde (judge) when mining disputes flared up. After he earned enough money, Manogue moved to Paris and was later ordained a priest.

MORGAN (Calaveras) Established in 1850 on Carson Hill when ore was found, the camp was named for Mr. Morgan, president of a mining company that recovered gold mainly by tunnel work. By 1871 the company had brought out about $1 million of ore. At the Morgan Mine a piece of gold that was in the shape of a tree limb was discovered. It was more than four feet long and valued at $300,000, believed to be the largest single piece ever recovered.

MORMON BAR (Mariposa) The bar is situated along the headwaters of Mariposa Creek where Mormons found gold in 1848. They stayed only a short time when the deposits were found to be superficial. They were soon followed by the Chinese. Later miners reaped rich rewards, making a good $50 per day and more. The diggings were revived for a time during the 1930s and realized about $1.2 million.

MORMON ISLAND (Sacramento) This site was not really an island, but was so called because it was a sandbar separated by a ditch, giving it the illusion of being an island. It was located near the North and South forks of the American River. Two Mormons who were returning from Coloma found gold on March 2, 1848. By April there were seven Mormons digging for gold, and they used tightly woven Indian baskets to wash out their paydirt. "In July, 1948, there were about two hundred men at work.... At the lower mines, the hillsides were strewn with canvas tents and bush arbors. Some of the miners used tin pans, some used closely woven Indian baskets; but the greater part had a rude machine known as the cradle. This required four men to work it properly. A party thus employed averaged one hundred dollars a day."[96]

Samuel Brannan, a director for the Mormon population, established a general store. He required those miners who purchased goods to tithe ten percent of their earnings to the church. The post office opened in 1851. Brannan made immense profits from his store in addition to his mining operations. He moved back to San Francisco where "he established a distillery on a grand scale, and here in '68 he received eight bullets and nearly lost his life in a quarrel for possession of a mill. Meanwhile he had given himself up to strong drink.... Domestic troubles led to divorce ... and Brannan's vast wealth melted gradually away. For the last year or two ... Brannan has lived at Guyama or on the frontier, remarried to a Mexican woman, a sorry wreck physically and financially."[97] Mormon Island is now under Folsom Lake.

MOSQUITO GULCH (Calaveras) This name was given the site for mosquitoes that drove the miners nuts. It was located on a fork of the Mokelumne River. The majority of miners who came here about 1851 were Germans. Within four years more than 40 arrastres were in operation, followed later by construction of a 10-stamp mill. By 1878 the gold was depleted and the place was deserted.

MOUNT BALDY DISTRICT (San Bernardino) During 1871 gold was discovered at San Antonio Creek in the San Gabriel Mountains. Placer mining was followed in 1894 by the discovery of lode mines. The Banks Mine was located and worked using hydraulics until 1900 when an injunction was brought against the company for polluting San Antonio Creek.

MOUNT BULLION (Mariposa) Once called Princeton, the name was changed to honor Missouri senator Thomas "Old Bullion" Benton, who also happened to be explorer John Fremont's father-in-law. Benton received his nickname for his political battles regarding currency. Located northwest of the town of Mariposa, Mount Bullion was once a thriving mining town with a post office in 1862. Mexican miners were the first to discover gold in 1848. Three years later the Merced Mining Company began operations at the Mount Ophir Mine, which recovered more than $250,000. Total production in the district was about $6.3 million.

MOUNT GAINES (Mariposa) Northwest of the town of Mariposa in the Hornitos Mining District, gold was discovered during the 1850s. The site was established as a supply camp for the mining population and named for John Gaines, a Texas man who came to search for gold. Placer mining was conducted during the 1850s, followed in the 1860s by hard-rock mining. One of the best producers was the Mount Gaines Mine, which yielded more than $1 million in gold until it closed in 1949. Total production of the region was nearly $3.5 million.

MOUNT GLEASON (Los Angeles) Named for George Gleason, who found gold there about 1869, the camp was located south of the town of Acton. Although a number of mines were active in the 1890s, not much was produced and the place was soon deserted.

MULETOWN (Amador) This camp took its name from the creek for the miners' mules. During the 1850s Muletown was a busy place. Stories were circulated that a man named Yancey was digging out about $100 a day from his placers, and of a Chinese miner who found a 36-ounce lump of gold who was so elated he took the first boat back to China. After the placers were worked out, William H. Fox & Company began hydraulic mining operations. When the gold was exhausted, so was Muletown until some copper was discovered, which gave the town some new life temporarily.

Muletown prospectors were often called to the Miners' Court to testify in cases. They finally got tired of leaving their claims for the hearings, and decided their disputes would be better settled by arbitration. Any party that felt aggrieved would enter into a fist fight with his opponent. If one person was much larger or stronger, the weaker man could have a friend do his fighting. But the fights escalated until supporters on both sides had at each other, and so it was back to the Miners' Court.

MURDERERS BAR (El Dorado) The bar received its name after prospectors killed three Indians who were only protecting their wives. The Indians retaliated by killing five miners. Situated on a fork of the American River, the very rich diggings that were discovered in 1850 were worked out by 1858. Approximately $2.5 million was recovered from the bar.

MURPHYS (Calaveras) This camp was one of the better known places near the southern mines. It was named for miner John Murphy who moved from Canada to search for gold. The post office was established in 1851. That same year a company was organized to bring water to the claims. Prospectors swarmed to the place, and the rich diggings yielded some of them more than $100,000. Another claim reportedly took out 37 pounds of gold one day, and another 63 pounds the following day. A 3-stamp mill was erected to crush the ore. Although the diggings were rich, they were exhausted by the 1860s. Total production came to about $20 million.

In 1859, Murphys experienced a fire as reported: "The fire commenced in the Magnolia Saloon.... The Magnolia was used as a fandango house and was tenanted by some disreputable Mexican women at the time, one of whom is suspected, acted the incendiary, in revenge for some harsh treatment she had received.... In fifteen minutes from the discovery of the fire, over thirty houses were in flames."[98] It would have been easy to abandon the town, but there was too much gold in the hills, so the site was rebuilt.

John Muir once wrote of his visit to Murphys, "According to all accounts, the Murphy placers have been very rich — 'terrific rich,' as they say here. The hills have been cut and scalped, and every gorge and gulch and valley torn to pieces and disemboweled, expressing a fierce and desperate energy hard to understand. Still, any kind of effort-making is better than inaction, and there is something sublime in seeing men working in dead earnest at anything, pursuing an object with glacier-like energy and persistence." He also quoted the last of the prospectors, "We have no industry left now, and no men; everybody and everything hereabouts has gone to decay. We are only bummers — out of the game, a thin scatterin' of poor, dilapidated cusses, compared with what we used to be in the grand old gold-days. We were giants then, and you can look around here and see our tracks."[99]

Murphys celebrates its Gold Rush Days with a Grape Stomp and Street Faire in October. There is also a historical walking tour. For more information call the Community Club at (209) 728-8978.

NASHVILLE (El Dorado) Located along a fork of the Cosumnes River, this camp was first called Quartzville. When the post office was established in 1852 it was changed to Nashville, a transfer name from Tennessee, honoring a Revolutionary War officer, General Francis Nash. This camp may have been the first place where arrastres were used. There were about 100 of them before the stamp mills were brought in. The Nashville Mine began operations about 1851 and realized about $150,000. The camp also served as a trading center for the nearby mines, which brought out more than $3 million.

NATCHEZ (Yuba) During 1850 there was some successful mining, precipitating the establishment of Natchez along a fork of Honcut Creek. It is a transfer name from Mississippi, for the Natchez Indians whose name was interpreted as "hurrying man," or "salt." When the gulches dried up, so did the camp.

NEEDLES (San Bernardino) Needles was a large railroad and supply center for the mines in the desert. On the Arizona side of the Colorado River are craggy peaks that rise from the desert floor. In 1854, Lt. Amiel Weeks Whipple named the peaks "The Needles" while leading a government party to survey a railroad. In 1883, the eastbound tracks of the Santa Fe Railroad met the westbound tracks of the Atlantic and Pacific Railroad, and their arrival caused the town to be founded. Railroad surveyors moved the town site to a better location a few miles north on the California side. The post office was opened in 1883 as Needles. Until the railroad arrived, steamers would haul ore across the Colorado River. The Old Trails Highway connected Needles to other communities in the area. When the road was paved, it became part of Route 66. Needles still owns part of the "Mother Road."

NELSON CREEK (Plumas) The creek is a tributary to the Middle Fork of the Feather River, where rich diggings were found in 1850. It was reported that some of the miners were extracting a good $500 a day for a period of time, but most were earning about $20 a day. The site was named for a miner named Nelson, and its post office opened in 1855. The placers didn't last long and the place was soon deserted.

NEVADA CITY (Nevada) John Pennington and Thomas Cross may have been the first miners to search for gold, but a prospector named Hunt discovered the rich deposits near Deer Creek in 1849. The town went through several names: Deer Creek Diggings, Caldwells Upper Store for store owner Dr. A.B. Caldwell, and finally Nevada City. The town received its post office in 1850.

The diggings were very rich because they were located in the gravels of an ancient river bed that got richer as the miners went deeper. One man found a lump of solid gold that weighed almost 400 pounds, and refused to sell it to someone who offered him $25,000 for it. During the 1860s lode mines were discovered, one of the most notable being the Lava Cap located in the Sierra Nevada foothills, which realized about $12 million. A cyanide plant was built in 1940 to recover the mine's concentrates, but it was a losing proposition. In one month, the Pittsburg Mine yielded more than $19,000. Other lodes in the region averaged between $1 and $5 million.

During 1856 a dam was built above town, but just before completion it rained so much the dam began to give way. During the pre-dawn hours on February 14, 1857, the dam broke, sending water roaring down Deer Creek and into town, washing away the bridges and buildings, in addition to damaging the mining claims.

During 1859 miners all but deserted the town when news of the Comstock Lode in Nevada spread. Business was so dull merchants became quite skillful in pitching quoits. One saloon keeper went to the extent of advertising his wants and desires as follows: "One hundred thousand square drinkers wanted at Blaze's, corner of Pine and Commercial. No drunkards tolerated on the place."[100] Then in September of 1852 a fire started in a hotel and spread through the town, destroying more than a dozen buildings, but determined residents rebuilt.

Benjamin P. Avery tried his luck at mining and wrote a letter regarding his experience in 1849: "I started from Mormon Island on a prospecting trip to Reading Springs [Shasta], in October 1849. Rode a little white mule along with pork and hard bread and blankets packed behind me. On the way from Sacramento to Vernon, a trading station just started at the Junction of the Sacramento and Feather rivers, I encountered a party on horse back who were coming from Deer Creek, and who told me big stories about 'pound diggings' in Gold Run. As 'pound diggings,' i.e. claims that would twelve ounces of gold per day to the man, were just what I was in search of.... Arrived at Caldwell's store, the only trading post on Deer creek at that time. I found it a square canvas shanty, stocked with whiskey, pork, moldy biscuit and gingerbread; the whiskey four bits a drink; the biscuit a dollar a pound.... I prospected with good success in a claim that had just been abandoned by the notorious Greenwood.... A few other men were carrying dirt from one big ravine back of Caldwell's on an ox cart, and washing it at the creek with good success.... All were closed mouthed about yield, and regarded me as an interloper.... I struck the conjunction of ravines in the little flat known afterwards as the site of Dyer's store. To my intense disgust I found that my ravine was occupied from one end to another by long-haired Missourians, who were taking their 'piles.'"[101]

Nevada City had some notable people. George Hearst, father of famous newspaperman William Randolph Hearst, spent some time mining near Nevada City, as did Herbert Hoover after he had graduated from mining school. In 1852 a Frenchman named Chabot invented what was the forerunner of hydraulic mining. He made a canvas hose to bring water to his claim. About a year later, local miner Edward Matteson made it better by attaching a metal nozzle to a hose to direct his water wherever he wanted it.

NEW MEXICAN CAMP (Mariposa) Situated west of Bear Valley, this camp was inhabited by Mexicans who were working the placers in 1851. They took out about $250,000 before the gold ran out.

NEW RIVER (Trinity) These diggings were located on a tributary to the Trinity River. The camp was called New River, but the post office established in 1890 was named Denny for a store owner. Supposedly rich diggings, miners actually realized only about $4–$8 per day. During the 1880s more placers were found, in addition to lode mines. One of the richest properties was the Mountain Boomer, which recovered more than $300,000.

NICOLAUS (Sutter) Nicholaus Allgeier established a ferry and trading post at the junction of the Feather and Bear rivers during 1843, and acquired a post office in 1851. The ferry transported miners traveling back and forth between the Lower Yuba and Feather River diggings. After the mining era, settlers raised hops, alfalfa, and cultivated fruit orchards.

NIMSHEW (Butte) This camp was named for the Nimshew tribe of Maidu Indians and bears their word for "big water." It was located northwest of the town of Magalia and had a post office established in 1880. The site served as a trading and supply center. A nearby drift mine produced more than $1 million.

NORTH BLOOMFIELD (Nevada) North Bloomfield was established about 1851 along Humbug Creek, and was initially called Humbug. This name came about because some prospectors who found gold went to Nevada City and told others of their discovery. When the newcomers arrived they found some gold, but not nearly the amount the former men had bragged about. When the post office was established in 1857, the name was changed to honor German surgeon Dr. Frederick A. Blume. Limited mining was conducted in the region until hydraulic mining began, which recovered nearly $4 million.

NORTH COLUMBIA (Nevada) Located near North Bloomfield, this camp was established in about 1853 with a post office in 1860. Placer mining was conducted by the Chinese, although most of the gold was retrieved using hydraulics. One of the lode mines discovered nearby, the Delhi, produced about $1 million. More than $3 million was recovered from the area.

NORTH SAN JUAN (Nevada) A veteran of the Mexican War, Christian Kientz, named the place after he discovered gold in 1853 because the nearby hills reminded him of the San Juan de Ulloa in Mexico. The post office was established in 1857. Gold extraction was slow until the Middle Yuba Canal Company began to supply sufficient water to miners. By the 1870s the population had reached more than 10,000. When hydraulic mining began, its actions ruined the landscape, cutting gashes in the hills more than 200 feet deep and a half a mile wide. This type of mining undermined the buildings and many of them were carried away with the operations. Revenues are included with You Bet.

NORWEGIAN GULCH/MINE (Tuolumne) This claim began as a pocket mine in the gulch during 1851 and operated until the 1930s, yielding about $200,000. Tradition says that Black Bart robbed a stagecoach of its gold in the gulch, which ultimately led to his being nabbed. The site was located near the camp of Melones.

OAHU (Sierra) Oahu was a very small camp near

Downieville where Hawaiian miners conducted placer mining. Most of them were realizing only about $8 a day. Oahu was thought to be Hawaiian meaning "gathering place," when in fact it has no meaning whatsoever.[102]

OLANCHA (Inyo) Olancha's name is derived from a Shoshonean village, but its origin is unknown. The place was a trading center and from 1870 served as a post office for the silver mines of Owens Lake. In 1863 Minnard Farley built a sawmill, stamp mill and amalgamating pans for the ore, but there was only minimal gold production. When nearby Cerro Gordo began to develop with the discovery of silver, Olancha was relegated to a stagecoach stop. After the mines played out, the town turned to agriculture and ranching for its economy. The Southern Pacific Railroad laid its tracks through town in 1910, supplying materials for construction of the Los Angeles Aqueduct, giving Olancha new life.

OLD DIGGINGS (Shasta) Just north of Redding, these diggings were discovered during the 1850s. Found to be quite productive, two stamp mills were built in 1858. The nearby Reid Mine yielded more than $2.5 million.

OMEGA (Nevada) Named for the last letter of the Greek alphabet, Omega was settled about 1850, and the post office was established in 1857. The camp became a trading center for the mines in the region. J.A. Dixon was one of the early miners in 1850 who first discovered the diggings. Before long more than 30 claims were being worked near the confluence of Diamond Creek and the South Fork of Yuba River. The placers yielded more than $3.5 million.

ONION VALLEY (Plumas) This was a short-lived camp that began in 1850 with the discovery of gold. It was located on a fork of the Feather River near the camp of La Porte. The abundance of wild onions growing in the valley gave the camp its name. The placers soon ran out and before long the site was abandoned.

OPHIR (Placer) First called Spanish Corral, the camp's name was changed to Ophir when its post office opened in 1872. A $3,000 gold nugget was discovered in Auburn Ravine by a Chinese miner in 1859, and a few years later a piece of pure gold weighing 107 ounces was found in the tailings of a gold mine. More than $5 million in gold was extracted from the claims.

OREGON CREEK (Sierra) A pocket of gold found nearby was valued at about $200,000, which brought in curious miners to check out other possibilities. The site was located near Forest City and was a tributary to the Middle Fork of Yuba River. One of the mines produced more than $700,000 and continued operating until the 1970s.

OREGON RAVINE (El Dorado) The ravine was located north of Georgetown and named for miners from Oregon who found rich deposits, took their profits, and returned home. During 1849 a miner found a lump of gold that was worth about $2,000. By the early 1930s more than $3.5 million had been recovered.

ORLEANS BAR (Humboldt) Orleans Bar was named by French miners who settled there during the 1850s. It is evolved from and is a modern name for Roman emperor Aurelius. The bar was located along the Klamath River near the confluence of the Salmon River, and served as a trading point for the miners. It was also the county seat until Klamath County was dissolved.

OROVILLE (Butte) Oroville was founded along the Feather River by Colonel John Tatham about 1849 and first known as Ophir. It was changed to Oroville (*oro* is Spanish for "gold") in 1854 when the post office was established. Cherokee Indians came here during the 1850s to work in the gold mines near town. A very rich load was the Cape River claim that in one month's time during 1852 had more than $1 million worth of ore extracted. Placer mining was further developed when a ditch was built to bring water. In 1857 one of the miners found a gold nugget that weighed more than 100 ounces. Ore from the surrounding mines yielded about $82 million. During the 1870s the gold began to run low and more than 10,000 Chinese came in to collect what was left.

Oroville is still a prosperous town. After the gold rush was over, residents learned that the town was right in the middle of a thermal belt, and made their living cultivating orchards and growing olives. One of the largest ripe-olive canning plants was established at Oroville. The town also greatly benefited when the Oroville Dam was built.

OSCEOLA (Nevada) This short-lived camp was named for Seminole chief Osceola, whose name

Butte County miners (Butte County Historical Society, Oroville, CA).

means "bread." Miners discovered a pocket of quartz worth more than $220 a ton. As a result a 24-stamp mill was built there in 1856. But after crushing about 120 tons, operations were suspended and the machinery was moved to another site. A short while later the site was abandoned.

OTTAWA BAR (Butte) The bar was discovered in 1850 by Delano and Company, but it didn't produce much. The company built a dam, then abandoned it because during the winters there was too much water. The site was located along a fork of the Feather River. When the company left, the Chinese took over and worked what was left of the gold, with some reports they were acquiring about $50 to $100 a day for a short period of time.

PANAMINT CITY (Inyo) Charles Alvord led a party of Mormons into the Panamint Mountains searching for a lost cliff of silver. He was abandoned by his companions when he was unable to locate the ore. One of the men went back later to search for Alvord, and found him camped on a slope of the mountains, hungry and exhausted. The following year Alvord returned to the search with a partner, who later went berserk and killed Alvord. It wasn't until 1872 that silver was found and initial assays showed a value of $2,500 per ton. The town came into existence after the Panamint Mining Company was established, and grew to a population of more than 2,500.

Panamint itself was a violent camp because of its inaccessibility, since there was only a sand trail for the pack mules until a wagon road was built. Most of the population was made up of those with notorious pasts and, as such, Wells Fargo refused to handle shipments because of the too-frequent hold-ups. It may have had the highest concentration of outlaws outside of the prisons. A Mr. Steward, owner of a mill and furnace, devised a way to ensure safe shipment of the ore. The silver was cast into 750-pound cannon balls, which were too heavy for any robber to lift. With the lowering prices in silver, the place declined.

PARADISE (Butte) Dice? Perhaps they had something to do with the origin of this town's name. Composed of gambling halls and saloons, one of the most popular spots was the Pair-O-Dice Saloon. When settler "Uncle Billie" Leonard, along with some friends, were riding around the valley and rested under some tall pine trees, he told his friends the place was some paradise. The town was a trading and supply center for the nearby mines, and acquired a post office in 1877.

Today Paradise has a population of more than 29,000, and is known as the Apple Center of California. The first Paradise Fair and Apple show was held in 1937 and lasted 5 days. The middle of Memorial Hall had a pyramid of 15,000 apples. In 1972 the Johnny Appleseed Days Festival presented the Guinness Book of Records an apple pie. Created from 900 pounds of the fruit, it was 10 feet in diameter and weighed 950 pounds.

Gold Nugget Days is held on the fourth weekend of April. For more information and directions, call the Paradise Ridge Chamber of Commerce at (530) 877-9356.

PARKS BAR (Yuba) First named Marsh Diggings for Dr. John Marsh, the name was changed to Parks Bar honoring David Parks, who opened a general store in 1848 and a post office in 1851. The camp was located near the Yuba River. These two men were smarter than most; they returned to their original homes after they made what they considered their fortunes. Marsh left with about $85,000 and Parker acquired about $40,000.

In 1855 Hinton R. Helper tried his luck at mining and garnered a resounding 93¢. That year he wrote a book about the hazards of those coming to California for gold: "I will say, that I have seen purer liquors ... larger dirks and bowie-knives, and prettier cortezans here.... In my judgment, the present condition and future prospects of California ... portend much poverty and suffering ... based upon fair estimates, I learn that since California opened her mines to the world, she has invested upwards of six million dollars in bowie knives and pistols."[103] By 1867 the bar had disappeared after hydraulic operations on the Yuba River carried debris there.

PEÑON BLANCO (Mariposa) Located northwest of Coulterville, the area was first mined by Mexicans in 1849. Their claims were relocated about 1863 by two miners who extracted more than $16,000 in a two-month period. The place was given a Spanish name meaning "large white rock," for a nearby rock formation used as a landmark. Moderate mining was conducted, but the site was gone by the early 1900s.

PENRYN (Placer) During the 1860s a number of quartz mines were discovered, which warranted a 10-stamp mill that was erected shortly thereafter. Its post office was established in 1873. During 1885 the gold began to decline and the mill closed down. One of the better producers was the Alabama Mine that was worked until about 1936, and yielded close to $1 million. Penryn is a transfer name from Wales, which means "nose," descriptive of a point of land.

PETTICOAT SLIDE (Sierra) This site was established near Poker Flat. Tradition says it received this unusual name after a few hurdy-gurdy girls ventured into camp and fell at the slippery entrance, sliding into the camp, much to the amusement of the miners. Hurdy-gurdy girls were women brought over from Europe and put to work in the saloons to dance with the customers at a dime a dance. Miners working the placers nearby were making about $8–$10 per day.

PICACHO (Imperial) Settled along the Colorado River, the camp was named for nearby Picacho Peak; the name is redundant, since *picacho* is Spanish for "peak." Mexican miners were the first to the area about 1857 who conducted placer mining. After the placers had run their course, hydraulic mining was attempted, but it was a total failure. One of the more successful lode mines was the Picacho Basin Quartz Mine, which produced about $2 million in gold. A narrow-gauge railroad was built to haul ore to the mills, but it experienced too many mechanical problems, and then a flood destroyed the tracks. Much of the site was submerged when the Laguna Dam was built in 1909.

PIKE CITY (Sierra) Pike City served as a supply point for the surrounding mines and may have been named for the Pike family. Its post office was established in 1877. The Alaska Mine was the main producer there between 1863 and 1961, yielding more than $1 million. The camp was located between the Yuba River and Oregon Creek.

PIKE FLAT (Nevada) The Pike Flat Company came in and formed this camp near Grass Valley. During the 1850s, it was recorded that some of the miners were getting $100 a day. The place was gone by 1867.

PILOT HILL (El Dorado) This was a mining camp near the American River where gold was discovered in 1849, with the miners' diggings yielding about $8–$60 per man. That same year the Pilot and Rock Creek Canal was built to bring water to the claims. One of the prospectors found a piece of gold that assayed out at more than $1,600. In 1854 Silas Hayes established a post office and became its first postmaster. Just southwest of Pilot Hill along the American River the Zantgraf Mine was opened about 1880. After the Montauk Consolidated Mining Company took it over, more than $1 million was recovered from the property. The waters of Folsom Lake later covered the site.

PINCHEMTIGHT (El Dorado) Located near Weber Creek, this site acquired its interesting name from folklore. A German shoemaker and saloon keeper named Ebbert took their profits by pinching the gold dust so tight to get as much as possible. The Pinchemtight Mine was also known as the Sailor Jack for a sailor who was a novice to mining. The old-timers told him of a claim that was rich in gold and the fellow went there to acquire his fortune. To the old miners' surprise, he actually did find a rich pocket of gold. By 1883 the mine was deserted.

PINE TREE (Mariposa) Gold discoveries in 1849 caused Pine Tree to be settled the same year. Placers were followed by discovery of the Pine Tree Mine which began operations the following year. It, along with the nearby Josephine Mine, yielded almost $4 million. The Josephine was later abandoned when it recovered only $8 per ton of ore that was treated at the Benton Mills.

In 1876 a horrendous fire occurred in the Pine Tree mine as reported: "Five Men Killed and Several Others Severely Wounded. A premature explosion of a blast of giant powder occurred in the Pine Tree Mine.... The news reached here about five o'clock on Thursday morning by a messenger, dispatched immediately after the accident occurred ... though it is hoped that the loss of life is no greater than it is stated, though the messenger did not wait until a thorough examination of the mine had been made before starting on his errand. This is the most terrible and distressing accident that we ever remember to have heard of occurring in the Mariposa mines, although deep quartz mining has been carried on in that county for the last twenty years and upwards."[104]

PINECATE (Riverside) Now the town of Perris, this site was originally a small placer camp that was established in the 1870s as Pinecate. The name was taken from an Aztec word meaning "black beetle," and its post office opened in 1881. The diggings were fair until discovery of the Good Hope vein in 1874 which was first worked by Mexican miners. It produced more than $2 million before 1896. During the 1930s attempts were made to reopen the Good Hope, but the efforts were unsuccessful.

PIUTE MOUNTAINS (Kern) During 1861 this area was prospected for gold, and some of the men found fair-paying diggings in Kelso Canyon high in the Piute Mountains. A post office was opened in 1875 and named for the Paiute Indians. The camp of Kelso was named for John W. Kelso, who freighted supplies to the miners. Some of the good producers were the Joe Walker Mine, that recovered about $600,000; Green Mountain, which yielded $1 million–$2 million up to 1890; and Bright Star (owned by the Bahten brothers), which also yielded about $600,000.

PLACERVILLE (El Dorado) A place of gold, lynchings, and name changes, Placerville started out as a small mining camp. At a nearby creek rich placer deposits were found in 1849 and prospectors from outlying camps came to Placerville hoping to find even more wealth. Because the creek was dry during the summer, the site was named Dry Diggings. Eventually the criminal element entered the picture, those who robbed the miners and Pony Express. The perpetrators were nabbed and quickly hung. As a result Dry Diggings became known as Hangtown. In 1850 the name was changed to Placerville for the type of mining conducted, which was washing and dredging of the ore. Its post office opened the same year and Thomas C. Nugent was appointed postmaster.

In April of 1856 a fire broke out and destroyed many of the buildings. Another fire started later the same year, nearly destroying all the wood structures. The residents quickly rebuilt, only this time they used brick and stone. Eventually placer mining gave out and residents became concerned about the town's economy. But the overland mail and stage coaches began running through the Sierra Nevada Mountains to Placerville bringing emigrants, and the place became a major commerce center. Placers in the region produced about $25 million. A nearby mine was the Linden, which recovered more than $130,000 in gold between 1882 and 1894.

Placerville hosts the Apple Hill Season, where a shuttle takes visitors through orchards, vineyards and forests in the region. The event is held every weekend between October 2 and October 31. Other events include the Annual Wine & Cheese Tasting and the Annual American River Festival. Information: Chamber of Commerce, 1-800-457-6279

PORT WINE (Sierra) This camp was established northwest of Downieville in 1850, and acquired a post office about ten years later. It was named by a party of prospectors who found a keg of port wine that had been concealed in the bushes by packers. They took advantage of it, drank the wine, and went in search of water, but what they found was a ravine full of gold gravel.

Placerville when it was known as Hangtown (courtesy of the Amador County Archives, Amador, California).

PUT'S BAR (Amador) Situated along the Mokelumne River near Lancha Plana, the site was probably named for a miner named Putnam who discovered gold there in 1853. Although it wasn't very rich, between 1855–1860 a few hundred Chinese miners worked the placers. Put's Bar fared better when it was discovered the soil was ideal for raising watermelons and cultivating vineyards.

RAILROAD FLAT (Calaveras) Railroad Flat was founded in a rich mining district that contained both placer and quartz mining. The largest producing mine was called the Petticoat. Named by miners for the female gender, specifically their wives, the property produced about $100 per ton of ore. Placer mining needed a lot of water, which was in short supply, so an engineer built a channel that brought water from the Mokelumne River and deposited it in a reservoir located on his land. Although Railroad Flat never amounted to much during its early years, it was almost totally wiped out when it received the deadly visitor known as the Black Fever.

The community might have been named by a peddler who sold merchandise to local prospectors. He was surprised at the number of tracks leading out from the mines. These wooden rails were used by miners so they could transport gold in an ore car which was pulled by mules to their destination. No real railroad ever made it to town. Railroad Flat is now a small community with about 500 people, general store, elementary school and a post office which was established in 1857.

RANDSBURG (Kern) Named for a place at Johannesburg, South Africa, Randsburg was established about 40 miles northeast of Mojave near the Rand Mountains. It bears an Afrikkans expression for "white water ridge." In 1895 three prospectors (Charles A. Burcham, John Singleton and Fred Mooers) wandered over from Summit Dry Diggins and found rich ore at Goler Gulch. It wasn't long before the entire hillside was covered with mining claims. The three prospectors named their mine the Yellow Aster, which was credited with $12 million. Total estimated production of the other mines in the area was about $40 million.

RATTLESNAKE BAR (Placer) Prospector John C. Barnett washed out a pan of gold at the bar in 1853, and when the news got out the place teemed with prospectors who staked their claims. But getting the ore out was a problem until a ditch was built to bring water to the diggings. The bar became a busy place, a stage line was established that ran to Sacramento, and Rattlesnake gained prominence as a principal town along the river. A miner by the name of Qua had a sluicing operation that netted him more than $30,000. Another group of men made about $6,000 in less than two months. In 1863 a fire occurred that destroyed most of the town and it was never rebuilt. It is now under Folsom Lake.

RED BLUFF (Tehama) Taking its name from the reddish cliff along the Sacramento River, Red Bluff was established by S. Woods in 1850 and was initially known as Leodocia. It became a center for miners headed to the Trinity gold fields, then evolved as a shipping point for agriculture and timber.

This town was once the home of William B. Ide, who was one of the leaders of the legendary Bear Flag Revolt. After learning that Mexican authorities were going to drive out the Americans from California in 1846, Ide formed a contingent of men and headed to the Pueblo of Sonoma. They captured General Vallejo and his brother, threw them in jail, erected a bear flag they had hastily fashioned, and proclaimed the state an independent republic. "These bold moves on the part of the settlers paved the way to occupation of California by armed forces of the United States. The Bear Flag, after flying for a few weeks, was replaced by the Stars and Stripes, July 9, 1846."[105] During 1848 Ide discovered a placer claim which rewarded him with more than $25,000. With his earnings, he purchased part of a Mexican land grant called the Red Bluff Ranch.

RED MOUNTAINS (San Bernardino) Not gold, but silver was king there, which was discovered in the early 1900s. The mountains were named by miner P.J. Osdick in 1921. Silver was discovered by accident when prospector Ham Williams got caught in a snowstorm in 1919 and was looking for shelter. He found an open pit, and while waiting for the storm to subside started picking at the rocks, one of which contained silver. He staked his claim, which was later renamed Big Kelley, and went on to yield $12 million.

REDDING (Shasta) Founded in 1872 by the Central Pacific Railroad, the town may have been named for an agent of the railroad, B.B. Redding, or for an early pioneer named Major Redding. It started out as a major trading and shipping center for the many gold and silver mines between the Sacramento River and Trinity County line. Redding's post office opened in 1872. When the mines declined, the town continued its prosperity by establishing itself as a shipping point for the fruit growing and farming interests.

RESCUE (El Dorado) During the 1880s Dr. M.A. Hunter moved to what was called Green Valley. After purchasing land and opening a store, he applied for a post office. Postal authorities wanted the name changed, so miner Andrew Hare asked that Rescue to be added to a list submitted. He suggested the name because a mine had rescued him from poverty. Hare had discovered a ledge he named Rescue, so called because of the gold found there and its profits allowed him to continue more extensive mining. Postal officials accepted the name and the post office opened in 1895.

Lotta Crabtree was an entertainer who began her fortune here (although others think it was at La Porte). She started out tap-dancing on an anvil at the blacksmith's shop, and the miners' practice of tossing gold nuggets to Lotta eventually made her a millionaire.

RICH BAR (Plumas) Rich Bar was discovered in the upper reaches of the Feather River by three German miners who traveled farther than other prospectors had ventured. One of the men was bringing water back to camp when he noticed some boulders along the creek, their cracks and crevices filled with gold flakes and dust. Tradition says they took out about $36,000 in dust and nuggets the first four days. Had the news not leaked out, the men would have acquired millions.

With the rush going full force, some of the men were taking out up to $1,500 in a single pan or just scooping it up, but others were near starving. In one incident, a man became so delirious he cut his throat. He was discovered in time, whereupon he accused a fellow known as Paganini Ned of trying to kill him. A mob then insisted on shooting the victim after discovering he did it himself. They argued that if a man was capable of attempting to take his own life, he would not hesitate to kill another. Fortunately for the man, authorities were able to restrain the mob.

Many of the miners' efforts at the diggings ended in failure, which affected the entire community. "The whole world (our world) was 'dead broke.' The shopkeepers, restaurants, and gambling houses had trusted the miner to that degree, that they themselves were in the same moneyless condition. Such a batch of woeful faces was never seen before."[106]

ROHNERVILLE (Humboldt) This site was named for a Swiss man, Henry Rohner, who established a general store, followed by a post office in 1874. It lived its life as a trading and supply center for the miners as well as a stopping point for the pack trains carrying freight and mail. The town was located near Fortuna.

A number of Chinese who lived here were hated by the whites, and were "gently" forced to leave, as the local paper noted: "Rohnerville is at last freed form the presence of the Mongolian in her midst. On Saturday the last two departed on the steamer, the 'Chester,' their buildings & property having been bought by residents of Rohnerville by subscription. This purified condition of Rohnerville is mainly due to the efforts of Martin McDonald, who 'rustled' continuously until he had secured money enough from the citizens to buy out the Chinamen."[107]

ROSAMOND (Kern) When the Southern Pacific Railroad came through, it made Rosamond one of its depots in 1877 which was named for the daughter of one of its officials. In addition to mining, many of the settlers were engaged in cattle ranching. During the 1890s the Lida Mine (later renamed the Tropico) was discovered on a precarious hill, located by Ezra Hamilton about 1882. He sold it to the Big Three Mining and Milling Company. It was later purchased by the Burton Brothers, who continued its operation until about 1956. The Tropico produced between $6 and $8 million. A nearby mine, the Cactus Queen, was worked for about six years in the 1930s, and produced nearly $5 million in silver and gold.

ROSES BAR (Yuba) Early miners in 1848 at the bar were Jacob Leese and Jasper O'Farrell. Their claim was thought to have produced more than $75,000 in less than three months. The site was located along the Yuba River, and named for a Scotsman, John Rose, who established a trading post there.

ROUGH AND READY (Nevada) Captain Townsend owned a mining company called Rough &

Tropico Mine, Rosamond (courtesy Art Naffziger).

Ready, from which the town took its name when the post office opened in 1851. He served under "Old Rough and Ready" Zachary Taylor during the Mexican War. One of Townsend's workers discovered a chunk of gold while hunting for game, and the company moved to the area, staked out claims, established a trading post, and a town was born. The first store was a tent constructed using the mainsail of a vessel owned by H.Q. Roberts, and Rough and Ready was on its way as a trading and supply center.

It wasn't long before the government imposed a mining tax on all the claims. The men were enraged and Colonel E.F. Brundage called a meeting on April 7, 1850, where it was proposed the people secede from the Union because of the taxes. A population of 3000 voted to go it alone as the Great Republic of Rough and Ready. Brundage was elected president and drew up a constitution similar to the U.S. With the Fourth of July nearing, the men were gearing up for a celebration by drinking tanglefoot whiskey. But how could they celebrate when they had seceded from the union? So they rejoined the good old U.S. In 1948, the Postal Department wanted the name changed because it was too long. Again, the citizens rebelled and Rough and Ready it stayed.

Secession Day is celebrated on the last Sunday in June, although it actually occurred in April, showing that residents are still mighty independent. The event includes a chuckwagon breakfast and chili cook-off. Information: Rough & Ready Chamber of Commerce, (530) 272-4320.

SACRAMENTO (Sacramento) The state capital was settled about 1839 with the arrival of John Sutter and bears a Spanish word for "sacrament." The post office was established in 1849. That same year Samuel Brannan opened the first general store. When gold was found in the region, it brought in a multitude of miners. Before becoming the capital, Sacramento served as a supply center for the Northern mines, and swelled to a population of more than 10,000. A Dr. Sullivan wrote of the conditions during 1849: "Hundreds are encamped in tents, through the rains and storms, scantily supplied with food and covering. Men are driven from the mines for want of food, and are begging for employment, asking only subsistence. Yesterday there were twenty-five deaths. The sickness does not arise from the severity of the climate but largely from overwork, scanty and bad food, disappointment and homesickness."[108] Then in 1850, the city experienced its first cholera epidemic. Dr. John Morse wrote that "three of the large gambling resorts have been closed, the streets are deserted, as frequented almost solely by the hearse, and nearly all business is at a standstill.... The daily mortality is so great (60 per day)."[109]

The city also became the terminus of the first railroad in the state built by Theodore D. Judah, which was a short line that went to the town of Folsom. Judah also planned the first transcontinental railroad through the passes of the Sierra Nevada.

SALMON RIVER (Siskiyou) Gold was found in the southwestern part of the county about 1851, but serious mining didn't begin for nearly ten years, as noted: "Early in the summer of 1861, faint and indistinct rumors were heard of rich diggings.... These fleeting shallows were soon crystallized into a well-defined story of unprecedently rich strikes having been made on Salmon River." The *Democrat* wrote: "Reliable men say that miners on that river are taking out gold by the pound — that they don't count by dollars and ounces at all — that they take out from three to ten pounds per day to the rocker — that there are hundreds who have fifty dollar diggings."[110] Two years earlier the country had gone wild over the Frazer River Mines, but it cost hundreds of men's lives and impoverished more. When the Salmon River excitement began, there was only a small interest in the beginning because the miners thought it was another Frazer River swindle. But with wonderful tales of riches, they changed their mines and headed to the diggings, which proved to be quite a bonanza. Total yields from the placers came to about $25 million.

SAN ANDREAS (Calaveras) San Andreas was first settled by the Spanish in 1848, and gold was discovered shortly thereafter. It was named for San Andrés, patron saint of Spain, although some think it is for St. Andrews, patron saint of Scotland. Its post office was established in 1854. By 1859 thousands of tents dotted the hillsides. The claims were rich because the site was located on ancient river channels and more than 1,500 miners were working their claims. San Andreas itself was located along the Mokelumne River. Since the diggings were mainly surface, the gold was depleted within a few years. A mining company called the Plug Ugly was extracting about $1,000 a day and a 10-stamp mill was built to crush the ore. In 1852 the famed San Andreas gold nugget was discovered and sold for about $12,000. When the easy gold ran out, the Chinese came in and cleaned up the remainders.

SAWMILL FLAT (Tolumne) Although this place was not a mining camp, it was important in that it provided timber for the mines in the region. Two sawmills processed the wood. The nearby mines brought out more than $2 million. Sawmill Flat saw its demise after a fire nearly destroyed Columbia, its neighbor and best customer.

SAWYERS BAR (Siskiyou) Dan Sawyer was an early miner in the vicinity during the 1850s. After placers were discovered along the Salmon River in 1857, the camp grew with a number of saloons, hotels and stores, and its post office was established in 1858. During 1895 a road was built that connected Sawyers with the camp of Black Bear. More than $20 million in gold was recovered from the bar.

During 1899 John Bailey, who was a bunko artist, brought in bogus nuggets and showed unwary prospectors the property and they purchased the worthless ground. Fortunately, he was caught. The local paper wrote that "John Bailey, the old man who is wanted by the Siskiyou authorities for bunkoing the gold miners of Sawyers Bar with bogus Klondike nuggets & incidently for feloniously borrowing the constable's horse there, has been apprehended at Crescent City & will be returned to await the pleasure of the Siskiyou officers.... The meek miners of Sawyers Bar could afford to lose a few dollars on worthless gilded quartz rather than let the story out but when he toyed with such valuable property as horses all the country was up in arms against him & particularily the Constable who owned the horse."[111]

SCALES (Sierra) West of Downieville, this site was settled by a saloon keeper named Scales. It is not known if the town was named for him or for gold that was discovered composed of scales in loose marl (a mixture of clay, sand and limestone). The post office opened in 1871. Most of the mining was conducted using hydraulics.

SCOTT BAR (Siskiyou) Discoveries of gold occurred in 1850 along a tributary of the Klamath River. The bar was named for miner Joseph Scott and its post office was established in 1856. The site enjoyed a thriving population with its rich and abundant placers. One of the miners talked about his discovery: "I was nearly down to bedrock.... At the next stroke of my pick it encountered ... something which stopped the pick, but which felt slightly yielding as though I had ht a bar of lead.... Picking it up, I knew by the weight that it must be gold."[112] And it was indeed, weighing 87 ounces, the largest nugget found in the region. Two of the nearby mines paid quite well. The Quartz Hill was producing between $10,000 and $12,000 a year, and the Joe Ramus experienced a yearly production of about $7,000, with a total yield of $500,000.

During 1851 rich diggings were found, and claimed, by two separate parties of prospectors. Arguments ensued and when it seemed things would get worse, someone suggested that ownership of the claim be settled legally. Attorneys and a judge were brought in from San Francisco to solve the problem. The winning party began working their claim as soon as they got the verdict. "At last, when the case was decided, the claim was opened by the successful party; and when they reached bed-rock, and were ready to 'clean up,' we all knocked off work, and came down and stood on the banks, til the ravine on both sides was lined with men. And I saw them take out gold with iron spoons, and fill pans with solid gold, thousands upon thousands of dollars. Ah! It was a famous claim, worth hundreds of thousands of dollars."[113]

SEIAD CREEK (Siskiyou) A member of the Whitney Survey named this creek in 1863, which was once home to the Shasta Indians. Seiad was interpreted as a Yurok word for "far away land." Gold was exploited along the creek during the 1890s, mainly drift and hydraulic mining.

SELBY FLAT (Nevada) The flat was settled just north of Nevada City by miners who reported they were recovering about $60 a pan of gold for a time. The placers ultimately went on to yield nearly $2 million.

In his diary, a miner wrote about the entertainment of the day: "There was a lively time over at Selby Flat Wednesday night. The landlord gave a ball at the hotel. All the women were there — seven of them — and about two hundred men.... It was a great sport for a while, but towards morning some of the men got too much gin aboard and a quarrel started about the right to dance with one of the Missouri girls. Pistols were drawn, the lights put out, at least a hundred shots fired; but, funny enough, only one man was hurt ... I went out through a window and did not wait to see the finish. It was too exciting for me."[114]

SHASTA (Shasta) In 1814, Alexander Henry and David Thompson came through the region and met Indians of the Walla Walla, Shatasla, and Halthwypum Nations. The Shastas called themselves weohaw, meaning "stone house," designating a cave in the region. After Major Pierson Reading found about 50 ounces of gold at Clear Creek Canyon in 1848, Shasta became a gateway to the gold fields and did a brisk business as a supply point.

First known as Reading's Springs, the name was changed to Shasta when the post office opened in 1851. The town was also known as the head of "Whoa Navigation," because more than 2,000 pack mules were transporting supplies back and forth to the Trinity mining camps. Merchants could sell a good $3,000 worth of goods before noon. During 1852 gold shipped from the camp was averaging about $100,000 a week, roughly $2 million that year. A road that was built to the camp of Weaverville bypassed Shasta, signaling its impending death.

Religious services occurred from time to time at the camps, sometimes with belligerence. One day the Reverend Hill began to preach to the miners. Someone in the crowd told him to leave and get out of town. "We don't allow any fanatical, psalm singing preachers in this burg." The reverend continued his sermon. "All right, get a move on and make it fast or I'll throw you over the balcony." Hill told the gent to go ahead. When the protestor rushed at him, he was brought up short by a blow between the eyes. "Now boys, if there are any more of you who object to hearing the word of God, let him come forward before the service proceeds. I do not like to be interrupted." [115] He had no problems after that.

SHAW'S FLAT (Tuolumne) Located south of Columbia, this camp was settled when placer mining began about 1848 and was named for Mandeville Shaw. Its post office was established in 1854. The flat was rich in placers. An early miner dug up a 26-ounce gold nugget, and was followed by others who were bringing out between $200 and $600 every few days. Total yield from the various claims came to over $6 million.

One story told of the miners experiences was about a fellow at one of the claims who was too lazy to wash his clothes: "You take, for instance, one of these first comers, he was even too lazy to wash his drawers. To save himself the work of scrubbing them, he just tied them to a limb that overhung a little stream and let them dangle in the water. He figured, you see, that the current would wash them for him overnight. And the next morning when he came to fish them out, lo and b'God! He found his drawers gold-plated."[116]

Those types of stories managed to bring in even more miners. John Mackay and Jim Fair started out by working a small claim at the flat. Until a water ditch was built they had to haul their ore for miles to the streams to wash out their gold. They later went to Nevada where they became among the famous for the Comstock in Nevada.

SHERLOCK (Mariposa) This camp was established after the Sherlock brothers accidentally discovered gold. Their rich claim yielded them about $30,000. Sherlock was located along the Merced River east of Bear Valley. The rest of the mines in the immediate area produced about $1 million.

SHINGLE SPRINGS (El Dorado) Starting out as a crude mining camp in 1849, the town began its growth a year later when Edward and Henry Bartlett built the first public house. They also operated a horse-drawn shingle machine, hence the town's name. The post office was established in 1853. The nearby mines were good producers. One recovered nearly $10,000 in gold, which precipitated the building of three stamp mills. In 1890 the Pyramid Mine was located and recovered about $1 million in gold. During 1865 the Placerville & Sacramento Railroad came through, and Shingle Springs became a supply and transportation center for the region.

SHIRT TAIL CANYON (Placer) Shirt Tail Canyon, which was near the North Fork of the American River, received its unusual name in 1849 when two prospectors named Tuttle and Van Zandt found a miner working his claim completely naked. When he saw the two men, the miner hastily put on his shirt. Tuttle asked the man, "What in the devil's name do you call this place?" The miner glanced at his bare legs and realized his ludicrous appearance and laughingly answered, "Don't know any name for it yet, but we might as well call it Shirt Tail as anything else."[117] No record exists of the man's name. Mining in the canyon was sporadic and not that successful. One worker was getting only about 4–6 ounces a day, although a few did manage to make $1,000 a day. After the miners left about 1861, the Chinese came in and worked what was left of the placers.

SICARDS BAR (Yuba) Pierre T. Sicard was a Frenchman and sailor who moved to the flat in 1848 and named the place for himself. In addition to gold mining, he also opened the first general store and began trade with the nearby miners. The site was located along the Yuba River. Some of the mines in the region were averaging $600–$1,000 per day. Hydraulic operations began in the 1870s but had only moderate success.

SIERRA CITY (Sierra) This place bears a Spanish word meaning "mountain range." The camp was established in the middle of the county when lode mining began during the 1850s, and acquired a post office in 1864. The most important mine was the Sierra Buttes, that produced between $15 and $17 million in gold, which made it the largest single gold producer in the county. It operated until 1938.

SKIDOO (Inyo) Situated in the Panamint Range, the camp was established by Matt Hovic, E.A. Montgomery and John Ramsey. Its post office opened in 1907. The first gold claim discovered in 1906 sold for a mere $60,000. It later produced more than $1.5 million. Skidoo was taken from a popular song, "Twenty Three, Skidoo," and means "get out," or "leave me alone." Total production from the surrounding mines came to more than $6 million.

Skidoo was the site of the last hanging in California. In 1908 a part owner of the Gold Seal Saloon, Joe Simpson, got into an argument with a resident named Jim Arnold, and in the altercation Arnold killed Simpson. Arnold was tried and summarily hanged.

SMITHS FLAT (El Dorado) Named for some fellow named Smith, the diggings were discovered in 1849. A few years later someone found a boulder full of gold that was worth about $6,000. Placer mines in the area realized more than $2 million.

SONORA (Tuolumne) Sonora was settled by Mexicans in 1848 who named it for a place in Mexico. Water for the claims had to be brought about 60 miles from the Stanislaus River. The cost of water was so high in the miners' opinion that they built their own ditch, but high water destroyed it in 1861. During 1851 a group of Chileans with American partners were digging in a gulch near town that turned out to be a bonanza. They worked the claim until they made their fortunes, then left. Later, Mexican miners tried their luck and recovered about $50,000, and left. The gulch was abandoned because it was thought all the gold had been taken out. About 1879 Joseph Bray and Jim Divoll worked the old claim and found pockets of pure gold that came to almost $1 million. As they continued working it, the *Union* newspaper took notice: "James G. Divoll shipped via Wells Fargo C. 900 lbs. of gold, with a value of $189,000. It was taken from a pocket mine on the north end of Washington Street, within the city limits of Sonora.... It is the biggest shipment ever made on this coast."[118] The gulch ultimately yielded more than $2 million in gold.

Judge Barry was a tough judge at Sonora. When he moved there he saw how wild the place was and helped to form a town government. Robberies, murders and assaults were the order of the day. He made law when it was necessary on which to base his decision. Of one proceeding in 1851, he wrote, "N.B. Barber, the lawyer for George Work, insolently told me that there were no law for me to rool [*sic*] so. I told him that I didn't care a damn for his book law, that I was the law myself. I fined him $50, and committed him to goal [jail] 5 days for contempt of court in bringing my roolings and discussions into disreputableness and as a warning to unrooly persons not to contradict this court." As to the plaintiff George Work, Barry ruled on his case after Jesus Ramirez stole Work's black mule. "This is a sute [*sic*] for mule steeling, in which Jesus Ramirez is indited [*sic*] for steeling one black mare mule, branded O with a 5 in it from Sheriff Work.... Jesus Ramirez not having any munny [*sic*] to pay with, I rooled that George Should pay the costs of court as well as the find [fine]."[119]

SOULSBYVILLE (Tuolumne) The first discovery of gold here yielded riches for the Platt brothers, who found ore accidentally while looking for stray cattle in the 1850s. Located east of Sonora, the camp was named for Benjamin Soulsby, who struck it rich nearby. A stamp mill was built to process the ore, and a post office was established in 1877. Soulsbyville was the first place in the United States where hard-rock miners from Cornwall, England, came to work the lodes in the late 1800s that yielded between $7 million and $20 million.

SPANISH DIGGINGS (El Dorado) During 1848, these diggings were found by Andrés Pico, who was the brother of the last Mexican governor of California. The first prospectors there were Mexicans from Sonora, followed by the Americans. First named Dutchtown, it was changed to reflect the Mexican miners.

One group of Forty-Niners wasn't having much luck finding ore, and other prospectors were laughing at them, but the men kept on digging. One of them said, "We got down through the gravel the boulder had been sunk in ... and it was all mixed in with black sand and with little flecks and bits of gold. We chattered like magpies scraping it up into our pans and stepping into the river to wash it out. And that when it was dark and we couldn't work any longer we got out the scales and weighed up our day's work by firelight, and we had twenty-six ounces of gold, or $416 worth, and one little chunk alone weighed $17 worth. We wouldn't have traded in our gravelly spot for any kingdom in Europe."[120]

During the 1860s the Grit Mine was opened, and someone found a nugget weighing out at 101.4 troy ounces. The nugget was put on display in the Ferry building in San Francisco, but it was later discovered that the piece actually weighed 201 troy ounces. Between 1852 and 1942, the mine yielded more than $2.6 million. The site was located a few miles northwest of Georgetown.

SULLIVANS (Tuolumne) This camp was named for an Irishman named John Sullivan who arrived in 1844 and opened a trading post in 1848. He also did some prospecting on the side. Discoveries by other miners included a Dutchman who found 40 pounds of gold in a few hours, and a nugget found by another man weighed more than 400 ounces, yielding over $7,000. Roughly $3 million was recovered from the area.

SUSANVILLE (Lassen) First known as Roop, the camp was settled by Isaac Roop in 1853, who later changed the name in 1857 to honor his daughter. Peter Lassen and a group of prospectors arrived about 1854 from their mining ventures at the Feather River. After their discovery of gold nearby, Susanville became a trade center. The region was so remote that during 1856 settlers held a meeting at Susanville to form their own government. They "cut themselves off from California," and christened their land the Territory of Nataqua. That didn't fly because Congress refused to recognize it. So Roop waged what was called the Sagebrush War. The Plumas County sheriff came to break up their intentions. After a few shots were exchanged back and forth, they came to an agreement with the result new boundary lines were drawn, and Susanville became part of Lassen County. Both sides repaired to the local saloon for drinks.[121] Just south of town the Diamond Mountain District produced nearly $50,000 in gold.

TECOPA (Inyo) Tecopa was founded east of Death Valley National Park as a gold and silver mining camp before 1872. The post office was established on May 24, 1877. J.B. Osbourne named the site for Paiute chief Tecopet, whose name means "wildcat." Although he was a pleasant person and got along well with the whites, he could get quite ornery. He was leader of the Southern Paiute Tribe, and owned a lead-silver mine that he later sold for a silk top hat. The Tonopah and Tidewater Railroad didn't make it to town until 1907 because of grading difficulties over the Amargosa River Canyon. It transported about six to eight carloads of ore a month from the surrounding mines. Gold mining was later supplanted when gypsum and talc were discovered.

TEHACHAPI (Kern) The Kawaiisu Indians (Shoshonean lineage) were the first to occupy the territory. During the 1860s the town was called Williamsburg, and later changed to Tehachapi, interpreted as "windy place," referring to the windy desert passes. When surveyor Lt. Robert Williamson came through in 1853, he wrote in his field notes, "We learned that their name for the creek was Tah-ee-chay-pah," and described a creek frozen over.[122] Other translations were "sweet water," and "many acorns." Miners came to the region during the Kern River gold rush in 1854. Gold was also found in the nearby mountains. The men were earning between $8 and $10 a day for their work. One of the mines produced only about $250,000 within a 50-year period.

TEMECULA (Riverside) Temecula's foothills were home to the Pechanga Indians until they were removed to a reservation in 1885. This town was part of a land grant given to Felix Valdez in 1845. The general store opened by John Magee in 1849 served as a trading center for gold seekers and settlers, and acquired a post office in 1859. It was the second post office in the state, behind San Francisco. In 1873 Simon Levi was postmaster until 1876 when he left Temecula and later founded the Simon Levi Company. Temecula is a Luiseno word meaning "sun shining through fine mist."

TIMBUCTOO (Yuba) First called Sand Hat in 1854, the name was changed to Timbuctoo. It came from either a song or for a black miner by that name, descriptive of a place most distant imaginable. Its post office was established in 1858. Timbuctoo mainly served as a supply center for the outlying mines until a fire destroyed it.

TODDS VALLEY (Placer) The valley was first settled in 1849 by Dr. F. Walton Todd, for whom it was named. He was a cousin of Abraham Lincoln's wife. Todd was soon followed by W.R. Longley, who opened the first store, and the post office was established in 1856. This site was located above the Middle Fork of the American River. Longley's store was a general stopping place for travelers and local miners. Gold was discovered just below Todd's land by two brothers from Illinois. After their claim yielded them each $20,000, they sold it to Alfred A. Pond and returned home. Total production of the claims was about $5 million.

TRINIDAD (Humboldt) Trinidad was first occupied by the Bruno de Hezeta Expedition in 1775 who named it Puerto de la Trinidad, for Trinity Sunday. The town itself was founded in 1850 with a post office in 1851 and named Warnersville for settler R.V. Warner, which was later changed. Trinidad served as a supply distribution point for the gold miners headed to the Trinity River and Klamath mines.

TRINITY CENTER (Trinity) Trinity Center was established by Moses Chadbourne, and became a supply and trading center for the mines in the region. Its post office opened in 1855. The name was taken from the Trinity River along which it was located. Miners were earning about $10–$20 a day in the nearby diggings. Bradbury wrote of exaggerated reports, "There is good news from Trinity River; gold is very plentiful and provisions scarce ... the richest gold deposits yet discovered were on Trinity River, where any man could dig $100 per day, and from that to $2,000. To every man who could be gulled by this story, some merchant would sell two mules, provisions for three months, mining tools, &c., &c."[123] When the placers ran out, those who stayed went into farming.

TUMCO (Imperial) This camp was about ten miles from Yuma, Arizona. It was initially called Hedges for a Swedish track walker with the Southern Pacific Railroad. He apparently deviated from his work and took some time to search for gold, which he discovered in 1880. Some excitement was created and a number of men came to check out the prospects. A 140-stamp mill was later erected to treat the ore from the mines in the area. When the post office was established in 1910, the name was changed to Tumco, an acronym for The United Mining Company. The camp enjoyed a population of nearly 3,000 and four saloons along Stingaree Gulch. Hedges later sold his property to a man named Borden (later the condensed milk baron).

TUOLUMNE RIVER/TOWN (Tuolumne) When Spanish explorer Alferez Morago visited in 1806, he called the stream the Rio de Dolores ("river of sorrows"). The town was founded in 1854 as Summerville honoring settler Franklin Summers. It was changed to Tuolumne in 1891 when the post office was established, thought to be from the Miwok word talmalamne, for "stone wigwams." Of course, there were no stone wigwams, but the expression probably referred to the Indians who sometimes lived in caves. Tuolumne existed as a supply center for the nearby mines as well as a railhead for the old Sierra Railroad. The Tuolumne River rises in the Sierra Nevada Mountains and enters the San Joaquin River. After the gold ran out, Tuolumne's economy was tied to the timber industry.

TUTTLETOWN (Tuolumne) Starting out as a stopping place for packers carrying supplies to the mines, the camp was first known as Mormon Gulch for the Mormon prospectors who arrived in 1848. They found abundant gold in the nearby streams, and one lucky miner extracted about 40 pounds of ore in a week's time. Another found coarse gold valued at about $10,000. The post office was established in 1857 and renamed Tuttletown for Anson A. Tuttle, a judge and mine owner. The camp was situated on a slope of Jackass Hill along Mormon Creek. By 1899 more than $2 million had been recovered from the local mines.

VOLCANO (Amador) Volcano began its life in 1848 after a group of soldiers found $500 in gold from a single pan of gravel. The town grew rapidly with a dozen restaurants, five hotels, and more than 45 saloons, with a post office in 1851. Some of the claims were so rich that the gold could literally be picked up by hand. One spot found to contain ore was located at a graveyard. The bodies were relocated so the miners could get the gold, which was valued between $500 to $1,000 a pan. So much for sanctity. By 1855 the placer deposits were depleted, so miners turned to hydraulic mining. Unfortunately the dirt washed down into the homes and buildings, undermining them, but more than $2.5 million in gold was extracted from the area.

During the Civil War, southern miners threatened to divert their gold for the Confederacy. But when Union volunteers brought in a cannon called "Old Abe," the sympathizers backed off. Unknown to them, the only ammunition available for the cannon was a pile of round river rocks. Eventually Volcano was left in ruins, the gold was gone, and it almost became a ghost town. But it was rebuilt, and today Volcano is a prosperous tourist town. Volcano received its name because settlers thought the site was the crater of an extinct volcano.

The first rental library was established at Volcano. In 1850 local merchants donated shelves for the books which were rented for 10¢ a day. Four years later the Miners' Library Association was formed.

Near Volcano is Daffodil Hill, a flower lover's paradise. Each spring between the middle of March and the middle of April, the hill explodes with daffodils. During the 1930s, the McLaughlin family began expanding their flowers, and today there are more than 300,000 bulbs and 500 varieties that burst into bloom. Each year, even more bulbs are added. Information: Amador County Chamber of Commerce, (209) 223-0350. The Daffodil Flower Farm may also be contacted at (209) 296-7048 for directions and hours.

"Old Abe" at Volcano used to thwart Southern sympathizers from directing their gold to fund the South in the Civil War (courtesy of the Amador County Archives, Amador, California).

WEAVERVILLE (Trinity) This camp was settled about 1850 and took its name from the creek that honored prospector George Weaver, who built the first cabin there. He found a small deposit which reaped him about $15 a day. Its post office was established in 1850. The mines were quite prosperous in 1852 with gold shipments amounting to about $30,000 a week. By 1856 more than $50,000 was being shipped out weekly. Large-scale placer mining was conducted using hydraulics, followed later by dredging. Total production came to about $35 million.

Weaverville was quite isolated until a wagon road was built in 1858 connecting it to the outside world. That year a newspaper wrote, "Weaverville now assumes its proper place among the foremost cities of our fair state of California. The opening of the luxurious new wagon road ends our condition of isolation."[124]

WHISKEYTOWN (Shasta) Local folklore has it that a miner was hauling supplies to his mine when the pack on his mule came loose. A whiskey barrel fell off, broke, spilled into a creek, and the miner dubbed it Whiskey Creek. The settlement next to the stream became Whiskeytown. The birth of this community began with the tents of miners. The arrival of businessmen transformed it into a supply depot and permanent settlement. Miners first extracted the gold by placer mining. In nearby Mad Ox Canyon, an 81-ounce gold nugget was found, and in Whiskey Creek two men found a 7- or 8-pound piece of gold quartz. Down at Mad Mule Canyon a nugget of crystallized gold was found. It was later exhibited at the Paris Exposition of 1878. Operations in the canyon ultimately recovered about $1 million. Eventually the surface gold started to decline, and other methods of mining had to be used. Placer production yielded more than $1 million in gold. The ore disappeared in the 1870s and the miners with it. When Shasta Dam was constructed in 1963, water filled the town, which was later turned into farmlands. Nothing is left except the post office, cemetery, and possibly a store. Whiskeytown is now a national recreation area.

YANKEE HILL (Butte) This site was probably named for Easterners who came to seek their fortunes, but it was settled by Mexicans and Chileans in the 1850s who discovered gold nearby where more than $1 million was realized. Yankee Hill later became a supply center for the mining companies. One of the most productive lodes was the Hearst, that yielded more than $1 million before 1910.

Nate Arnold and his family moved to Yankee Hill, and since there wasn't much entertainment, he decided to purchase a piano for his wife's pleasure. But when he got it to the cabin it wouldn't fit through the door or window. It spent its life in the front yard where it was used for parties that Arnold occasionally held and invited the entire town.[125]

YANKEE JIM'S (Placer) This camp was located on the northern side of Devil's Canyon, and settled about 1850 with the arrival of wagoners Nicholas F., George W. and Benjamin F. Gilbert. The post office was established in 1852. Yankee Jim was no Yankee; he hailed from Australia, and his name was Jim Robinson. He had a penchant for "raw brandy, a light hand for other people's property, and a Cockney accent."[126] He built a corral hidden among the trees and spent much of his time stealing horses, getting away with it for a long time. He was finally nabbed and promptly hanged. Upon discovery of Jim's corral, the ground was found to contain a fortune in gold. Some believe Yankee Jim was an Irishman named Jim Goodland.

YOU BET (Nevada) You Bet received its interesting name from a saloon keeper named Lazarus Beard who took up residence there. He called to his assistance William King and James Toddkill, who frequently repaired the saloon. They had the pleasure of drinking free whiskey while trying to come up with a name, but were always careful to suggest one they were sure Beard would object to, so as to protract the deliberations and enjoy the accompanying whiskey as long as possible. If Beard used one expression more than another, it was "you bet," which was a great favorite with him, and in a joking way the two suggested that as a name for the town. To their surprise it met with favor and was adopted, and their free whiskey was stopped.[127]

Although gold was discovered in 1849, You Bet wasn't established west of Dutch Flat until 1867 when hydraulic operations began, which badly scarred the foothills of the Sierra Nevadas. This type of mining recovered about $60 million from the region. After passage of the California Debris Commission Act of 1893 all hydraulic operators closed down.

YREKA (Siskiyou) Discovery of gold caused Yreka to be founded 1851 when Abraham Thompson found ore at Black Gulch. A few days after the discovery, two freighters named Theodore F. Rowe and Burgess Brothers brought their goods to Yreka from Scott Bar and started a flourishing supply business. First called Thompson's Dry Diggings, then Shasta, and Butte City, it was changed to Yreka in 1853 when the post office opened. The name was interpreted as "north place," "north mountain," or a Shasta word for "the white," descriptive of the snow on Mount Shasta. During the camp's early days, it was described as "A pole leaning against a tree and covered with canvas [which] made an excellent store.... Brush when artistically arranged by the hand of experience, made a couch so soft and springy as to woo the tired miner to sweet repose and happy dreams.... This commingling of tents, brush shanties, and open camping places probably bore as striking resemblance to Washington's camp at Valley Forge as has ever been produced since the 'times that tried men's souls.'"[128] A prerequisite for any mining camp was saloons and whiskey. Sam Lockhart boasted of a meager attempt at architecture by building his saloon with shakes nailed upon poles and covered with canvas and brush. But eventually the tents were replaced by wood and brick buildings.

Yreka became a major trading and supply center for a large mining district. In 1852 a disastrous flood occurred in the region which made the road impassable. Supplies could not get in or out. Food ran low, and even the basic staples like flour and salt were unavailable. What salt was available was sold for $1.00 an ounce, too little to even cure the settlers' meat. Then during 1854 the town suffered a fire that destroyed a large portion of the business district. The bucket brigade could do little since water was scarce and there was no other way to fight the conflagration.

At one point in time a group of horse thieves began their operations near the mines. One of the robbers stole one of a matched pair of mules and claimed it belonged to him. His case was brought up before the local alcalde, George Vail, who ruled in favor of the thief. A few days later, the real owner brought the other mule into town. The stolen animal heard the braying of the other mule and responded. As they kept calling to each other, the alcalde decided there had been a gross injustice. He said, "Old man, that is your mule. That is better evidence than was ever given in a court of law." He turned to the thief and said, "If I ever hear of you attempting this game again, I will head a crowd that will hang you higher than Haman."[129]

Daguerreotype of miners at Thompson's Dry Diggings, 1852, renamed Yreka (photograph courtesy of the Siskiyou County Museum, Yreka, CA, #P4306–22).

YUBA CITY (Sutter) Yuba was thought to be derived from *uva*, the Spanish word for "grape." Others suggest it is from a branch of the Maidu Indian Tribe, the Yu-ba. When John Sutter moved here, he named the river Yuba for the Indian village. This site was deeded by Sutter in 1849 to Samuel Brannan, Pierson Redding and Henry Cheever, who had the town surveyed and platted the same year, followed by a post office in 1851. Yuba became a trading center for the outlying mines. The first store was opened by Tallman H. Rolfe and Henry Cheever, who advertised heavily in the local papers: "Notice to Miners. Rolfe & Cheever, having established a store at Yuba City, will keep constantly on hand a large and general assortment of dry goods, groceries, provisions, etc., which will be sold low for cash or gold dust."[130] After the gold rush days were over, the town became an agricultural community.

COLORADO

ALICE (Clear Creek) Alice was settled in the 1880s after placer gold was located along the Fall River. Most of the production came from the Alice Mine, named for the wife of the owner, Mr. Taylor. A 200-ton stamp mill was erected, followed later by hydraulic machinery used to recover the ore. Roughly $60,000 was taken from the mine. In 1897 the Alice was sold and worked until about 1889 when it closed. The only other recorded production was from the North Star-Mann Mine, which yielded about $116,000.

ALMA (Park) Alma was established in 1872 as a supply camp for the South Platte River gold mines, and acquired a post office in 1873. During the first three years of placer mining, about $19,000 was taken out. Between 1879 and 1880 silver was discovered, bringing in a horde of prospectors. The claims needed a smelter, which was later built to treat the ore. Between 1904 and 1942 more than $570,000 had been recovered. There are three versions of its name origin: Alma James, the wife of a store owner at the town of Fairplay; Alma Jaynes, daughter of a settler; or Alma Graves, wife of mine operator Abner Graves.

ALPINE (Rio Grande) Silver was discovered about 1872, but not much development occurred until 1875 when the Tilden Mine was located. Shortly thereafter, Alpine was established and named for the Alps in Europe, Celtic for "rock mountain." Until a smelter was built there, the ore had to be freighted to Pueblo. The Tilden Mine yielded about $36,000 during its early operations. The new smelter failed to work properly and was later sold. Miners thought Alpine would become a boom town when they heard rumors that a railroad might come to town, but it bypassed the site in the 1880s and Alpine began to decline.

ANIMAS FORKS (San Juan) This site bears a Spanish name for "soul." About 1875 a mill was erected to serve the Red Cloud Mine near Mineral Point, and a camp was built around the structure. Located at an altitude of 11,000 feet on Wood Mountain, Animas Forks experienced bad storms with snow drifts nearly 25 feet deep, so most of the miners abandoned their claims during winter. Adding to the danger was the cutting of trees to build cabins, equipment, etc., which caused a horrific snow slide down a bare mountainside that destroyed many of the buildings. The camp began to decline in the early 1900s, but got a new breath of life when a railroad was extended through the region. After the mill was dismantled in 1916 Animas Forks was history. Ore production is combined with that of Silver City since they were both located in the same district.

APEX (Gilpin) Apex was established about 1890 with discovery of the Mackey Mine by Dick Mackey, and a post office opened in 1894. Mackey later sold his property to a Mr. Mountz and partners. Mountz was left high and dry when his cronies stole their working capital, but he continued to try to work the mine. In a last-ditch effort he put dynamite into the claim, which yielded him more than $1,000 a ton of ore. Gold values are included with Gold Dirt.

ASHCROFT (Pitkin) Silver deposits were discovered in 1880 by prospectors Charles Culver and W.F. Coxhead, who had come from the boom town of Leadville. Their mine yielded about 14,000

Miners with pack mules at Alma, ca. 1888 (courtesy South Park Historical Foundation, Fairplay, CO).

ounces of silver and then the vein ran out because it was too shallow. During this time period the town was established, even thought its growth was hampered by Indian attacks and a road to the camp was almost inaccessible. In order to reach it, travelers had to cross Taylor Pass where wagons were disassembled and hauled down cliffs, then reassembled. Ashcroft lasted only a few years and folded when the nearby town of Aspen began to grow.

Horace Tabor, famous for his Matchless Mine in Leadville, and his mistress Baby Doe, moved to Ashcroft where Horace built his lady love a lavish home, paneled with gold-encrusted wallpaper. When Baby Doe arrived, Horace proudly pronounced a holiday and offered free drinks to everyone in town. He went broke after squandering all his money.

ASPEN (Pitkin) The first exploration occurred about 1879. A number of lode silver mines were found in the Aspen and Smuggler mountains, notably the Durant, which began producing about $150,000 a month. The town was founded about 1883 as Ute City, then renamed for the abundant quaking aspen trees. About the same time the Spar and Emma mines were located. The Emma produced more than $500,000 from a single chamber, probably a geode. A smelter was constructed by Jerome B. Wheeler, and the ore was brought down via tramway for treatment. Wheeler also had a hotel and opera house built for the town. The silver was hauled out by pack train at a high cost, but when the Colorado Midland Railroad was built in 1887, financial burdens were greatly alleviated. Aspen became a boisterous camp of more than 15,000 people. In 1894 one of the world's largest silver nuggets was found at the Smuggler Mine, weighing a little over 2,000 pounds and consisting of 93 percent silver. Mining declined with the Panic of 1893. Total production for the district came to about $10

Animas Forks (courtesy David Nestor).

million. Aspen survived though, becoming a premier ski resort town.

During 1899 a young fellow walked into the store of feed merchant Owen Thom looking for a job. Thom asked him if he had any money at all, and the youth pulled out a rock. Thom was astounded with all the gold it contained and asked the boy where he got the rock. "Oh I stopped to sleep the other night about 10 miles south of town. On a big cliff where I bedded down, I saw this pretty stone."[131] Thom realized the rock might just be a piece of a big strike. He asked the boy if he could relocate the spot and was told that he could, but he was really hungry, so Thom gave the boy a hefty amount of cash for food and clothing. He waited for the boy to return, but the fellow never showed up again, having bought himself a new set of clothes and a sumptuous meal at the hotel. Thom had the rock assayed, which turned out to be rich in gold, but no one ever found out its location.

BIEDELL (Saguache) In 1881 Biedell was established on Crystal Hill and a post office opened two years later. Rancher Mark Biedell was an investor in the nearby mines who built an ore mill and attempted different methods to process the low-grade ore. He finally sold it in 1899. A big producer was the Buckhorn, which recovered about $80,000 in gold. In 1886 a feud began between owners of the Buckhorn and Alexander properties. The camp declined after all the litigation, although the disagreement was eventually settled. By that time Biedell was already in its death throes.

BLACK HAWK (Gilpin) Black Hawk was born as a milling and smelter town. The post office opened in 1880, named for a mill that was brought to town by the Black Hawk Company from Illinois. Black Hawk was a Sauk chief from the Midwest. The camp was located in Clear Creek Canyon and became very prosperous with its facilities serving the hard-rock mining industry in Colorado. This came about because Colorado's smelters could handle only high-grade ore, much of which was shipped long distances at great expense. Professor Nathaniel P. Hill introduced new milling methods from Europe, which proved quite successful. That, plus the building of the Colorado Central Railroad in 1872, eased shipping costs. Visitors to the camp described it: "Narrow and dingy as is this mining town its people are making a brave effort to give it a look of comfort.... Scarcely a tree or shrub is to be seen, or even a flower, except it be in some parlor window."[132]

BONANZA (Saguache) James Kenny, who was with General George Custer's cavalry, discovered silver along Kerber Creek near the Cochetopa Hills in 1880. He named his claim the Rawley and worked it profitably for about ten years until he sold it for about $100,000. His find brought prospectors swarming to the area, and Bonanza was founded bearing a Spanish name meaning "prosperity." The Empress Josephine Mine was located and yielded more than $7 million in ore. Most of the mining operations were small. At one point in time, some of the claims were producing more than 2,000 tons of silver per month. But much of it was low grade and the expense of freighting and smelting proved too expensive. By 1882 the town had lost much of its population. Total gold production was about $350,000.

BOULDER (Boulder) Named for the nearby large boulders, the town served as a supply and trading center. Prospectors arrived in 1858 and a post office was established the following year. The miners discovered what was called "shot gold" along Boulder Creek. After a silver strike at Caribou along the stream, Boulder boomed. The mines in the nearby hills recovered more than $35 million in gold and silver in addition to other minerals.

BOWERMAN (Gunnison) Bowerman was a wannabe mining camp, named for settler and miner J.C. Bowerman. Its post office opened in 1903. Local newspapers claimed this camp would rival Cripple Creek. Bowerman hit pay dirt, but kept it a secret and told his wife to keep her mouth shut. Tradition says she didn't, which brought in a small rush. Bowerman never shipped out any of his ore, and the miners left. Within a year the camp ceased to exist.

BRECKENRIDGE (Summit) During 1859 rich gold placers were discovered at Georgia Gulch on the north side of Farncomb Hill, and it wasn't long before more gold was found in other nearby streams. The camp was established and named for the U.S. vice president at the time, John Cabell Breckenridge, and had a population of about 8,000. Its post office opened in 1860. A blockhouse was also built for protection against possible Indian attacks. During the winter of 1899 Breckenridge was snowed in so bad it would be 79 days before anyone saw the outside world again. The only way to get supplies was via snowshoes to the neighboring mining camp of Como. Georgia Gulch contained one of the richest placers in the area. Water for the diggings was brought in via ditches and flumes from more than 50 miles away. Its production was estimated at nearly $3 million in gold.

A trapper named Sam found lots of gold during the 1860s, but kept his claim a secret. Sam regularly purchased his supplies at Breckenridge, and folks got curious when he began paying with gold dust. After he failed to appear one winter it was thought he had simply died in the mountains. But a miner named Zeb stumbled upon Sam's cabin in 1890, and discovered a mine dump and tunnel behind the structure. Hurrying back to town with some specimens of rock, Zeb showed them to friends. They headed back to the cabin, but Zeb couldn't exactly remember its location, and Sam's secret riches died with him.[133]

BUCKSKIN JOE (Park) This camp was named for Buckskin Joe, who found gold there in 1859. His real name was Joe Higganbottom, and he received that nickname because of the buckskin

Buckskin Joe (courtesy South Park Historical Foundation, Fairplay, CO).

clothes he wore. It was formerly known as Laurette for two women who lived at the camp, Laura and Jeanette Dodge. Records show that a man named Harris was the first discoverer of ore, and Buckskin Joe was right behind him. Joe supposedly sold his claim for a gun and horse. Folklore has it that Harris was hunting and shot at a deer, missing it. He recovered the bullet and there found gold-bearing quartz. Harris worked his claim in secrecy so word would not get out, but as in all mining areas somehow it did leak. Gold obtained from the lodes between 1860 and 1867 was valued at about $710,000. During the summer of 1871 silver ore was discovered on Mounts Lincoln and Bross and mining activity increased. Output of lode gold was about $2.7 million.

Silver Heels was a dancing girl at the time a smallpox epidemic broke out. When the women and children were removed to Fairplay for their safety, Silver Heels stayed. She nursed all those who got ill and comforted the dying. After the smallpox was quelched, residents donated about $5,000 to thank her, but Silver Heels had already departed. Many years later a veiled woman was seen visiting the cemetery. No one ever saw her face, but many believed it was Silver Heels, who contracted the smallpox that left her face badly scarred. Nearby Mount Silver Heels was named in her honor.

A Methodist preacher named Father Dyer arrived at Breckenridge about 1861. He delivered mail between Buckskin Joe and Cache Creek, more than 40 miles away, in addition to other outlying mining camps. On Sundays he preached at the mining camps. Because Dyer's mode of travel was a pair of snowshoes (Norwegian skis), he later became known as the "Snowshoe Itinerant."

BUENA VISTA (Chaffee) Located at the junction of Cottonwood Creek and the Arkansas River, Buena Vista began as a shipping point and trading center for the area miners and farmers. It was founded in 1879 with a post office the same year, and bears a Spanish name for "beautiful view." Buena Vista also became the county seat after resident thieves stole into Granite and swiped the county records. A smelter and ore sampling works were erected in 1881, and later a railroad came through with Buena Vista as its terminus until the line was completed to Leadville. Ingham once wrote, "Tents greet the eye along the streets in every direction. Large canvas covered buildings, tents filled with merchandise and mammoth tents transformed into warehouses crammed full of freight supplies en route to Leadville and the Gunnison country are everywhere seen."[134]

The town was a wild and lawless one, boasting more than a dozen saloons. Nearby was the post office, run by a woman who apparently sympathized with the less desirables. She put a box at a back window with a missing pane, and set their letters in the box so they could pick up their mail without being seen.

Buena Vista hosts its annual Gold Rush Days in August. Information: Buena Vista Area Chamber of Commerce, (719) 395-6612.

CAÑON CITY (Fremont) Before establishment as a gold center, this place was once a camping spot for the Ute Indians and explorer Zebulon Pike. A member of Pike's party wrote in his journal that the site was "surrounded by the grandest & most romantic scenery I ever beheld — what a field is here for the naturalist, the mineralogist, chemist, geologist and landscape painter."[135] Located near the Grand Canyon of the Arkansas River, Cañon City saw a surge of miners during 1859, serving as a supply point and trading center with a post office established in 1860. At one time Mayor Joaquin Miller tried to change the name to Oreodelphia, but the prospectors couldn't pronounce it, so he lost out. Cañon City became home to the Colorado State Penitentiary in 1868.

CARIBOU (Boulder) George Lytle was an early miner who located the Caribou Mine during the 1860s. The camp was established almost 10,000 feet up on a bald mountain, where prospectors suffered from tremendous winds and huge snow drifts during winter. Despite those obstacles, Caribou managed to keep a population of more than 3,000. By 1871 most of the rich lodes in the region had been found. In 1875 alone, more than $500,000 in gold and silver had been recovered.

Promotional techniques were used to encourage investors. Ore from the mines was displayed at territorial and national fairs and exhibits. In 1876 ore samples won awards at the Centennial Exhibition in Philadelphia. During 1873 another scheme was hatched to entice even more investors when President Grant visited Central City. He found the entrance to the Teller Hotel paved with silver bricks from the Caribou mines.

When the Panic of 1893 hit, most of the lodes closed except the richer gold mines. The value of silver produced was about $6 million. In 1900 a fire came through Caribou and decimated the site.

CARSON (Hinsdale) Christopher J. Carson was the first to discover gold and silver, and later located the Bonanza King Mine near the summit of the Continental Divide. Carson was formed about 1880 near the mine at about 11,000 feet. More than 100 men worked their claims, but shipping the ore at that altitude required a makeshift road to be built which was hazardous at best. The St. Jacob's Mine was a good producer, recovering more than $450,000 in ore. After demonetization of silver in 1893, activities slowed down.

CENTRAL CITY (Gilpin) John H. Gregory was the first prospector to find gold at nearby Clear Creek. He hailed from Georgia and moved to Colorado intending to work as a muleskinner. Gregory was considered a foul-mouthed, dirty person, but he "at least had an eye for 'flake' gold and sense enough to hold his tongue until he was ready to stake out his claims."[136] Central City became the supply center for the area's gold camps, and received its name because it was the center of trade. The post office was established in 1869. Placer mines nearby were recovering up to $35,000 a week in gold dust. In 1863 the Bobtail Mine opened and a tunnel was constructed so the ore could be brought out to the surface, hauled out by bob-tailed oxen. More than $84 million was recovered from the vicinity.

When prospectors learned that Horace Greeley was coming to Gregory Gulch that ran along town, they wanted to make sure he was impressed. So they salted a sand bar where they led Greeley so he could pan for gold. He was surprised at the amount of gold: "Gentlemen, I have washed with my own hands and seen with my own eyes, and the news of your rich discovery shall go forth over all the world as far as my newspaper can carry it."[137] And he did, creating a wild rush to the area.

One of the newcomers was George Pullman, who reaped $150,000 in a short period of time, not by mining but by lending money, charging an exorbitant 20 percent interest per month. He later went on to invent the famous railroad Pullman Car. In 1874 a fire that began in a Chinese laundry, caused by a defective flue, sent most of the town up in flames.

CLEAR CREEK CANYON (Clear Creek) George Jackson, who was a cousin of famous frontiersman Kit Carson, arrived at the canyon about 1859. He camped on a sand bar of Clear Creek during winter and while putting water on his campfire noticed a glint of gold. He took his drinking cup and pulled up a gold nugget, followed by another ounce of gold dust. Jackson kept a diary in which he wrote: "After a good supper of meat — bread and coffee all gone — went to bed and dreamed of riches galore in that bar.... Dreamed all sorts of things ... I've struck it rich. There's millions in it!"[138] Marking his find, he left and returned in the spring to work his claim. Jackson recovered about $2,000 in gold. He told one friend, who supposedly had a mouth as tight as a beaver trap, and lo and behold, in came the stampede. Total production of gold came to about $347,000, although other estimates came to more than $10 million.

CREEDE (Mineral) In 1899 Nicholas Creede and his partner, George L. Smith, discovered a vein of silver so rich that within a few months they had sold it to railroad magnate David H. Moffat for $75,000. Named the Amethyst, it rewarded Moffat with $2 million the first year. As a result the town swelled with more than 10,000 money-hungry gold seekers. The post office opened in 1891. A year later Richard Davis wrote of the burgeoning camp, "It is like a city of fresh cardboard, and the pine shanties seem to trust for support to the rocky sides of the gulch in which they have squeezed themselves. In the street are oxen teams, mules, men, and donkeys loaded with ore, crowding each other familiarly, and sinking knee deep in mud. Furniture and kegs of beer, bedding and canned provisions, clothing and half-open packing cases, and piles of raw lumber heaped up in front of the new stores — or those still to be built — stores of canvas only, stores with canvas tops and foundations of logs.... It is more like a circus tent which has sprung up over night and which may be removed on the morrow, than a town."[139] By 1897 over $4 million in silver and gold had been recovered, followed by more prosperous years until 1910 with the mines yielding about $1 million annually.

During 1892 famed bunco artist Soapy Smith came to Creede and conducted his soap scam, wherein he told would-be buyers that for $5 a person could get a bar of soap and if they were lucky one of them might have a hundred-dollar bill in it. The bars were wrapped with bills of various denominations and then wrapped again with blue paper. Soapy's shills purchased the first few bars that contained $100 in them, which brought the unwary to grab up the rest of his merchandise. Soapy also had other scams, but residents finally got tired of his shenanigans, and in 1890 he was kicked out of Creede and went to Leadville where he was kicked out again. Soapy went on to Alaska continuing his "art," and was eventually killed in 1898 by a group of vigilantes.

Old mine on a precarious cliff at Creede (courtesy Peter MacGregor).

Creede holds its Days of '92 Mining Events in July. Contact the Creede & Mineral County Chamber of Commerce at 1-800-327-2101.

CRIPPLE CREEK (Teller) Levi Welty and his family were the first people to settle in Cripple Creek. A shack was built over a spring to keep animals from polluting it. While constructing the shack a rolling log hit one of the boys, injuring his leg. During this commotion Levi accidentally discharged his rifle, nicking his hand. Then a pet calf tried to jump the stream and broke his leg. In the aftermath Levi said that it was some cripple creek.

Starting out as a cow camp, what some called a $300 million cow pasture, Cripple Creek became one of the greatest gold camps in the world, even more so than Virginia City or the Klondike. One of the most famous mines was the Independence, owned by Winfield Stratton. He nearly lost the property after agreeing to sell it to a representative of mine syndicate owner E.R. Pearlman. Because this area was mainly underground mining, getting the ore out was slow and costly. In debt, Stratton gave Pearlman a 30-day option on his property when offered $155,000 for it. Giving his mine a last look, Stratton accidentally hit the wall with an object and exposed a huge vein of gold. Prior to that lucky strike, Pearlman had decided there wasn't enough gold to make it worth his while and asked Stratton to release him from the lease, which Stratton gladly obliged. The Independence Mine turned out to be a "gold mine" when Stratton sold it for $11 million in 1859. It ultimately realized nearly $30 million. During the 1890s the mines were producing more than $250,000 worth of gold every month, and by the 1900s over $100 million had been recovered.

One of the biggest discoveries near Cripple Creek about 1914 was the Cresson Mine. While an engineer was tunneling through the mine, he drilled through what was a huge geode filled with gold hanging from the ceiling and walls, and nuggets littering the floor. In less than a month more than $1 million had been taken from the pocket. Another mine was the Elkton, said to have produced about $13 million.

Cripple Creek's Annual Donkey Derby Days is held on the last full weekend of June, with the World's Championship Donkey Race and other festivities. Thought of as pets, the donkeys are free to roam the city and protected by city ordinance. They are rounded up and brought into town in May, then sent back to pasture in October. Information: Cripple Creek Welcome Center, toll-free at 1-877-858-GOLD, or the Woodland Park Chamber of Commerce, (719) 687-9885.

CRISMAN (Boulder) This camp was named for Obed Crisman, who built the first ore mill there. G.A. Kelley established his store in 1876 and the post office opened the same year. Prospecting was concentrated in the nearby gulches. A good producer was the Yellow Pine Mine, located in Sunbeam Gulch. But the most important was the Logan, discovered in 1874 and thought to be the richest in the county at that time. The gold was so pure it was taken directly to the Denver Mint in a strongbox. The mine was nearly abandoned earlier because initially not much gold was found, but A.S. Coan, superintendent of the mine, requested that one more blast be made. The first blast uncovered more than $1,000 in high-grade gold. The mine ultimately produced more than $200,000.

CRYSTAL (Gunnison) Crystal served as a trading center and mill town, with a post office established in 1882. The mill powered tools used in the mines and a nearby stamp mill. Crystal was located at a high altitude and became snowbound quite often in the winter. The year 1899 was exceptionally bad as the *Denver Post* wrote in a letter it received: "Crystal is snowbound. The stage road between the city and Marble, six miles below, is under ten to fifty feet of snow and the mail carrier has only been able to come through once in the past week and then on snowshoes. Supplies will be exhausted in a short time and unless the road can be opened, several of the mines will probably be forced to close down and the miners go out on snowshoes. As soon as the storm ceases and weather settles, a force of fifty men will begin work to open the road for a train of pack horses to bring in supplies and it is believed that by tunneling through snow at points most exposed to avalanches it will be possible to open a terminal station."[140] With the demonetization of silver, Crystal lost its luster and faded away.

DECATUR (Summit) This camp was named for Indian fighter Stephen Decatur, who was also a member of the Territorial Legislature. The only mine of importance was discovered in 1879 by J.M. Hall. Called the Pennsylvania, it recovered a little over $2 million in gold.

DURANGO (La Plata) Spanish explorers who visited the La Plata Mountains during the 18th century may have found gold, but mining did not begin until 1873 when placers were discovered along the Animas River near Durango and the La Plata River. Before the miners moved in, Durango served as a watering point for stage coaches and wagon trains. The town bears a Basque name meaning "watering town." Durango was established as a mining camp in 1881 when the Denver and Rio Grande Railroad decided to build a track to Silverton and establish Durango as the hub of its rail system to transport ore from the mountains to the smelters in town. Gold production in the region exceeded $100,000 by 1914.

Durango features a number of guided tours, from cultural and historic to local agriculture and manufacturing. There is also a day-long train ride up and down the valley to the town of Silverton, with a layover there for lunch and shopping. Information: Durango Area Tourism Office, toll-free at 1-800-525-8855.

ELDORA (Boulder) Located near the mining camp of Caribou, Eldora was given high hopes in 1869 when silver was first discovered in the Spencer Mountains, and more than 50 claims were staked. The first claim yielded about $70,000, and was later sold to a Dutch syndicate for $3 million. In addition to silver there was some placer mining, but their early production is not known, although it was estimated at about $200,000. Two stamp mills were built near the mines, and a drainage and transportation tunnel constructed to drive the ore to the mills. After the rich placers were depleted, what was left turned out to be low grade and most of the mines shut down. By 1917 Eldora ceased as a mining camp.

EMPIRE (Clear Creek) Gold was found about 1860 by prospector Henry D. Cowles and his companions. The camp was established and named for the Empire State by miners who hailed from New York. A post office was established the following year, and the population rose to about 1,500. Situated at the confluence of Lyons and Bards creeks, Empire was the last of the mining towns between Idaho Springs and the Continental Divide and was relatively isolated. As a result food supply prices rose, especially butter and eggs, going for $3 a

Main street of Fairplay (courtesy South Park Historical Foundation, Fairplay, CO).

pound and $3 a dozen, respectively. Because of lack of water for the claims, by 1866 most of the men had left for better diggings. Total minimum gold production of the district was a little more than $3 million.

EUREKA (San Juan) Eureka was settled about 1872, and the following year the Sunnyside Mine was located. Most of the miners working the property lived here. In 1948 the mine and its mills were sold for $225,000.

FAIRPLAY (Park) When gold was discovered in 1859, a number of districts were established, one of which was the Tarryall Mining District. New prospectors were not welcome, so they left and headed west to establish their own district along the South Platte River, which they called Fair Play Diggings. The name was chosen because the miners agreed that there would be fair play for all. It was changed to Fairplay in 1880 and served as a supply center. Gold placers discovered nearby were valued at about $1 million. Total yield of the district was a little over $4 million.

The town has memorials erected to two burros named Prunes and Shorty. Prunes is said to have worked at every mine in the Fairplay district. His last owner was a prospector named Rupert M. Sherwood. After the burro could no longer work, Prunes was allowed to wander the streets of Fairplay at will. When Sherwood died in 1930 he was buried next to his beloved burro. Prunes has a brass plaque charting his 60-plus years of service to miners and his collar is embedded in the memorial wall. Shorty's memorial is smaller. He wasn't the hard worker like Prunes, only the town mooch, but loved by all. He had a dog friend named Bum, who died of a broken heart when Shorty passed away, and was buried next to him.

Burro Days is a large event the last weekend in July, commemorating early mining days. Featured is a Pack Llama Race on Saturday, and on Sunday the famous World Championship Burro Race. Auto tours in the area travel through Mosquito Pass (named when someone opened a ledger and saw a squashed mosquito), Horseshoe/Four Mile, Weston Pass, Buckskin Gulch, and Placer Valley, showing a look back in the history of old mines and artifacts. Fairplay's museum depicts life during the gold rush days, with about 35 buildings that house more than 60,000 artifacts. Information: South Park Chamber of Commerce (719) 836-3410.

FULFORD (Eagle) This place was the consolidation of three mining camps: Fulford, Polar City and Gee's Addition. Formed in 1892 with a post office the same year, the site was named for rancher Arthur H. Fulford, who was killed in an avalanche a few years earlier. A man named Nolan discovered rich gold along Brush Creek. By 1896 the Polar

Star and Cave mines had been located, which warranted a 25-stamp mill. Mining continued until about 1918, but the ore values are unknown.

GEORGETOWN (Clear Creek) Georgetown is located at the foot of the Continental Divide, walled in by the high mountains. It was established about 1859 and named for George Griffith, who was the first person to find gold nearby. He later served as county clerk. Tradition says that he sat down to rest, then as he braced himself with a rock to get up, he saw a glint of gold. When it was learned that the Anglo-Saxon Lode had ore assaying up to $23,000 a ton in silver, stampeders headed for Georgetown. The post office opened in 1866. Estimates of silver recovery in the area range between $90 million and $200 million. The gold yielded a total of about $2.9 million.

Although Georgetown had its questionable characters, it was fairly peaceful, moreso than most mining camps. But it did have its moments, such as a sign over a saloon that read, "We sell the Worst Whiskey, Wine and Cigars." A sign painter, apparently unhappy with the owner of the saloon, thought his sign would detract from business. But it brought a lot of attention and so much more business that the owner kept the sign as it was.

A unique railroad track was laid near town in 1882 by the Union Pacific Railroad. Called the Georgetown Loop, it was the most difficult railroad project ever attempted. With a little over 4 miles of track, it had to climb more than 360 feet between Georgetown and Silver Plume. Built in a spiral fashion, with the tracks literally folding over themselves to gain altitude, it had 14 curves, many of them at an 18-degree angle. The spirals were built so the train could gain momentum to make the steep grades.

The Clear Creek County Railroad & Mining Days is held the last weekend of May, and features a number of tours in addition to a Pack Burro Race. Information: Chamber and Tourism Bureau of Clear Creek County at (303) 567-4660.

GILMAN (Eagle) Initial discoveries of silver during 1879 brought in the prospectors, who later found gold in the underlying quartz. During this time more than 100 claims had been staked. The camp was established on a steep slope 1,000 feet up Battle Mountain and its post office opened in 1886. The camp was named for mining entrepreneur Henry M. Gilman, who died a short time later. A local paper wrote, "He fell dead in the road near the town not long after the town had received his name."[141] Two of the most productive mines were the Ground Hog and Iron Mask. Owners of the Ground Hog recovered what was called wire gold. Because of its rarity, the ore was valued at $750,000. Also located was the Eagle Mine, the tailings of which were valued at about $84 million.

GLADSTONE (San Juan) The establishment of Gladstone was a result of Olaf Nelson's discovery of the Gold King Mine in 1877. The camp was named for the prime minister of England, William E. Gladstone. The property went on to produce between $8 million and $9 million in gold. After Nelson died, Cyrus Davis and Henry Soule purchased the mine from Nelson's widow for a paltry $15,000. As other properties were developed, a wagon road was built to enable transportation of the ore to Silverton. A 20-stamp mill was erected there, then changed to handle 80 stamps when even more discoveries were made. In 1899 the Rocky Mountain Construction Company built the Silverton, Gladstone and Northerly Railroad from Gladstone to Silverton, a boon to miners, enabling the ore to be shipped quickly and less expensively. Total production for the area came to about $70 million. In 1907 a fire started at the Gold King Mine, killing six men.

GOLD DIRT (Gilpin) Gold was found about 1859 at Gamble Gulch, followed by the erection of a number of stamp mills. The Perigo Mine, discovered in 1860, had very rich surface gold. However, the easily worked ore was soon exhausted and within six years only a few companies were still operating. After 1868 the camp was almost deserted until 1879 when the Perigo Mine once again became active. Total amount of gold in the region was about $723,000.

GOLD HILL (Boulder) This camp was founded in 1859 soon after gold was found in the region, and is believed to be the first mining camp in the state. The discovery brought in more than 1,000 miners, and that same year the first quartz mill in the state was brought in by ox team. Placer deposits yielded about $100,000 the first year. In 1894 a huge fire occurred, followed by high winds that fanned the flames, destroying much of the town. Not long after that disaster Gold Hill suffered another setback when a flood finished destroying what was left, but the town was rebuilt. By the time the gold had played out, more than $13 million had been realized.

Abandoned ore track in San Juan County (courtesy David Nestor).

GOLDEN (Jefferson) Golden was settled at the mouth of Clear Creek Canyon by Thomas L. Golden, George A. Jackson and James Saunders in 1858, but it was only their temporary camp. George West, who was affiliated with a Boston company, was the person who actually established the town with a post office in 1860. Golden had the honor of being the state's territorial capital from 1862 to 1867 until Denver was made the permanent capital. By the 1870s the town had five smelters to process the region's gold and silver. Placer deposits and lode mines amounted to about $60,000. Silver ore recovered was more than $1 million annually.

Golden is the home to the Colorado School of Mines. It was founded as Jarvis College in 1869 by a bishop of the Episcopal Church of Colorado, the Reverend George M. Randall, who wanted to provide technical training in mining.

GOTHIC (Gunnison) Gold and silver were discovered here in 1879, and the camp was formed with a post office the same year. News of the rich ore got out, and within a week more than 100 miners had arrived, which later grew to nearly 8,000. The camp took its name from the mountains, which resembled gothic architecture. The Sylvanite Mine that was discovered by James Jennings contained wire silver. Located in a pocket (or geode), the ore was worth over $15,000 a ton. In addition to mining, Gothic also served as a supply and smelting center for the outlying camps. By 1884 most of the silver was depleted. Gothic later became home to the Rocky Mountain Biological Laboratory, used for advanced studies in plant and animal life.

GRANITE (Chaffee) Named for the nearby Precambrian granite, this camp was established about 1860 when placer gold discoveries were made along the Arkansas River and Cache Creek. Although lode mines were discovered, it was the placers that brought in the revenue. The Yankee Blade Mine recovered about $60,000 in gold. When the placers got too deep, hydraulic mining operations began. Granite was county seat until the town of Buena Vista stole the county records.

HAHN'S PEAK (Routt) Prospector Joseph Hahn came here in 1862 seeking gold, which he found near Willow Creek. As the first winter was drawing near, Hahn left his claim with the intention of coming back the following spring. But it would two years before he returned with partners William Doyle and George Way, followed by other prospectors. In 1866 the camp was established high up on Hahn's Peak at about 8,200 feet. The men staked numerous placer claims and planned to leave again for the winter, but three men stayed behind. Way left camp one winter day to get supplies and never returned. It's not known what happened to him. The men who were left at camp used all their resources until they were close to starvation. So they headed out on snowshoes to another camp, but Hahn died en route. Not much is known of the mining history; however, placers were worked intermittently and between $200,000 and $500,000 in gold was recovered. There was once a small camp near Hahn's Peak called Bugtown, so named because of the bugs (mosquitoes) that caused typhoid and diphtheria epidemics, wiping out the camp.

IDAHO SPRINGS (Clear Creek) After George Jackson found gold quartz near Chicago Creek in 1859, he drew a detailed map of his claim and wrote in his diary, "Dug and panned today until my belt knife was worn out, so I will have to quit and use my skinning knife. I have about a half ounce of gold; so will quit and try to get back in the Spring."[142] Before Jackson returned, he told a trusted friend of his find, and of course news like that was never kept secret long, so other prospectors made a beeline to stake their claims. Shortly thereafter, John Gregory hit a rich strike. The ore was freighted to Blackhawk for smelting, but transportation costs were exorbitant until the Colorado Central Railroad built its line in 1877. A tunnel was later built through the mountains to Central City, where the ore was hauled out and loaded onto railroad cars. Idaho Springs built its own smelters and in 1903 had more than 25 of them. By 1918 the mines began to decline. Gold recovery estimates came to about $38 million.

IRWIN (Gunnison) Dick Irwin was a prospector from Canada who arrived in 1879 and discovered silver in Ruby Gulch. Called Ruby wire silver, the ore was sent by burro to Alamosa for shipment by rail to the Denver smelter. Irwin was established the same year, as was the post office. The following year the *Gunnison Review* wrote, "The only brass band in the whole Gunnison Country, is at Irwin, and the sweet music discoursed by it every evening has a charming affect, echoing through the gulches and over the distant hills and peaks."[143] Not so charming when winter set in. Those who stayed in the camp had to deal with 50-foot snowdrifts, and the miners had to dig themselves out of their cabins. With the arrival of spring, prospectors headed to Irwin were surprised to find no buildings. One of the men saw a chimney with smoke rising from it and in front of the building was a man standing by a hole in the snow. One of the newcomers asked the resident where the post office was located and he was told at the next hole. Other mines in the region also producer wire silver ore valued at about $3,000 a ton.

JAMESTOWN (Boulder) Although this site had been settled in 1806 by George Zweck who was unsuccessful in his prospecting, it would be a few years before any ore was found that brought in hundreds of miners. The boom lasted about three years and then lay idle for a time. A camp was formed in 1865 with a post office two years later. First called Elysian Park, the name was later changed. Two prospectors named Indian Jack and Frank Smith found gold in 1875 and named their claim the Golden Age. They did not realize their mine contained 50 percent gold and sold it for a mere $1,500. The big productive mine in the region was the Buena, which recovered about $1 million in gold. Other claims yielded more than $5 million in gold and silver. Much of the camp was wiped out by a flood in 1894.

LAKE CITY (Hinsdale) Now the seat of Hinsdale County, this town was originally a supply center and shipping point for gold and silver. Ore was first discovered by explorer John Fremont and his party about 1848, but there was no development until 1871 when Henry Henson and J.K. Mullen relocated the ore. They were unable to work their claims because it was on Ute Indian territory. When the land was opened by the Brunot Treaty in 1874, serious mining got underway after Enos Hotchkiss found gold near the lake. The Golden Fleece Mine was later discovered, but the ore seemed insufficient until one of the owners threw in a stick of dynamite in disgust and opened up a rich lode. More than 400 people lived at Lake City and its post office opened in 1875. The establishment of the Denver & Rio Grand Railroad brought Lake City into the forefront as a major shipping center. After 1893 the mines began to close, but the town survived with cattle ranching and agriculture.

Near town is Cannibal Plateau, so named because it is believed to be the site where infamous prospector Alfred Packer ate five of his "friends" during the winter of 1874. He went on trial at Lake City where he was found guilty of murder and sentenced to hang. But Packer got another trial and escaped the noose, receiving a prison term instead.

LAMARTINE (Clear Creek) The first veins were discovered in 1861 but not developed until 1868 after a smelter was built at the camp of Blackhawk. Gold and silver were discovered by prospectors named Bougher, Chavanne, Cooper and Medill, who staked their claim and named it the Lamartine. The camp grew around the mine and its post office opened in 1889. Mining development picked up after a railroad was built to the region in 1870. The Lamartine kept the camp stable until the ore began to run out during the early 1900s. Owners sold their mine to a man named Himrod for $280, who spent thousands trying to redevelop the property, but he was unsuccessful. More than $18 million in gold and silver was recovered, most of it from the Lamartine.

LEADVILLE (Lake) Leadville was first called California Gulch because prospector Abe Lee said that he had California in his pan when he discovered gold in 1859. One settler described the camp: "It was strung along through the gulch.... There were a great many tents in the road and on the side of the ridge, and the wagons were backed up, the people living right in the wagons. Some of them were used as hotels."[144] Eventually the gold ran out and many of the miners left. Sometime later it was discovered that tailings from the mines were full of lead carbonate containing silver, bringing more than 6,000 avid miners to the site. By 1878 the mines had yielded about $2 million in silver, and the following year about $9 million, then again in 1880 nearly $12.5 million. Total gold output of the district was about $61 million.

Horace Tabor and his wife, Augusta, arrived at Leadville to establish a store. Tabor grubstaked a couple of prospectors and was rewarded with a dividend of $10,000. In 1878 he purchased the famous Matchless Mine in addition to other claims, which netted him about $10 million. Tabor's downfall began in 1880 when a divorced young woman

Leadville (courtesy USDA–Forest Service).

named Elizabeth McCourt Doe moved to town. Tabor couldn't keep his eyes off the comely lady and they began an affair that scandalized the town. He divorced his wife and married Elizabeth, who was nicknamed Baby Doe. Tabor sold some of his property for about $1 million, then squandered it away. His situation went from bad to worse with the demonetization of silver. Tabor eventually became a postmaster of Denver until he died in 1899. Before Tabor passed away he told Baby Doe to hold on to the Matchless. She was found frozen to death in 1935 in a broken-down shack next to the mine she believed would bring her great riches.

The highest incorporated city in the West was once the wildest, richest lead and silver town in the West, as an old newspaper once wrote: "On all sides was a conglomerate mass of diversified humanity — men of education and culture, graduates of Harvard and Yale and Princeton, mingling with ignorant and uncouth Bull-whackers, men of great wealth mixing with adventurers of every degree without a cent in their pocket with which to pay for their night's lodging at the bit corral down the street; men of refinement jostling against cheap variety actors and scarcely less masculine actresses, dance-hall herders and others with callings less genteel; representatives of the better element in all the callings of life — hopelessly entangled in throngs of gamblers, burro-steerers, thugs, bullies, drunkards, escaped convicts, dead beats and the 'scum of the earth.'"[145] Another journalist wrote, "All roads leading to town are alive. Freight wagons coming down loaded with ore, and freight wagons going up loaded with every conceivable thing from mining machinery and railroad iron down to baby-wagons and pepper casters."[146]

At one time Leadville was one of the largest producers of molybdenum until the American Metals Climax Molybdenum Mine closed. One of Leadville's more colorful visitors was a Soapy Smith, who had earlier been ejected from Creede. He and his buddy sold soap bearing the name Sapolion to residents. He was run out of town when the people objected to him, so Soapy went on to Alaska where ended up getting shot by the vigilantes and was laid to rest in Skagway.

The Annual Leadville Boom Days is held in August, depicting its heritage and mining days. There is also a Leadville Mining Tour Class in June. Information: Leadville–Lake County Chamber of Commerce, toll-free at 1-800-933-3901.

LEAVICK (Park) This site was named for prospector Felix Leavick; it grew up around a stamp mill that served the nearby mines. It had a population of about 200 people, and even boasted a baseball team. In 1873 the Hilltop Mine was discovered at more than 13,000 feet in the mountains, but it wasn't developed for about 10 years. That same year the Last Chance Mine was located at the 12,000-foot level between Mounts Sherman and Sheridan, but the only way to haul the ore was in buckets using a 13,000-foot aerial cable tram. When the Panic of 1893 hit the miners left.

MAYSVILLE (Chaffee) Maysville's existence was ensured since it served mainly as a toll-gate camp for the Monarch Pass. It was founded on the ranch of Amasa Feathers, who set up shop when gold was located in the 1870s. First called Crazy Camp, it was renamed after a town in Kentucky (for Virginia politician John May). The post office opened in 1879, and before long two smelters were built to serve the surrounding mines. By 1893 the gold had been depleted and Maysville went down hill.

MINERAL POINT (San Juan) Located in the San Juan Mountains, Mineral Point was very isolated and named for a 60-foot knob where the camp was established. Its post office opened in 1875. During this period of time the Old Lout mine was located, but nothing was found after the first shaft was sunk. The Old Lout Company was about to pull out when one of the miners fired one more shot of dynamite which uncovered a rich body of high-grade ore. The first carload was valued at more than $8,000, and during the first month of operation the company recovered more than $86,000.

MONTEZUMA (Summit) This silver camp was founded along the Snake River in the 1860s and named for the famed Aztec emperor. Its post office opened in 1871. H.M. Teller and D.C. Collier were early discovers of silver, but transportation difficulties inhibited development until a railroad was built to haul the ore to a smelter at Pueblo. The Madonna Mine produced about 50 percent of the region's ore output. By 1882 most of the large deposits had been located, and were in operation until 1898 when they closed down. Total gold production was a little more than $400,000.

MONTGOMERY (Park) Thought to be one of the oldest camps in the state, Montgomery was established in 1861 a few years after gold was discovered. It received its post office in 1882. In addition to the usual businesses and saloons, the camp also had the largest theater in the area. The camp grew rapidly with more discoveries, but relapsed in the 1860s. It saw renewed life in 1881 after silver was discovered on Mounts Lincoln and Boss. With all its amenities, Montgomery didn't last long, and most of the residents moved on to the mining camp of Buckskin Joe. The town went under water with the construction of a reservoir.

MOSQUITO (Park) Gold found in the region during 1860 resulted in the establishment of Mosquito. While residents were trying to decide on a name, a secretary opened a book and saw a smashed mosquito between the pages, hence the name. The post office began operations in 1862. Numerous placers were worked until the Civil War when many of the miners left, but later they returned and went on to locate lode mines. One of the most important was the London Mine, which operated for about 50 years and recovered more than $4 million in gold. Before the camp's demise, a pass was cut through Mosquito and into the booming town of Leadville.

OHIO CITY (Gunnison) Although gold was discovered during the 1860s, not much occurred until a rich silver boulder was found at the mouth of Gold Creek in 1879 that brought in the gold seekers, followed by formation of the camp in 1881. First named Eagle City, it was changed to honor prospectors who hailed from Ohio, a Seneca expression meaning "beautiful river." Numerous stamp mills were erected but development costs of the properties ran into the millions. In 1882 the Denver, South Park & Pacific Railroad built its track through town, enabling cost-effective shipment of the rich ore. The Carter Group that purchased many of the claims was well-managed and processed ore at about $1,500 a week using its 20-stamp mill that operated for almost 20 years. A very productive mine was the Raymond that yielded nearly $7 million in gold, and the Gold Link Mine recovered more than $1 million in silver and gold.

OURAY (Ouray) A.J. Staley and Logan Whitlock were early discoverers of silver in the region while they were fishing. W. Begole and Jack Eckles later opened the Mineral Farm Mine, so called because the silver was simply dug up like potatoes. In 1874 the headwaters of the Uncompahgre River and Poughkeepsie Gulch were prospected, resulting in many claims staked out.

The town is surrounded by 14,000 foot peaks on three sides. It was founded in 1875 and served as a supply center and trading point. Named Uncompahgre City ("hot water spring," or "red water canyon"), it was later changed to Ouray for a highly respected Ute chief, whose name means "arrow." Ouray was instrumental in preventing bloodshed between the whites and Utes when the Indians were trying to claim western Colorado. After the Meeker Massacre in 1879, Ouray was distraught because all his efforts towards peacemaking came to naught. He died near Ignacio, a broken man.

During 1884 there was much development, but low-grade ore and expensive transportation allowed only the best mines to ship ore. The American Nettie Mine was a very rich lode discovered in 1889 that produced over $4 million in gold. The Camp Bird Mine, owned by Thomas Walsh, was located about 1896 and recovered nearly $4 million within six years. Walsh later became a U.S. senator and was also one of the people who purchased the Hope Diamond for his wife. By 1916 the Camp Bird Mine had yielded more than $27 million in both gold and silver.

There are daily tours of one of the old mines. Reservations are recommended. Call (970) 325-0220.

PITKIN (Gunnison) First known as Quartzville, this site rose on the heels of Leadville and was established as a supply point when prospectors arrived in 1878. A year later the name was changed for Colorado governor Frederick W. Pitkin when the post office opened. Rumor has it that two elderly prospectors searched for silver with no luck until one of them hit a rock and found wire silver. After that, the growth of Pitkin was so swift that newcomers had no room to board because the accommodations were always full. One of the miners griped, "We have three women, eight children, three fiddlers, 180 dogs, two burros and one cat — and need a newspaper and sawmill."[147] He at least got his wish about a newspaper because it went into business by 1880. The silver panic of 1893 forced the closure of the mines.

QUERIDA (Custer) Bearing a Spanish name meaning "beloved" or "darling," the camp was settled in 1887 near Tyndal Mountain by David Livingstone, nephew of the noted explorer of Africa. Its post office was established in 1880. One of the richest mines was the Bassick, which produced both gold and silver. John M. True located it in 1887 but did not develop the claim because he didn't think it was worth the effort. E.C. Bassick, a former sailor, took over the claim and his first shipment was valued at more than $12,000. During

Present-day Ouray (courtesy Bruno F. Smid).

its first year the Bassick had recovered ore valued at about $500,000 in silver and more than $1 million in gold. Then in 1904, another large body of ore was found that yielded about $2 million in gold.

RED MOUNTAIN (Ouray) Named for the nearby red-colored peak, this camp was settled about 1880 with a post office in 1883. Threatening to rival Leadville, the town grew with more than 100 businesses and a population of about 10,000 with the discovery of silver. In 1881 John Robinson located the Yankee Girl Mine that rewarded him with more than $8 million. Two years later the National Bell Mine was discovered, its ore showing values up to $14,000 a ton. As other properties were developed, transportation became a problem since freighters were the only mode of hauling the ore at an enormous expense. So about 1888 Otto Mears built the Silverton Railroad to the mine sites. More than 20,000 tons of ore were shipped out annually.

Transportation was not the only problem for the miners; frequent snow slides killed a number of them. During winter the only way for supplies to reach the mines were stages that were fixed as sleighs, and they were not always reliable. One of the driver's rigs got out of control and plummeted down Red Mountain Pass.

The Silver Panic began Red Mountain's decline and the railroad ceased running. Gold and silver valued at nearly $30 million was recovered before everything shut down.

RICO (Dolores) Joseph Fearheiler and Sheldon Shafer may have been the first prospectors in the region during 1861, but no claims were filed until 1869. Three years later a prospector named Darling found some low-grade ore which brought in a few miners. Although a smelter was built in 1873, the ore didn't pan out as expected and the men left. Finally, in 1879 silver ore was located, prospectors returned from other camps, and the Chestnut (silver) and Johnny Bull (gold) mines began operating. Rico was established and given the Spanish word for "rich."

Because of the camp's isolation, hauling out the ore was problematic with the nearest railroad located more than 200 miles away. In 1880 the Rio Grande Southern Railroad reached town, the Grandview Mining & Smelting Company built its

Remains of Romley (Library of Congress, Prints and Photographs Division, Historic American Buildings Surveys, 8-Rom, 3–1).

facility, and stamp mills were erected. One of the richest lodes was the Enterprise, discovered in 1887 by David Swickheimer. After running out of money trying to develop his property, Swickheimer was going to be forced to sell. But his wife purchased a lottery ticket, winning $5,000 which saved their mine. They later sold the property to an English company for $1.2 million. It was estimated that Rico's gold mines recovered a little over $2 million.

ROBINSON (Summit) Gold placers were found in McNulty Gulch near the Tenmile Range during 1861. But the ore was depleted after about $300,000 was recovered and not much mining occurred until the 1870s. A businessman from Leadville, George B. Robinson, grubstaked a group of prospectors in 1878 who located the Robinson Mine two years later, causing the formation of the camp. Robinson never realized his fortune because he was accidentally killed shortly thereafter. The camp flourished when the Rio Grande Railroad extended its line through Robinson. By 1890 most of the mines had run their course. Total silver output in 1881 was estimated to be about $2 million, and gold production was a little more than $1 million.

ROMLEY (Chaffee) The discovery of ore near Romley occurred during the 1860s. The Mary Murphy Mine was the best producer, and the camp was built around its operations followed by a post office in 1886. The name was coined from Colonel B.F. Morley, who supervised operations of the property. The town of Alpine constructed a smelter for the ore but it did not work properly and was later sold. The mine received its name from Mary Murphy, a nurse from Denver. A prospector who initially found the lode had been cared for by Murphy at one time, and he named it in her honor. To bring the ore down from the mine, a tramway more than 4,000 feet long was built to Romley. It had the capacity to carry 96 buckets of ore, each weighing up to 200 pounds. The Mary Murphy produced about 75 percent of the ore output of the region, nearly $4 million, in addition to some silver. After further development, it recovered more than $60 million by the 1920s.

RUSSELL GULCH (Gilpin) In 1859 William G. Russell and his party found gold in the gulch, and were followed by more than 800 prospectors. That same year about $20,000 in gold was recovered. Other strikes began yielding about $35,000 a week. Within a few years the ore was exhausted and the site abandoned until Prohibition, when the deserted mines were used as warehouses for bootlegger hooch.

SHAVANO (Chaffee) Taking its name from the mountain which honored a Tabeguache Ute chief ("blue flower"), the camp was established after a toll road was built in 1880 from the town of Maysville. Mining promoters offered free lots to anyone who would grade one half the width of his own street. J.H. Akin and E.W. Carpenter discovered a chunk of rock that contained more than 115 ounces of silver. News of this find brought in other prospectors, but the boom lasted only a short time because the silver ran out and Shavano with it.

SHERMAN (Hinsdale) The first gold and silver ores were discovered about 1877 in Cuba Gulch at the confluence of Cottonwood and Lake Fork Creeks, and the camp formed the same year. However, it wasn't until about 1880 that any of the mines were sufficiently developed. Promoters thought the camp would rapidly grow, and in 1880 the *Denver Daily News* wrote: "Sherman is going to make the boss camp of that section."[148] But it never stood a chance.

SHERROD (Gunnison) Located near Missouri Gulch at about 12,000 feet, Sherrod was settled after W.H. Sherrod along with others discovered gold during the early 1900s. Its post office was established in 1904. Some of the claims were averaging ore valued at more than $1,500 a ton. The material was shipped out via the Colorado and Southern Railroad's spur line. Because of Sherrod's elevation, massive snow slides in the winter, and lack of adequate gold, the town folded by 1907.

SILVER CLIFF (Custer) Silver Cliff came to life about the time of Leadville's rise in 1879 after R.J. Edwards found silver in the region. He named his claim the Racine, which assayed out at 24 ounces of silver per ton. Trying to keep his secret was useless, and the camp soon burgeoned with a population of more than 10,000. Much of the ore was hornsilver, which was soon exhausted, but in Silver Cliff's first year more than $2 million had been recovered. Edwards later sold his mine to James Craig, who realized about $2.5 million. The *Silver Cliff Miner* wrote in 1880: "No town in the country with the exception of Leadville has grown with such astonishing speed as the gushing, energetic little city of Silver Cliff."[149] But it was to be short-lived because the demonetization of silver in the 1890s brought Silver Cliff to a screeching halt.

SILVER HEELS (Park) Located in Silver Heels Mountain, the camp was named for Silver Heels, a dancing girl who went to Buckskin Joe's to aid the people who were stricken with smallpox. When silver was discovered, the *Mining and Industrial Report* wrote in 1897 that "no new gold camp in the state has brighter prospects than Silver Heels."[150] A stampede was on and the camp grew rapidly. Although the claims were said to be very rich, for some reason the town just up and died, and was heard from no more.

SILVER PLUME (Clear Creek) The first discoveries of silver were made in 1868 by Owen Feenan. After Feenan got ill, two of his best friends took over his claim and kept it working for him, or so he thought. When he was well enough Feenan headed out to his mine, the Pelican, only to find his "friends" had jumped his claim. The property was later sold for about $50,000. In 1870 the camp was formed and given the name Silver Plume for the streaks of silver in the rocks that resembled plumes (or feathers). Its post office opened in 1875. Another property located next to the Pelican was the Dives, and their veins were so close that litigation ensued because each property owner believed the other was tapping into his seam. The Dives Mine was later sold for $50,000 at a sheriff's sale, and the new owner consolidated the Pelican and Dives, then sold them for $5 million.

Love was not lost on the prospectors either, some with sad endings. One English miner named Griffin fell in love and was planning to get married soon. He had earlier discovered the Seven-Thirty Mine that contained gold and silver. Every night residents would see Griffin sitting on a chair on the porch playing his violin. One evening he wasn't there, and concern for Griffin caused a few people to check up on him. They found him dead with a bullet in his head. Apparently his fiancée changed her mind.[151]

Old mine at Silverton (courtesy David Nestor).

SILVERTON (San Juan) Prospectors came searching for ore but were hampered because they were trespassing on Indian land, so mining didn't get underway until the Brunot Treaty was signed in 1874. Silverton was founded between the Uncompahgre and San Juan mountains, and received a post office in 1875. It later became known as the "Silver Queen of Colorado." The building of a railroad to the region assured Silverton's permanence as a supply and trading center. Many of the mines were located in the mountains so an aerial tramway had to be built to get the ore down to the smelters.

Silverton experienced horrendous winters, and snowbound miners had to wait long intervals for their provisions, as a local newspaper wrote: "Last Tuesday afternoon our little community was thrown into a state of excitement by the arrival of jacks, as they came into sight about a mile above town. Somebody gave a shout, 'turn out, the jacks are coming,' and sure enough there were the patient homely little fellows filing down the trail.... It was a glad sight, after six long weary winter months of imprisonment to see the harbingers of better days, to see these messengers of trade and business, showing that once more the road was open to the outside world."[152] The total amount of gold produced was not recorded, but estimates range between $8.2 million and $20 million. Silver recoveries also ran into the millions.

At one time Silverton and Ouray went head to head in a feud. When someone wrote that the mines at Ouray were producing like mad, Silverton's paper responded with, "The mines are not as mad as the men who believe that they are anything more than a flash in the pan." Ouray replied that it was only natural that Silverton should throw mud for "they have a plentiful supply of it, all of their roads being between six and seventeen feet deep in the stuff."[153]

Silverton holds its Step Back in Time Historical Celebration in June. Information: Silverton Chamber of Commerce, (970) 387-5654.

SUMMITVILLE (Rio Grande) Early discoveries in the San Juan Mountains during 1870 began with James and William Wightman and their party.

But the hundreds of prospectors who followed them to the gulch experienced disappointing results and left. The camp was later established as Summitville because it was located at an elevation of 11,000 feet. Someone made a statement that the camp was "reached by a good wagon road in summer, and saddle and snow shoes in winter."[154] The year 1874 brought in a number of stamp mills, one of which spit out a gold brick valued at $33,000. At one time it was thought a railroad would come through, but that was just rumor and many of the miners abandoned the camp. The placers recovered only $10,000, and by 1893 Summitville was deserted. There was a slight resurgence when the Little Annie Mine yielded about $2 million in gold and silver.

SUNSHINE (Boulder) Early prospectors Hiram Fuller and George Jackson located a gold claim called the American Mine and worked it until they recovered about $17,000. Fearing their luck wouldn't last, they sold it to Hiram Hitchcock for another $17,000. Fuller and Jackson should have kept their mine because it realized more than $195,000 in less than two years. Ore from the mine was shipped out by pack trains to a smelter at Blackhawk. In 1874 D.C. Patterson discovered a lode near the town of Boulder, followed by the formation of the camp with a post office in 1875. It was named Sunshine because Patterson felt so lucky that he discovered a bonanza. Others believe it was named for Sunshine Turner, the first baby born at the camp. Hitchcock's mine ultimately yielded more than $1 million.

TARRYALL (Park) Early prospectors at this site in 1849 never had a chance to make their fortunes because the Indians killed them. The second group arrived ten years later with better luck. Finding gold flakes nearly the size of watermelon seeds, one of the men said they should tarry all, thus the name of the camp that was established in 1861. It had a population of about 200 and for a short time was the county seat. By the 1870s the camp was on a backward slide. Nearly all the gold in the region came from placers, which yielded a little more than $1 million.

One of the claims had a deep pit known as the Whiskey Hole for the miners who spent the majority of their earnings on whiskey. Those who wanted a drink had to pan gold from the claim in order for the owner to pay them so they could hit the saloons. When the owner went back east one winter, his claim was jumped, giving the miners enough gold to keep them in hooch until the following spring.

TIN CUP (Gunnison) Jim Taylor made one of the first discoveries in the county during 1881 while searching for a stray horse. Tradition says the camp received its name when he got water out of a stream to drink from his tin cup and saw specks of gold in it, or it may have been for the tin cups used in saloons. There was sporadic mining until about 1879 when the Gold Cup Mine was discovered, in addition to a number of placer claims, and a year later two smelters were built. The Gold Cup recovered more than $7 million in gold before it closed in 1917, and the nearby silver mines yielded about $1 million. Until a branch of the Denver and South Park Railroad was extended to this region, all the ore had to be shipped out by wagons to the railhead at Alpine. By the end of 1880 Tin Cup had a population of about 4,000 and began serving as a supply center for the mining population. The camp was prosperous until the Panic of 1893. There was a slight resurgence, but by 1917 Tin Cup had turned to rust.

TOMICHI (Gunnison) Placer miners dotted the banks of Tomichi Creek during the 1880s and the community was established, bearing a Ute word meaning "dome-shaped rock," or "boiling." Its post office was established in 1880. Mining was fairly successful from 1882 to about 1895. Snowstorms were not strangers to Tomichi, and often supplies were delayed. In 1899 while the miners were working their claims, a tremendous snow slide came down on the town, killing about five people, destroying mining equipment, and totally wiping out the town.

VICTOR (Teller) This town was literally built on a gold mine and first named City of Mines. The lode was discovered about 1891 while a basement was being excavated for a hotel. Called the Gold Coin, it produced more than $4 million its first few years of operation. There was gold everywhere. James Burns and James Doyle staked off the Portland Mine, which reaped them more than $65 million in gold. Later, two druggists who knew absolutely nothing of mining literally threw a hat in the air and began digging where it landed, striking it rich. In 1894 the Strong Mine was located and yielded more than $20 million. With all that wealth to be had, the Woods Investment Company began promoting the town, bringing more than 35,000 to Victor. Ore was found under houses and even dug up from back yards. In 1899 the site was almost leveled after a fire broke out and destroyed more than a dozen blocks of homes and businesses, with a loss of more than $1 million. But the town was rebuilt.

During 1904 the Western Federation of Miners went on strike against mine owners. A professional agitator named Harry Orchard was hired by the union to rouse the troops. Only Harry planted a bomb at the railroad instead. More than a dozen people died, and there were untold wounded among the strikebreakers.

WALDORF (Clear Creek) In 1868 the Big Stevens Mine was discovered near the top of the Continental Divide and a camp was established, followed by a post office in 1908. Waldorf was also a stop for the stagecoaches that that traveled over Argentine Pass. The Argentine Central Railroad was constructed over the pass, which hauled out the ore in addition to serving as a tourist train. Most of the mining claims were purchased by the Waldorf Mining and Milling Company. Millions of dollars in gold and silver were recovered.

WARD (Boulder) The first discoveries of gold in 1860 by Calvin Ward were the beginning of a small camp, which developed about a year later. Cy Deardorff later located the Columbia and Utica mines, yielding him more than $5 million and $1 million, respectively. C.H. Merrill purchased Ward's claim (Miser's Dream) for $15, who in turn sold it to Sam Breath and W.A. Davidson. They developed the property, then sold it for $300,000, and the new owners realized $500,000. The region around Ward boasted more than 50 mines, each of which produced more than $5,000. Much of the ore had to be stored because of difficult transportation problems until the Denver, Boulder and Western Railroad built a line to town in 1897. Between $15 million and $35 million in gold was recovered

Horace Tabor, who made his fortune at Leadville, moved to Ward and worked in the mines for a short period of time until he went to Denver and became postmaster.

WHITE PINE (Gunnison) Located at the head of Tomichi Creek, White Pine was established in 1879 with a post office the following year. It gained a population of more than 3,000 who worked the nearby silver mines. There were rumors of gold discovered during the 1860s by two prospectors who either died from a winter storm or were killed by Indians. Traces of a sluice box were found along the creek, but no one was able to locate their claim for many years. Ore from the mines was hauled by wagon during the summer, and ox-drawn sleds in the winter to the railroad for shipment. A local newspaper wrote one summer in 1885 after a bad winter, "The first wagon over the road between Sargent's and Whitepine since last fall was the stage last Friday. It was quite a treat to our people to once more see a wheeled vehicle."[155] By 1901 the camp was deserted.

WILLIAMSBURG (Fremont) Bolus Mitchell and George Williams came to the region about 1871, found a deposit of silver, and within a few months the place was overrun with miners who had high hopes. When the camp was founded, it was named for either mine owner John Williams, or coal operator Morgan D. Williams. What ore was found assayed at about $328 a ton. Williamsburg declined rapidly and was later relegated as a railroad station known as Switchville.

IDAHO

ATLANTA (Elmore) Named by Southern sympathizers, Atlanta was established on the Middle Fork of the Boise River after gold was found in 1864. Much of the ore was located on Atlanta Hill, but the extreme inaccessibility of the area hindered operations. Supplies and stamp mills had to be hauled in by pack trains over a narrow trail. Problems with milling caused a loss of much ore until more efficient processes were put in place. The greatest activity was during the 1870s when the Monarch Mining Company shipped ore valued at $700,000. Mining declined after the 1880s. Total production was nearly $6 million.

BAY HORSE (Custer) Settled along Bayhorse Creek, the camp was first called Aetna, but was renamed after a prospector lost his bay-colored horse. While looking for the animal, he accidentally came across an outcropping, and it wasn't long before other prospectors arrived to stake their claims. Until a toll road was built in 1879, all goods had to be carried in by pack-mule trains. Silver discovered in the area was located in heavy iron ore deposits, and miners had to find someone to raise capital to get the silver out. There were also rich veins of pure silver, but as they went deeper and the ore became more refractory, mills and smelters were built. Then a big vein was hit when the Ramshorn Mine was discovered, which yielded nearly $10 million in silver by 1898. There was a short resurgence during the 1920s when another $2.5 million was recovered. After that the mines closed down.

BELLEVUE (Blaine) This camp was established after the Minnie Moore Mine was discovered in 1880, and the post office opened the same year. Located along the Pioneer Mountains, it received its name for the beautiful scenery. Bellevue lived its life as a shipping point for the surrounding mines, and a smelter was built there to treat the ore. The Minnie Moore was later sold to a British interest for $500,000, and ultimately yielded $8.5 million. When silver prices dropped in 1893 the mines closed down.

BLACKFOOT (Bingham) This name was applied to an Indian tribe after they traveled through a burned prairie, coating their moccasins with ashes. A band of Crow Indians who saw the travelers called them siksika, meaning "black foot." About 1880 the Utah and Northern Railroad built its line through Blackfoot, which served as an outfitting center for the gold mines in the county. The American Falls Canal Company built a ditch for irrigation in 1879, enabling farmers to grow sugar beets, and the first sugar mill was established, one of the best industrial resources for the region after the mining era.

BURKE (Shoshone) This remote and geographically constricted camp was named for miner and politician J.M. Burke. Gold lured prospectors to the canyon in 1885, but the ore failed to live up to their expectations. What saved Burke from immediate extinction was the Couer d'Alene rush, in addition to rich deposits of silver discovered nearby. Stamp mills were erected and a railroad came through in 1887. More than $8 million in gold was recovered.

Burke was squeezed so tightly in a narrow canyon there was barely enough space for buildings, much less streets. When the hotel was built, it also included a tunnel through the middle of it in order to accommodate the railroad tracks. In 1923 the whole place went up in flames.

CHALLIS (Custer) Challis was founded in 1878 and grew up as a trading center for the nearby mining camps of Bay Horse and Yankee Fork. It lies in a valley surrounded by mountains, and was named for surveyor Alvan P. Challis.

CLAYTON (Custer) Clayton was established at the mouth of Kinnikinnic Creek and served as a smelter camp beginning about 1870. It was named for Clayton Smith, who owned and operated a house of ill repute. The Salmon River Mining and Smelting Company operated its facility during 1881 and 1882.

DELAMAR (Owyhee) Located about five miles from Silver City near a narrow creek bottom, this site was originally called Wagontown, named for the freight wagons. It was changed to honor mining capitalist Joseph R. DeLamar who promoted low-grade lodes in 1886. The post office was established in 1889. He later had a 10-stamp mill brought in, which was producing about $3,000 a day from the nearby properties until the camp reached its peak in 1890. Roughly $5 million in silver had been recovered by 1898. About all that remain are old workings of early day miners.

DEWEY (Owyhee) Holding the honor as the oldest settlement in Owyhee County, Dewey was founded in 1863 as Boone and became a placer center. A year later the *Boise News* wrote that the place, located on the creek between high and rugged hills, had narrow, crooked and muddy streets which "resembled the tracks or courses taken by a lot of angleworms, and might have been laid out by a blind cow."[156] In 1896 Colonel William Dewey purchased the townsite and had a 60-room hotel built. Its post office opened the same year, taking Dewey's name. He also purchased the Trade Dollar Mine and erected a 20-stamp mill. The property recovered more than $12 million in gold. Although Dewey had built an elaborate water system to ensure that no fires would burn the town, he had less luck with his hotel when a fire consumed it in 1905.

DREAM GULCH (Shoshone) The gulch was named for the Reverend Floyd M. Davis' dreams of discovering riches. He was a circuit rider, but did little preaching because he was more interested in gold, which he eventually discovered there. But none of the ore was profitable. The gulch was located near the town of Murray.

ELK CITY (Idaho) Shortly after discovery of gold near Pierce, prospectors found more ore along the Clearwater River, which attracted more than 2,000 people the first year. The town was established in 1861 and named for the magnificent animal. Spring thaws presented problems with the placers having either too much water or not enough, so the miners dug ditches to supply dependable water for their sluicing. By 1872 the best ground had been worked out and Chinese miners took over operations. Then in 1892 hydraulics and dredges were brought in which operated with great success. Lode deposits were developed in 1891, but their exploitation was hampered by the remoteness of the area. During the years 1861 through 1863, between $270,000 and $1.5 million in placer gold was shipped from the district. Although gold veins were found in 1870 at the Buster property, very little was mined until 1902. The Buster became the largest lode producer in the district and recovered about $300,000 in gold between 1907 and 1909. Early gold production was estimated between $5 million and $10 million.

Elk City, ca. 1905 (Kennedy Collection, Idaho Goldfields Historical Society, Elk City, ID).

ERA (Butte) Located along Champagne Creek, Era was established with the discovery of silver in 1885, and took its name from the mining company. Within two years Era was enjoying a population of more than 1,000. The region's mines yielded nearly $1 million.

FEATHERVILLE (Elmore) This site started out as Junction Bar. Located at the confluence of the Feather and South Fork Boise rivers, Featherville was originally a stop on a stage line until it became a mining camp. It took its name from the Feather River, a transfer name from California, so called because the Indians decorated themselves with feathers that were found along the stream. There are no accurate records of placer production. Between 1922 and 1927 dredging operations recovered nearly $600,000.

FLORENCE (Idaho) Placer mines were discovered in the nearby draws and gulches during 1861 by prospectors from Elk City. One of the discoveries occurred after a miner washed dirt that was laced with gold from the base of an old, uprooted tree. As a result the town of Florence was established with a post office in 1862. The previous winter a huge snowstorm had left 10 feet of snow. Supply wagons could not reach the miners, and by January of 1862 very little food was left. What was available cost a fortune. The majority of the men lived only on flour and the inner bark of trees. In May the freighters were finally able to get their provisions a few miles short of the camp, where they were packed in. The snowbound men were charged 40 cents a pound for their supplies. The height of the rush lasted until about 1864, when placers were bringing out about $40,000 a day.

Florence was home to Cherokee Bob, who killed two soldiers at Walla Walla, Washington, then hightailed it to Lewiston. After moving to Florence, he took up with a lady of the evening. One night she was tossed out of a dance hall; Bob came to her defense, and ended up getting gunned down.

GIBBONSVILLE (Lemhi) Gibbonsville was founded in 1872 along the North Fork of the Salmon River when gold was discovered in Anderson and Dahlonega creeks. The site was named for Colonel John Gibbon, who was involved in the pursuit of Chief Joseph and his Nez Perce band. Placers were worked extensively, followed by discovery of a number of lode mines. The ore was initially crushed with arrastres until a 30-stamp mill was built by the American Development, Mining and Reduction Company in 1895. After a disastrous fire in 1907 mining was sporadic. Total production was about $2 million in gold.

GILMORE (Lemhi) Gilmore became a trading and shipping point in 1873 for the nearby mines. It was named for a stage company owner, John T. Gilmore, and acquired a post office in 1903. Since the claims were so isolated, the ore had to be hauled to the mills by mule teams. When the Gilmore and Pittsburg Railroad pushed through in the early 1900s, the ore became less expensive and easier to ship to the smelters. After the silver had run its course, the railroad pulled up its tracks. The Viola lode brought out about $1.5 million in silver. Total production was approximately $11 million.

GRAHAM (Boise) This former town was named for Matt Graham, who developed the silver mining industry in the eastern part of the county. Graham spread the news of his discovery on Silver Mountain, which aroused interest and others came to check out the prospects. The camp was then established to support the prospectors and a mill was erected. A road at a cost of $15,000 was built and Graham had high hopes, but the ore proved to be insufficient. The mill ran about eight hours, then permanently shut down and the miners hastily departed. Graham had attempted to acquire British capital to develop the mines, but an attachment was put on the claims for unpaid debts, and the camp's reserves were sold at a sheriff's sale.

HAPPY CAMP (Elmore) Located in the Boise Basin, this short-lived camp was formed about 1863. Tradition says it was named for the happy prospectors who found gold, and averaged about $25 a day.

HEATH (Washington) Tom Heath and Jim Ruth discovered silver along Brownlee Creek in 1875, but the camp wasn't formed until the 1890s. Prospectors crushed their ore with a small stamp mill they had freighted in. A smelter was later built, but unprofitable ore caused the place to become a ghost town.

IDAHO CITY (Boise) This town was established during 1862 in the Boise Basin, followed by a post office two years later. Idaho City boasted a good 40 saloons and more than 35 stores. Sherlock Bristol was one of the first to find gold in the nearby hills. Since he and other prospectors spent the winter there, they dug a well near the claims so they wouldn't have to make numerous trips to a stream for water. While doing so, one of the men hit pay dirt that turned out to be worth about $100,000. Word got out and a stampede ensued. The ground

Jail and pest house at Idaho City (courtesy David Decareaux).

was so rich even the buildings were razed to get the gold.

Supplies were hauled in from Walla Walla, Washington. Herman Reinhart and Phillip Scheible were freight haulers who traveled the route from Walla Walla to Idaho City. Reinhart wrote a journal of his freighting experience: "We loaded in 11,000 pounds.... We forded the river near the Indian Reservation. We had to double at the crossing, for it was high and rocky and steep banks, and eleven yoke had all they could do to take my heavy wagon over, and out of the river banks.... On the fourth day we commenced to go up the Blue Mountains ... and we all had a fearful time, such rough roads and hard pulls.... We crossed our teams over Snake River and kept on up to the We[i]ser River; here we had to ferry that river, and it took time and money too. We had to swim all the cattle ... and some cattle do not like to take to water.... We at last got to Idaho City, to where our goods were to be delivered.... The mines were rich and money plenty and the miners gave it a chance. Wages were high, and the working men spent their money drinking, gambling and dancing."[157]

During 1865 the town experienced a fire that destroyed most of its buildings. Then in 1876 another fire struck. The *Idaho World* wrote, "Another Great Conflagration, Idaho City Again in Ashes, Nearly the Whole City Burned.... The fire sped — a wild, destroying, irresistible, unconquerable element ... every building was soon reduced to a mass of cinders and ashes." Within two months the city was completely rebuilt, mostly of brick, and mining continued. Later the paper printed, "Messrs White, on the bar above town, have cleaned up about $13,000 since the clean up of $24,000 noted two or three weeks ago. We understand a claim on Thomas' Gulch, cleaned up on day this week, after a days run, $11,000. This is 'the biggest thing' yet."[158]

Idaho City was a wild place with its saloon-lined streets. One of the local papers wrote in 1863, "Several parties were found in the streets Tuesday morning. Some with fractured skulls; some with bunged eyes and swollen faces, indicating very clearly that there had been a muss somewhere during the night. Blood was freely sprinkled about the town on woodpiles and sidewalks. As the puddles of blood were distributed over a large district, it was impossible to locate the fight."[159]

The region was extremely rich. Between 1863 and 1869 more than $44 million was recovered. Most of the gold production came from the placers during the first few years of mining. Nearby was the Gambrinus lode mine, which yielded about $265,000.

JOHN DAY CREEK (Idaho) This site was only a rest stop for the miners on their way to the Salmon River Diggings after freighter John Day established the John Day House. Its post office opened in 1882. What was left of the buildings completely burned down during the 1920s.

KELLOGG (Shoshone) First called Milo, the camp's name was changed to Kellogg in 1891 for Noah Kellogg, the first person to find ore about 1885. A carpenter by trade, he was unemployed and asked O.O. Peck and Dr. J.T. Cooper to grubstake him. Supplies in hand, he and his mule headed for the hills. He made camp for the night, and when he woke the next morning discovered his jackass was gone. Kellogg found her high up in the mountains. Tired from climbing, he sat down for a rest and lazily picked up a rock, which turned out to contain silver. He returned to town with his samples. While there, he complained about his beast, "That jackass was always runnin' off.... Then to my utter amazement I seen he was standin' on a great outcroppin' of silver that was reflectin' the sun's rays like a gigantic mirror. We wasted no time in filing claims." Kellogg later killed the animal. "That jackass run off one too many times so I strapped a stick of dynamite on his back, drove him down off a mountainside where even a goat could not keep its footin.' A few seconds later there was an explosion an' I fell on my knees and gave thanks 'cause never again would I have to chase that ornery jackass."[160]

Kellogg returned to claim the Bunker Hill and Kellogg mines. To ship out the ore, a wagon road had to be constructed to the headwaters of a fork of the Coeur d'Alene River. The silver was so pure it produced more than $300 million.

In 1879 Tom Irwin found gold, built his cabin and worked his claim. For some reason he left three years later and never returned. But before Irwin departed, he dynamited the hill to cover his mine. Nobody knows how much gold he recovered. But in 1991 someone relocated the mine (Crystal Gold), and it is now open for tours. Email: goldmine_Idaho@hotmail.com.

LANDORE (Adams) Named for a town in Wales, this was a very short-lived mining camp. After T.G. Jones staked his gold claim in 1898 Landore was established. It grew slowly with a stage depot and a small newspaper. A few years later a smelter was built to process the local ores, but it did not succeed and Landore disappeared.

LEADORE (Lemhi) Established as a trading center for the surrounding mines, Leadore was settled in the early 1900s near the confluence of Texas and Eighteen Mile creeks. Its post office opened in 1911 and the town was named for the silver lead in the region. A smelter was later built. Leadore had high hopes when the Gilmore & Pittsburg Railroad came through, but within a short period of time the mines ceased operating. The town had to be content with becoming a center for agriculture and ranching supplies.

Folklore says gold may be hidden nearby. A stagecoach containing $37,000 in gold was stolen by two outlaws. During a pursuit by the sheriff and his posse, the thieves barricaded themselves in cliffs above the smelter, but they were discovered and both were killed. Supposedly the gold was hidden in the cliffs, but it has never been found.

LEESBURG (Lemhi) Prospectors who discovered gold along the Napias River in 1876 were quite successful. One of the men was getting about $1,000 in each pan, and others were realizing nearly $30,000 per week. Ditches were built to bring in water for sluicing the gravel. One prospector wrote of their riches: "We hear a whooping and a shouting and looking up the hill we saw Mulkey waving his hat and hollering like the Indians were after him. We scrambled up the steep, rocky bank to see what the matter was. A big tree had blown over in the previous winter's storms and the roots had dragged up a lot of quartz with them. The stuff was loaded with gold, so rich we were all very much excited and breakfast got all burned up."[161]

During one winter some of the men were hesitant to leave camp. Running short on provisions and needing medical supplies, a newspaper correspondent wrote, "We have two men with ax cuts in their feet; two frozen men; one gunshot wound, and any quantity of coughs and colds, together with some rheumatism, and there is no physician nor a particle of medicine in the camp."[162] A few of the miners decided to make a trail, which took them nearly a month to accomplish, but it did allow a pack animal to get through with provisions.

Frank Phillips moved to Leesburg and established

a store, then conceived a unique way to deliver supplies to the outlying claims by way of his burros. Those miners who had no animals would load the burros up with supplies and take them to their diggings. The burros would then be turned loose and immediately wend their way back to the store.

Leesburg enjoyed a population of about 7,000. Gold placers were worked for about 14 years and yielded more than $5 million. Lode gold was later discovered, but lasted only a short time with a small output of about $250,000.

MULLAN (Shoshone) Mullan was initially a way station on the Mullan Road during the 1880s. It was located on the South Fork of the Coeur d'Alene River, and later became a supply point for the miners who found silver in 1884. Its post office was established two years later. The Lucky Friday and Morning were two of the more prominent silver mines. John D. Rockefeller later purchased the Monday claim for more than $3 million. About $6 million in silver was recovered.

MURRAY (Shoshone) Murray was named for miner George Murray, and established along Prichard Creek after gold was found about 1883. It also became the county seat until the nearby town of Wallace wrested away the honor. In 1898 Wallace supporters stole into Murray and took the county records. One Murray resident said, "Murray was a better town than Wallace when they stole the county seat away from us. Stole it, that's what they did; just took the safe and the records and went off. Wallace is afraid if we get that road fixed the camp will come back. By gosh, someday we'll show 'em."[163]

While the placers were declining, rich deposits of silver were located in 1885 along the South Fork of the Coeur d'Alene River. Virgil and Wyatt Earp owned a few claims in the region, and were also proprietors of a saloon at the town of Eagle. A railway to the region was completed in 1887. The succeeding decade was marked by strife between the miners' unions and mine owners, and several times troops were called in to restore order. Approximately $4.7 million in gold was recovered between 1884 and 1905.

OROGRANDE (Idaho) Nathan Smith found placer gold during the 1860s, but Orogrande wasn't settled along Loon Creek until about 1899. It served as a trading and supply center and the post office was established in 1900. At that time it had about five stores and as many saloons. The placers lasted only about a year, then were sold to Chinese miners who reworked the diggings. Following the placers, lode mines were located. One of the better producing properties was the Gnome which yielded about $240,000 in gold from 1932 until 1937 when it was shut down.

PEARL (Gem) This camp was established about 1867 after gold was discovered on Willow Creek. The site was first named for a prospector named De Lava, and later changed to Pearl for a nearby mine. The local claims produced only about $80,000.

PIERCE CITY (Clearwater) In 1860 E.D. Pierce and his party of twelve men discovered gold at Orofino Creek. Despite the Nez Perce Indians' forceful objections, the news brought a stampede. The local paper printed a letter written by Pierce: "We found gold in every place in the stream ... I never saw a party of men so much excited. They made the hills and mountains ring for joy."[164] The town was founded at the headwaters of Orofino and Orogrande creeks and the post office opened in 1863. While other mining districts were collapsing, Pierce City continued to be active. There was an estimated total production of between $5 million and $10 million in gold before 1875. In later years the region was reactivated by large-scale dredging operations. During the early 1900s lode mines were developed and yielded about $250,000 in gold.

Pierce City experienced racial tensions during 1885 after a white merchant was murdered with a hatchet. Two businesses were in competition with each other. D.M. Frazer owned a store and Lee Kee Nam owned the other. When Frazer was shot and "terribly chopped to pieces in his own store," most believed Lee Kee Nam was the culprit and he was arrested. While the sheriff was transporting his prisoner to the town of Murray, a lynching party waylaid them and summarily hung Lee Kee Nam.[165]

Violence continued, and Governor Stevenson issued a proclamation expressing hope that Congress would deliver Idaho from the presence of Chinese immigrants. The Pierce City incident demonstrated that racial intolerance and the violation of basic legal rights was very much alive.

PIONEERVILLE (Boise) This camp was established by the Splawn-Grimes prospecting group about 1862 and the post office opened two years later. Moses Splawn was told by an Indian about a place where there was gold, so he and his partners came from Walla Walla, Washington, where they had initially begun their prospecting venture. The

Indians living here attacked the party, which scurried back to Walla Walla, ending their prospecting dreams. During 1863, gold was found by two men as noted: "In the fall of 1863, Branstetter and A. Saunders rocked out from $50 to $75 a day near Pioneerville."[166] Marion Moore arrived some time later and founded the town, which included construction of a stockade for protection against the Indians. Moore and other miners headed for the hills and found gold along Prospect Creek. The placer diggings were fairly successful, bringing a little more than $60 a ton. Lode mining followed and two of the biggest producers were the Golden Age, which yielded about $200,000, and the Mammoth Mine that recovered $472,000. Hydraulic mining operations later washed away the camp.

QUARTZBURG (Boise) Quartzburg was established about 1864 and its post office opened ten years later. The first discoveries of placers brought in the prospectors, who turned to lode mining about 1863. W.W. Raymond erected the first stamp mill to treat the ore. One of the largest lodes was the Gold Hill Mine, that was consistently productive and profitable. A horrendous fire swept through town, destroying all but one structure. The mill also went up in flames. A total of $8 million was realized from the district.

ROOSEVELT (Valley) Located near the Thunder Mountains along Monumental Creek, Roosevelt was formed with the discovery of gold about 1902. It was named for either Theodore Roosevelt or his daughter, Alice. This same year Colonel Dewey discovered the Dewey Mine along nearby Mule Creek, which became a principal producer. He later brought in a 10-stamp mill. A road was also built to the region in order to haul in mining equipment. During its early days, someone noted that Roosevelt was "a log town with one street and no society."[167] The Dewey Mine was in operation until about 1906 when a main tunnel collapsed. Then in 1909 a landslide caused by the unstable rhyolite on the hill dammed up Monumental Creek and the water buried Roosevelt. Total production estimates were about $400,000.

RUBY CITY (Owyhee) Gold was discovered during the 1860s when Michael Jordan and his group traveling from Placerville camped near Jordan Creek. One of the men idly checked out the creek and noticed a glint of gold. The rest of the men scrambled to stake their own claims. They went back to Placerville to stock up on provisions and returned with about 2,000 anxious miners right behind them. Although there was moderate placer mining, it was the rich silver ledges that set off the real stampede during 1863. Shortly thereafter, the camp was established and named for the red-tinted silver ore. In an eight-day period more than $15,000 in bullion was recorded by the assayer. Wells Fargo shipped out a good 200 pounds of bullion valued at about $14,000, in addition to gold dust and bars worth $13,000. Total values for the mines are combined with those of Silver City, since it absorbed Ruby City about 1867.

SALMON (Lemhi) Established as a supply and trading center in 1866 for the region's mines, Salmon was given an appropriate name for the abundance of fish in the Salmon River. It received a post office in 1867. State governor George L. Shoup was one of the camp's founders.

SAWTOOTH CITY (Blaine) Situated along Beaver Creek, this camp took its name from the mountains for their rugged-looking peaks. It was settled about 1878 with discovery of two lode mines, but their yields are unknown.

SHOUP (Lemhi) Much of the ore mined in the vicinity was lode gold, although there were a few placer diggings along Boulder Creek. After discovery of a lode mine in 1881 the town of Shoup was established with a post office in 1883. It was located along the Salmon River at the mouth of Boulder Creek. The site was named for George L. Shoup, the first state governor. Sam James and Pat O'Hara found gold while camping along the Salmon River in 1882. Disappointed with so little yield, they climbed up a small cliff and found what would be called the Grunter Mine. Other lode mines were developed in rapid succession. Getting supplies to the mines was difficult, since the nearest road was more than five miles away, so provisions had to be shipped via the Salmon River on barges. After a trail was made passable, the goods were brought in by pack animals. The properties yielded about $1.4 million. Shoup still exists with a small general store, gas station and a few houses.

SILVER CITY (Owyhee) This site was located about five miles south of Murphy. Gold and silver were located in 1863 along Jordan Creek and the town was settled in 1865. While placer mining, prospectors followed the diggings to their source, which were later developed into lode mines on War Eagle Mountain. The placers lasted about two years

Silver City (courtesy David Decareaux).

before the gold was depleted, but Chinese miners came in gleaned what was left. In 1865 the Poorman and Orofino mines brought in so many prospectors that it caused violence and disorder. As a result, the military was brought in to quell the riots. Before 1875 the Poorman Mine produced about $4 million, but it was closed after the secretary of the company absconded with much of the capital. Also found at the Poorman deep in the ground were solid ruby-silver crystals that were exhibited at the Paris Exposition in 1866 and won a gold medal.

Production of silver and gold came to about $12.5 million. Then in 1899 the Black Jack and the Delamar mines started another boom. The Delamar was named for Joseph R. DeLamar, mining capitalist and a former sea captain, and was producing close to $500,000 in bullion a month. By 1914 the ore had run its course, but not before realizing more than $23 million. Silver City suffered a crushing blow in 1935 when the county seat was moved to Murphy, but continued to cling to life.

THUNDER CITY (Valley) Thunder City was established about 1900, and took its name from the mountain. Indians believed that when their gods were angry they caused thunder to roll down the mountain (most likely rock slides). The camp served as a way station and outfitter for the mines on the mountain, and grew with a population of nearly 5,000. Its demise came after a railroad made the town of Cascade one of its stations.

TRIUMPH (Blaine) Initial discoveries were made during 1864, but the area was not developed until 1880. This camp was established after the Triumph Mine was located and received its first post office in 1889. The Triumph became the largest producer in the district. Early production records are not known, but after 1932 the region recovered more than $1 million, mainly from the Triumph.

WARREN (Idaho) This camp was established about 1862 and named for gambler and prospector James Warren. Its post office opened in 1885. Warren brought in a group of prospectors who discov-

Dilapidated Warren (courtesy Barbara Alger).

ered gold at what was called Warren's Meadows. The Rescue Mine was later located and went on to yield about $13,000 in gold. Rich placers also brought in a horde of miners, and by 1900 more than $15 million had been recovered.

WEIPPE (Clearwater) Lewis and Clark stopped here in 1805 and were given food and shelter by the Nez Perce Indians. The camp started out as a trading outpost and stopping place for miners on their way to the gold fields at Pierce during the 1860s. Wellington Landon laid out the town about 1874. During the Nez Perce War of 1877, all the buildings were destroyed and the settlers took refuge at Fort Pierce. Weippe is derived from *oyipe*, meaning "a very old place," "beautiful place," or "gathering place," descriptive of the meetings or potlatches held by the Nez Perce at certain times of the year.

YANKEE FORK (Custer) During the 1870s placer gold was found along the Salmon River near the confluence of Yankee and Jordan creeks. Three prospectors named Baxter, Dodge and McKein came to Yankee Fork about 1876 in their quest for wealth. And riches they did find in the General Custer Mine, loaded with both gold and silver. A few years later the Lucky Boy Mine was located, also rich in both ores, so a 30-stamp mill was built in 1880 to accommodate them. The General Custer yielded about $8 million, and the Lucky Boy $2.5 million. Placer gold revenues were $50,000.

YELLOWJACKET (Lemhi) This camp was named because prospectors' horses were constantly getting stung by yellowjackets. Gold was discovered along Yellowjacket Creek in 1869 by Nathan Smith and Long Wilson, who were followed by others. What the stampeders didn't know was that one of Smith's friends had salted the prospects which caused an outrage after its discovery and the miners left. But the gold was there, which soon required a 30-stamp mill that was brought in by a slew of pack mules in 1893. The equipment had to be hauled by cables and tramway. It took 80 mules several weeks to bring in the machinery. Yellowjacket was in full swing during the 1890s, but began to decline in 1900. The property produced about $450,000.

MONTANA

ADOBETOWN (Madison) This site was located in Alder Gulch and bore a Spanish name for the sun-dried bricks with which the houses were built. Adobetown enjoyed a population of more than 175, and its post office was established in 1865. It may be the oldest placer gold camp in the gulch. The miners recovered about $350,000.

ALTA (Ravalli) Alta was settled at the headwaters of the Bitterroot River, with a post office in 1898. The name is Spanish for "high," descriptive of its altitude. Tradition says that Barney Hughes was the first to discover gold along nearby Hughes Creek, although placer diggings were known as early as the 1870s. Gold and silver were recovered up until the early 1900s, estimated at about $500,000.

ANDERSONVILLE (Fergus) During 1880 a mixed-blood Indian named "Skookum" Joe Anderson, along with David Jones, came from the Black Hills in 1879 and discovered gold nearby. The camp was founded in 1881, and the post office established a year later. This place did not last long though, for as soon as the community of Maiden began its growth, Andersonville was abandoned in favor of the more lively camp.

ARGENTA (Beaverhead) After placer deposits were discovered in the Bannack area about 1862, more gold was found near Argenta and mined on a small scale during the 1870s. The camp was established near Rattlesnake Canyon and grew to a population of more than 1,000. The post office opened in 1871, given the Latin word for "silver." W.S. Keyes came to Argenta in 1866 and built a smelter, claiming it to be the first serious attempt at smelting in the west. But it proved to be economically impractical, so Keyes moved on to Nevada where his process was more successful. Eight furnaces were built at Argenta during its early years, but the ore bodies were soon depleted, and by 1875 the district was almost deserted. It would be 50 years before any activity resumed. During 1926 gold deposits were discovered, bringing total production of the region to a little over $1 million.

BALD BUTTE (Lewis and Clark) A number of businessmen from Helena purchased a claim on the butte, but later sold it to an English firm after disappointing returns. For three years the company continued operations but came up with empty pockets. Then one day an employee named Tatem accidentally found traces of gold ore, which blossomed to gold valued at about $3 million.

BANNACK (Beaverhead) John White was the first to discover gold in 1862, and his find brought about the establishment of the camp, which later became the first capital of territorial Montana. Placers along Grasshopper Creek were very rich, but the following year lode deposits were located and the Blue Wing District was formed. The region's mines produced gold and silver valued at about $7 million by 1910. Bannack is now a state park.

During 1877 Chief Joseph was taking his people to Canada after the Big Hole Battle. News that that Joseph was heading towards the camp brought people living in close proximity to come here for protection. During this time a church was being built. The preacher, Brother Van, and resident John Poindexter slipped out of town and brought back the cavalry. The residents didn't hide from the Indians; instead, while Van and Poindexter were gone they

Three miners next to a sluice box at Argenta, ca. 1905 (courtesy Beaverhead County Museum Photo Archives, Dillon, MT).

used their time to finish building the Methodist Church. Fortunately, Chief Joseph and his band never did try to attack the community.

Numerous camps had their questionable characters and Bannack was no exception. Henry Plummer was a crafty fellow who became the camp's sheriff. To all appearances, he was a well educated person of means. What the citizens didn't know was that he and his friends were stagecoach robbers and killed more than 100 men during their reign of terror. When Plummer's other life was discovered, a vigilante group broke into his house and summarily hanged him.

BASIN (Jefferson) Although gold was discovered in the 1860s, the camp was not settled until 1880. Early placer diggings were soon depleted, but rich outcrops of gold and silver were later found on the nearby slopes and headwaters of the Boulder River. Two miners named Allport and Lawson founded the camp, a post office opened in 1880, and before long more than 5,000 were living there. A smelter was built to process ore from the mines, two of the most prominent being the Katy and the Hope. A stamp mill was built near the Hope Mine, and a concentrator at the Katy. The concentrator burned down in 1896, taking seven workers with it, forcing the Katy to close, and then it filled with water. The high cost of smelting and a drop in the price of silver caused many of the properties to shut down. In 1893 a fire razed most of the site. Total production of gold was close to $4 million.

BEARMOUTH (Granite) Bearmouth served as a supply and transportation center for the mining camps in the Bear Gulch region. It was also a shipping point for ore headed to the smelters. This interesting name was for the camp's location near a gulch that resembled a bear's mouth, or for prospectors' numerous encounters with bears. Its post office was established in 1875. At nearby Clark Fork River, John Lannen established a cable-operated ferry to move supplies and freight. One day the cable snapped while the ferry was hauling two men and pack mules carrying large barrels of whiskey. One of the men drowned, but naturally most of the barrels were recovered. Bearmouth later became a station on the Northern Pacific Railroad.

BEARTOWN (Deer Lodge) Four prospectors discovered gold in 1865 and the camp was built in Bear Gulch at the confluence of Bear and Deep creeks. Until a small railroad was built to the region, all supplies and equipment had to be brought in by pack train. The wealth of ore prompted a miner to write a letter to an employee of the *Montana Post*: "Friend Ben, You may 'toot the old horn and blow the bazoo' as much as you please over Bear Gulch. It is enormously rich — claims are selling at $2000, and scarce at that.... The stampede is greater than ever known to any mines before."[168] The gulch was indeed rich, and tradition says that almost $30 million had been recovered there. The town grew with a population of nearly 5,000. By 1869 most of the ore had been depleted, or so it was thought. Additional placers and lode mines were located. In 1878 a good wagon road was constructed and Beartown was connected to the outside world. It was estimated that between $3 and $7 million was recovered by 1918.

BLACKFOOT CITY (Deer Lodge) Also known as Ophir City, this camp was founded about 1865 when placer deposits were discovered. The post office opened a year later. A man named A.K. McClure came to the camp and wrote, "Half the cabins are groggeries, about one-fifth are gambling saloons, and a large percentage are occupied by the fair but frail ones who ever follow the miner's camp."[169] Freighters made their fortunes delivering supplies to Blackfoot until a railroad came through in 1882. Storeowner William Clark also made a killing when he discovered the miners were running out of tobacco. He drove 250 miles from Bannack with a wagon full of tobacco, and returned home with a 300 percent profit. After 1875 the placers were pretty well exhausted and Chinese miners came in to glean the remainders. Early production was estimated between $3 and $5 million.

BRANDON (Madison) Brandon was settled in the 1860s after Alfred Cisler discovered the Buckeye Mine, and later the Broadgauge. A quartz mill was erected to serve the mines, and yielded Cisler nearly $100,000 in gold.

BURLINGTON (Silver Bow) The discovery of the Bluebird Mine in the 1880s drew miners to check out more silver prospects. As a result the camp was established and enjoyed a population of about 600. It may have received its name from the Chicago, Burlington & Quincy Railroad that built a short line to the area. The Panic of 1893 forced the mines to close and prospectors went off seeking greener pastures. Burlington continued to survive, acquiring a post office in 1901, and later became a dairy center.

BUTTE (Silver Bow) The first mineral discoveries were gold placers during 1864 in Missoula Gulch at a point now within the Butte City limits. The ore was located by William Allison and G.O. Humphrey. That same year more placers were discovered along Silver Bow Creek and German Gulch. As a result, Butte was founded and named for a sentinel-like peak and its post office opened in 1868. The placers produced about $1.5 million their first three years. When the gold began to decline about 1874, silver lodes were developed. Mining reached its peak about 1887. Total production was estimated at more than $60 million. Copper later took over the gold and silver industry.

The M&M Cigar Store, bar and restaurant was once the place where hordes of thirsty miners enjoyed their leisure during the 1890s. It was originally opened by Sam Martin and William Mosby (M&M) and continued operating until 2001, when it closed because of bankruptcy. In 2005 new owners took over and is now open for business. The building was restored as close to possible to its original structure, including its swinging doors. It is *the* place to go in Butte. They serve a great "garbage omelet." The best time to go is during St. Patrick's Day, when hundreds of thousands of visitors converge on the place. For those interested in gambling, the M&M has a poker room and keno machines. The store is located on Main Street.

CABLE (Deer Lodge) Cable has a historical marker that states:

> Atlantic Cable Quartz Lode. This mining property was located June 15, 1867, the name commemorating the laying of the second Trans-Atlantic cable. The locaters were Alexander Aiken, John E. Pearson and Jonas Stough. They were camped on Flint Creek and their horses drifted off. In tracking them to this vicinity the men found float that led to the discovery. Machinery for the first mill was imported from Swansea, Wales, and freight by team from Corinne, Utah, the nearest railroad point. The mine was operated with indifferent success until about 1880 when extremely rich ore was

Cable, MT (used with permission from the Montana Bureau of Mines and Geology).

opened up — a 500 foot piece of ground producing $6,500,000 in gold. W.A. Clark paid $10,000 for one chunk of ore taken from this mine in 1889 and claimed it was the largest gold nugget ever found.

Cable took its name from the mine; a post office opened in 1868. Another good producer was the Southern Cross Mine that yielded more than $600,000 in gold. Some placer mining was also conducted, with production close to $40,000.

CASTLE (Meagher) During 1882 silver outcrops were located, but it would be two years before any major mining occurred. Lafe and Isaac Hensley settled down to farm, but one of them came across a piece of silver float and their farming dreams went by the wayside. They worked their new claim which they dubbed the Yellowstone Mine. The Hensleys later built a smelter, but freighting costs for coke to run the machinery proved too high, and they shut down operations within one year.

When the camp was established it was named for the mountains, the rocks of which resembled a castle. Castle prospered quite well for a number of years with its numerous gold and silver properties. Ore was hauled out by ox teams to Livingston for treatment until three smelters were built at Castle, although transporting the ore was still expensive until a railroad was built. The Cumberland Mine, also discovered by the Hensley brothers, proved to be one of the biggest producers in the region. More than $1 million in silver was realized before it shut down.

Publisher Shelby E. Dillard once wrote, "The man who thinks Castle is not destined to be the greatest lead and silver camp in the land has a head on his shoulders as peaked as that of a roan mule." When the gold rush to the Klondike generated interest, Dillard followed with: "Castle is the lead and silver Klondike of Imperial Montana. There is no one in Castle inoculated with the virus of Klondike gold." He later wrote, "We had rather be a livid corpse in Castle than to own Alaska and all the sordid gold in the auriferous Klondike Country!"[170] After the price of silver dropped, Castle began to crumble.

CLANCY (Jefferson) Clancy was settled a few miles from Helena in 1865 with the discovery of ore nearby, spending its life as a trading and supply center. The camp was named for a local character known as "Judge" Clancy. Placer gold was worked until silver lodes were located about 1872. The ore was so rich it was hauled to Fort Benton and then on to Swansea, Wales, for smelting. The most prominent silver mine was the Legal Tender, which brought out about $1 million. In 1902 a fire burned down most of the business district; a few years later a second fire occurred, destroying what was left.

COLOMA (Missoula) Coloma was established in the Elk Creek District at the headwaters of McGinnis Creek after gold was found in 1865 by a group of prospectors from Last Chance Gulch. A 10-stamp and a 20-stamp mill were erected to crush the ore, which was then transported to Helena or Butte for treatment. Lode deposits were discovered about 1897. The Mammoth and Comet mines accounted for most of the early lode production, recovering about $200,000. It was estimated that between $1 million and $2 million came from early placer workings.

COOKE CITY (Park) The first miners to the region may have been Peter Moore and George Houston, but their search proved fruitless. In 1869 four trappers fleeing from an Indian raiding party discovered an outcrop that would later become the Republic Mine. By 1872 more claims were staked, although the land was still part of the Crow Indian Reservation. Cooke City not only served as a mining camp, it was also a receiving point for goods shipped by boat up the Missouri and Yellowstone rivers before the railroads arrived. The camp was named for mining promoter Jay Cooke, Jr. A smelter was built in 1875 to treat the ore until 1877, when Chief Joseph and his Nez Perce band retreating to Canada came through Cooke and burned down the mills. Miners were unable to get funds for their claims because capitalists refused to invest their money until the reservation was made public domain in 1882. Activities then increased, but high freight rates prevented extensive development, and by 1887 the mines closed down. More than $1 million was recovered.

DEWEY (Beaverhead) Dewey began as a timber camp until gold was located in the 1870s. It was named for rancher D.S. Dewey with a post office established in 1878. The camp served mainly as a supply and milling center for the mines at Quartz Hill. Its three stamp mills were supplied with water from the Big Hole River.

DIAMOND CITY (Broadwater) After a group of Confederate soldiers captured during the Civil War were released, they headed to Montana. There, they discovered gold in 1864, staked claims, and built their cabins. The camp received its name because paths between the cabins resembled a diamond. The diggings were very rich and before long the place grew to a population of almost 10,000. Placers along Confederate Gulch proved to be exceedingly rich, with a total yield estimated at $12 million. At Montana Bar miners were recovering nearly $1,000 a pan. In order to get water for the placers, a seven-mile long flume was built to the diggings. After 1869 placer mining declined rapidly, and by 1880 only a few hardy souls remained.

DILLON (Beaverhead) Dillon was established during the 1880s near some of the earliest gold and

Freighting outfitters and hotel at Dillon, c. 1880 (courtesy Beaverhead County Museum Photo Archives, Dillon, MT).

silver strikes in the state. It served as a supply and commerce center, which was further improved when the Utah and Northern Railroad came through. After the mining era was over, Dillon prospered with development of the cattle industry and agriculture, and later became the county seat.

ELKHORN (Jefferson) During the 1870s Peter Wyes made the first strike in the area, later to become the Elkhorn Mine. But the property was not fully developed until A.M. Holter purchased the claim. The camp was formed south of Helena and named for the abundance of elk. Ranchers and farmers also benefited from the town's establishment because miners and residents purchased all their meat and produce from them. After the Northern Pacific built its line to the area, supplies were easier to acquire. The economy of the camp depended mainly on the wealth of the Elkhorn Mine that produced more than $14 million in silver. An additional 70,015 ounces of gold was realized from the Elkhorn and other mines, amounting to a little over $1 million. Demonetization of silver caused the mines to close. In 1901 the Elkhorn was reopened by Frank, Henry, and John Longmaid, who made a tidy sum reworking the tailings.

EMERY (Powell) The discovery of gold in 1872 brought about the establishment of Emery. During the next 20 years the placers yielded about $75,000. The initial lode discoveries in 1887 are credited to either Joseph Peterson or Samuel Beumont. The best producer of the lode mines was the Emery that was located by John Renault in 1888. It was sold to the Deer Lodge Consolidated Mines Company, which later lost the property when it went too far in debt. John White took over the claim and worked it until 1948. The Emery was valued at $1 million. Other mines in the region recovered about $675,000.

EMIGRANT (Park) Emigrant took its name from a peak in the nearby mountains. It was founded about 1862 and received a post office in 1885. Thomas Curry and others found gold at Emigrant Gulch nearby, but the Crow Indians invaded their campsite, stole all their food, and ordered all the white men to get off their land. Curry went to Virginia City, but returned the following spring to stake his claims. Unfortunately, most of the diggings produced very little. There was a slight resurgence in the early 1900s when placers brought out about $300,000.

FARLIN (Beaverhead) This camp came into existence about 1875 after O.D. and W.L. Farlin discovered the Indian Queen Mine. It was located north of Dillon along Birch Creek. The mine was

Gathering of miners at Indian Queen Mine, Farlin, 1903 (courtesy Beaverhead County Museum Photo Archives, Dillon, MT).

Reminders of long-lost Garnet (courtesy Thomas Pritchard).

developed in 1902 and found to contain copper, silver and gold. A smelter was built during the 1890s by the the Beaverhead Mining & Smelting Company, and was later taken over by the Western Mining Company. The camp experienced growth during this time and gained a population of about 500. Residents had to get their mail from the town of Apex until the post office was established in 1905. But it operated for only four months. During 1904 the mine was sold to the Indian Queen Mining and Smelting Company, and operated until 1923 when it closed. Farlin was shortly abandoned after that. Most of the ore was copper, but silver production was about 43,000 ounces, and gold nearly 300 ounces.

FRENCH BAR (Lewis and Clark) In 1865 miners discovered placer gold along White Creek and French Bar. Most of the gravels were worked using hydraulic methods. Mining continued for about 20 years, yielding nearly $1.5 million in gold. During 1947 a dredge was brought in by the Perry Schroeder Mining Company, but it had a low recovery rate and shut down after a few months. The nearby ravine of Magpie Gulch recovered about $330,000, and the Avalanche Creek placers nearly $100,000.

GARDINER (Park) Joe Brown and his prospector friends discovered gold near Bear Gulch in 1865. Brown built a ditch to bring water to the diggings and also erected a small quartz mill. Within six months they had recovered nearly $1,800. Gardiner wasn't established until much later because of the Crow Indians' hostility. The camp was located at the northern entrance to Yellowstone Park and received a post office in 1880, honoring an old trapper named Johnston Gardiner. It served mainly as a freighting center for the local mines. Disagreements between mining stockholders and difficulty in hauling the ore to the mills caused the camp to decline. In 1903 Teddy Roosevelt traveled to Gardiner and dedicated a natural arch through which travelers passed on their way into Yellowstone.

GARNET (Granite) Both placer and gold lodes were discovered near Garnet. Worked along Bear Creek, those placers were very productive. The camp was founded about 1862 and received its name for the garnet rocks scattered about the area. Garnet experienced severe winters because of its location, and many of the miners left until spring. Those who stayed behind worked on repairing equipment. During one winter the supplies were getting critical, so one of the men dug a maze of underground tunnels through snow to the nearby camp of Bearmouth. Placer mining continued until about 1917, and at various times the camp had about 4,000 men seeking their fortunes. The plac-

Glendale, ca. 1880s (courtesy Beaverhead County Museum Photo Archives, Dillon, MT).

ers produced between $5 and $7 million. Total lode gold recovered was nearly $2 million.

GERMAN GULCH (Deer Lodge) After hearing of gold discoveries at Bannack and Alder Gulch, a group of German prospectors led by Edmund Alfeildt came to this area in 1864 and struck pay dirt. They decided to tell no one about their find, but a miner named Charles Bass followed them and the secret was out. The camp's location was reached via a very steep trail and until it was improved upon provisions cost the miners a fortune. After the placers were depleted, mining companies from Europe came in and conducted large-scale hydraulicking. It was estimated that the gold recovered was valued between $5 million and $13 million.

GILT EDGE (Fergus) Located near Ford's Creek, this camp was established after a post office opened in 1894. Buildings from a nearby failed mining camp were hauled over to Gilt Edge and a mill was erected at the foot of the Judith Mountains. Although about $1.2 million in gold was recovered, the remaining ore turned out to be marginal. The Gilt Edge Mining Company had investors believe there was a wealth of gold, but it had problems paying the miners, so officials issued worthless checks to various businesses. Eventually the company was seized by the sheriff and everything was shut down permanently.

GLENDALE (Beaverhead) Glendale was founded in the Hecla Mining District and served the mines when a smelter was built in 1875. Kilns were constructed at nearby Canyon Creek to provide charcoal to run the smelter. After processing, the ore was sent to the American Smelting & Refining Company in Omaha, Nebraska, for refining. When the silver became low grade, in addition to the silver panic in 1893, Glendale began its decline. Then in 1900 the smelter was torn down. The region's mine produced more than $15 million in silver and gold.

GLOSTER (Lewis and Clark) This camp was established about 1884 with a few stores and two saloons. Located a few miles from the town of Marysville, the Gloster Mine was discovered in 1880 and produced for about ten years, yielding a good $600,000 a year in gold and silver.

GOLD BUTTE (Tooele) Before the white men arrived, Blackfeet Indians discovered gold in the Sweet Grass hills during the 1860s. Gold Butte was founded on their reservation during the 1870s, but it didn't receive a post office until about fifteen years later. Four prospectors found gold in the fall of 1884, but since it was getting late in the season they camped there during the winter and waited until spring to begin their work in earnest. After a miner brought his gold to Fort Benton to buy supplies, the

River Press wrote: "The gold is found from the grass roots down.... If water was plenty, it is thought that the placers would pay at least $10 a day to the man."[171] Apparently they weren't that great because the prospectors abandoned the site in favor of better strikes elsewhere.

GOLD CREEK (Powell) Gold Creek was thought to be the site of the first gold discovery in the state made by Francois Finlay. Others believe Orrin D. Farlin was the first in 1862, or it may have been James and Granville Stuart, as noted: "No one has been able to establish exactly who discovered the first gold in Montana or where. Francois Finlay, a half-blood, better known as Bentesee, is said to have brought a teaspoon of gold dust into Fort Connah to Angus McDonald and in 1850 reported that he had gotten it at the present site of Gold Creek, near Garrison."[172] When a railroad exploring party came through, one of the men washed some flakes from the stream and named it Gold Creek. In 1883 the iron spike was driven here when the Transcontinental line extended its line. Three years later a post office opened, but the camp didn't last very long since the gold placers yielded only about $3 in a handful of gold pans. Gold discovered at other streams in the county yielded more than $2.3 million.

GRANITE (Granite) This mining camp was located in the Flint Creek District, sprawled out on a big hillside. The first property discovered was the Hope Mine in 1864, followed by other rich lodes. The Granite Mountain Mine, located by a prospector named Holland, proved to be one of the richest and most productive. Unfortunately, Holland let his claim lapse and it was taken over by Charles D. McClure, who realized more than $25 million in silver and gold. The cost of shipping ore to the smelters ran high, but was alleviated when the Northern Pacific Railroad was completed to Granite in 1883. After the price of silver fell the mines were shut down, and in less than two days the camp was empty. For about three years Granite lay as a ghost town until the mines were reopened and recovered about another $1 million.

GREENHORN (Lewis and Clark) Not much is known of this place except that it was a mining camp with a post office in 1871. Washington postal authorities received a telegram in 1883 from the territorial governor of Montana: "Vigilantes at Greenhorn, Montana have removed postmaster by hanging ... office now vacant."[173] Apparently nobody had the courage to take them up on the offer because the post office closed that year.

HASSEL (Broadwater) Placer gold was discovered in 1866 on Indian Creek and the camp that was established nearby was first known as Saint Louis. It was changed to Hassel, honoring miner Joseph H. Hassel. A stamp mill was brought in from Fort Benton by landowner Tom Reece, who had a gold mine called the Jawbone on his property. He realized about $50,000 from his claim. By 1880 most of the placers had been worked out, yielding about $5 million.

HECLA (Beaverhead) Hecla was established in the Pioneer Mountains. During 1872 William Spurr

Granite (courtesy Thomas Pritchard).

discovered a lode, but for some unknown reason he deserted his claim. It was relocated by James A. Bryant and Jerry Grotevant the following year. Tradition says that Grotevant was chasing his loose horses and went to throw a rock at one of the animals. Looking more closely, he discovered it had silver in it. That little rock later led to the Trapper Mine. When the news got out, the area was swarming with miners. The Trapper recovered more than $11 million of the area's total silver output of about $15.5 million. Gold recoveries came to nearly $250,000.

HEDDLESTON (Lewis and Clark) Gold was discovered in the vicinity during 1889, followed by the formation of a small camp. It was named for William Heddleston, who located the Calliope Mine along with his partner, George Padbury. They first worked their claim using an arrastre, which rewarded them with more than $10,000 in gold. A mill was later built, but total recovery of the few nearby mines came to only $120,000.

HELENA (Lewis and Clark) The capital of Montana began its life as a gold camp. In 1864 John Cowan and a group of prospectors on their way to Alder Gulch dug a few prospect holes while they were resting. They discovered placer deposits at what was called Last Chance Gulch, later changed to Helena. These diggings proved to be among the richest and most productive in the state. Before five years had passed, it was estimated that between $7 million and $16 million had been recovered from the gulch. Other nearby sites were also successful: Grizzly Gulch produced about $5 million, and in Nelson Gulch a little more than $2 million. The region was so rich that some of the buildings were undermined to get at the ore. By 1900 most of the placers and lodes were mined out. Estimated placer gold recovered before 1868 ranged between $10 million and $35 million. Total production through 1959 was at least $26 million.

HELMVILLE (Powell) This small camp was established after placer gold was found during the 1860s at nearby Nevada Gulch. The site was named for rancher Henry Helm and its post office opened in 1872. Very little ore was found, so only the hardiest miners stayed, and they realized only about $1,000 a year.

HUGHESVILLE (Judith Basin) This camp was established in 1879 after E.A. Barker and Patrick Hughes located silver and gold in the Little Belt Mountains. The post office opened in 1881. It wasn't long before others came to seek their riches. The Wright and Edwards lode was discovered, in addition to gold placer deposits. Ore was hauled by bull team over wicked roads to Fort Benton and then transferred for shipment to the smelters at Swansea, Wales. A newspaper reporter visited the camp and wrote: "From the very mouth of the gulch up to Gold Run the way is very bad, yes, outrageously bad for ten miles. Freighters who try it once declare that it is the last trip unless they are paid five cents a pound to compensate them for the risk."[174] About 1893 a railroad line was brought to the region, enabling easier transportation. However, with a drop in the price of silver and discoveries elsewhere, Hughesville began its decline. More than $2 million in ore was recovered, most of it from the Wright and Edwards Mine.

IRON ROD (Madison) This site got its start along the Jefferson River after the Iron Rod Mine was located in the 1860s, and a post office was established in 1869. The camp never grew much and contained only a hotel, store, and one quartz mill. Silver recovery from the Iron Rod was approximately $600,000. The camp survived for a time treating ore from other nearby mines and acting as a station on the Northern Pacific Railroad.

JARDINE (Park) Not far from the town of Gardiner, this camp was established and named for a Bear Gulch Mining Company employee, A.C. Jardine. In 1865 prospector Joe Brown discovered placer gold in Bear Creek. The ore was crushed using a mule-driven arrastre until it was replaced with a small stamp mill. Placers were also found along the Yellowstone River, but contained only small amounts and mining ceased by 1886. Later prospecting brought the discovery of rich gold lodes, and the small stamp mill was enlarged. Jardine became quite prosperous and its population warranted a post office, which was established in 1898. During the early 1900s activities were interrupted by an extended period of litigation. Operations were again suspended in 1942 because of Federal restrictions on gold mining, but increasing war demands for arsenic led to the reopening of the mines which operated until 1948 when fire destroyed the cyanide plant. Total gold production was about $4 million.

KENDALL (Fergus) The first discovery was made in the region during the 1890s. Harry T. Kendall, mine developer and rancher for whom the future camp would be named, took possession of the Goggle Eye Mine. It received this odd name because of the way the discoverers stared wide-eyed at its wealth of glittering gold. Kendall later sold the mine and received more than $400,000 for it. The Goggle Eye ultimately produced more than $6 million. Kendall was established with a boarding house, followed by a stamp mill. Numerous thefts had occurred during shipment, so much of the gold was stored in nail kegs or some other container to disguise what the freighters were hauling. Mining was continuous up until about 1920. When the ore played out Kendall ceased to exist. The district produced more than $9 million.

LANDUSKY (Phillips) Powell Landusky and Robert Orman struck it rich quite by accident in 1893. While getting a drink of water from a stream, one of the men broke off a piece of rock loaded with gold. Realizing they were near the Fort Belknap Reservation and unsure of the land boundaries, Landusky and Orman worked their claim at night, which they named the August, since it was discovered in that month. When the news got out that some of their ore assayed at $13,000 a ton, wild-eyed miners hurried to stake their own claims. One of the local papers wrote, "The boys have enough to pay the national debt after making themselves rich."[175] Landusky and Orman recovered more than $2 million. The camp itself was built on a mountainside in the Little Rocky Mountains in 1894. Some of the mineral-rich properties within the reservation's boundaries were later withdrawn from Indian lands. Total production was nearly $8 million.

LIBBY (Lincoln) During the 1860s prospectors found placer mines along Libby Creek, but they were attacked by the Kootenai Indians. Because of continued Indian threats, it would be about 20 years before serious mining was attempted by B.F. Howard and Tom Shearer, who relocated the claims. Some of the miners realized more than $75,000 within a few years. When the camp was established, it was named for the daughter of settler George Davis. Shortly thereafter, the Libby Placer Mining Company was formed, which brought in hydraulics to get the gold. Total production of the district was about $400,000.

A ferry was built across the Kootenai River so miners could travel back and forth, but spring freshets brought problems as the *Libby News* wrote: "During high water in 1894, there was a wreck at a point not far above the falls and the water was nearly up to the rails. A freight train and three of eight box cars were thrown into the water.... The cars and men floated down the stream and just before the falls were reached, one of the brakemen tried to swim ashore but the other one and the hoboes went over the falls and that was the last seen of them."[176]

LUMP CITY AND GULCH (Jefferson) This region was the scene of extensive gold mining. The camp and gulch were named for lumps of gold found by prospectors Fred Jones and William Sprague in 1864. Placer mining began along Prickly Pear Creek, which was quickly filled with gold-hungry prospectors. The ore was shipped out by freighters to Utah, where it was forwarded to Wales for processing. In 1866 Thomas G. Merrill located the Liverpool Mine, which yielded him about $1.5 million. By 1910 most of the claims were idle and the camp was abandoned. Then in 1933 a dredge was brought in to glean what remained of the gold. Other placers and lode mines produced over $2 million.

MCCLELLAN GULCH (Lewis and Clark) Gold was discovered in the gulch about 1864, and initial yields were so rich that a miner recovered about $500 in one pan. A small camp was formed called Pacific City, but later changed to McClellan. The placers ultimately yielded about $7 million.

MAIDEN (Fergus) Perry McAdow discovered gold about 1879 after he was grubstaked by a man named Joe Anderson. Placer gravels were later discovered in the Judith Mountains, and by 1883 the camp had grown to almost 6,000, many who came from the dying camp of Andersonville. Maiden may have been named for an Indian called Little Maiden, or for the daughter of a visitor, Mrs. James H. Connely, whom the prospectors nicknamed Little Maiden. It may also have been derived from a prospector named Maden. When it was discovered the camp was located on a military reservation, the people were told to leave. But citizens submitted a petition and the land was excluded from Fort Maginnis. The first lode mine discovered was called Skookum Joe, later renamed Spotted Horse. The next best producer was the Maginnis Mine, and between the two properties about $10 million in gold was realized. In all, the area mines produced more than $18 million.

MAMMOTH (Madison) Placer gold discovered about 1870 generated interest in the area and a

Neihart and narrow-gauge railroad tracks (courtesy of the Cascade County Historical Society).

mining district was formed. Shortly thereafter the Mammoth Mine was located and became the biggest producer of gold under the authority of the Mammoth Mining & Power Company. More than $14 million in silver and gold was recovered from the Mammoth and nearby mines.

MARYSVILLE (Lewis and Clark) The first placer mining occurred along Silver Creek about 1864. These operations continued until the early 1900s and accounted for a good 75 percent of placer production in the region, which came to approximately $3.2 million. In 1876 lode mining began. An Irishman named Tommy Cruse located the Drumlummon Mine and subsequently founded Marysville about 30 miles from the state capital. He sold the property in 1882 to English capitalists for $1.5 million which erected stamp mills and a cyanide plant. Cruse later converted to Catholicism and donated a large amount of his wealth to the Gothic Cathedral located at Helena. During the 1890s the Drumlummon became involved in lengthy litigation and the mine was worked only intermittently. It produced about $15 million in gold. Then during 1911 the property was sold again. Combined with the Drumlummon, other lodes produced between $30 million and $40 million.

NEIHART (Cascade) Neihart was once a trading center for the Little Belt Mining District. It was named for James L. Neihardt who discovered one of the richest silver deposits in the Little Belts about 1881. The post office began operating the following year. Queen of the Hills was the first lode mine located at Neihart, and was soon followed by other silver strikes. The ore was sent to Fort Benton where it was taken by steamboats for transshipment to Swansea, Wales. Then about 1885 the Hudson Mining Company built its smelter at Neihart. The camp became idle for a few years due to a lack of high-grade ore. When a branch of the Montana Railroad was constructed near the camp, Neihart came alive again, and as a result other mines were developed. The Broadwater Mine recovered about $5 million, and Queen of the Hills produced about $435,000. The silver panic of 1893 was the final blow to the camp. Gold that was discovered in the area was estimated at nearly $800,000. In later years sapphire mining supplanted silver.

PIONEER (Deer Lodge) The year 1852 brought about the first discovery of gold along Gold Creek. Although the diggings were shallow and not very rich, they still generated interest. New claims discovered about 1858 were found to be even more promising. Significant development commenced in 1868 after a 16-mile-long ditch was built to bring water to the placers and the camp grew to about 4,000. By 1871 dredgers were operating in the nearby creeks and gulches, in addition to hydraulic mining. The *Montana Standard* wrote, "Up Pioneer Gulch the dredge will go, sluicing the territory crosswise like a loaf of bread."[177] More than $1 million was recovered from Pioneer Bar alone. By 1920 most of the ore had been gleaned and only one hydraulic plant was in operation. Gold production was estimated between $5.6 million and $20 million.

POLARIS (Beaverhead) Located northwest of Dillon, this camp was established in the 1880s after the Polaris and Silver Fissure mines were discovered. A smelter was built by the Silver Fissure Mining Company to treat the ore and acquired about $60,000. During 1885 a group of prospectors located the Polaris Mine near Billing Creek. It was sold in 1892 to the Polaris Mining Company, which recovered about $250,000. But financial problems plagued the company and the property went into litigation. In 1922 the smelter was destroyed by fire.

PONY (Madison) Tecumseh Smith, whose nickname was Pony, made the first locations of gold during 1866. He staked out his claim and returned two years later with a friend. Pony worked his property, achieving moderate success which interested other prospectors who came to check out other possibilities. George Moreland accidentally discovered a rich silver lode while moving his pick among wild strawberry plants. The number of claims warranted a 10-stamp mill which was brought in about 1877. About three years later Henry Elling and W.W. Morris purchased some of the lode mines, which yielded them about $5 million in gold. The ore was hauled out for processing at a very high cost, which was alleviated when the Northern Pacific Railroad extended its line during 1900. Total gold production was about $7 million.

RADERSBURG (Broadwater) Radersburg was founded about 1865 after John Keating located the Keating Mine, which was loaded with silver. It was followed by the discovery of the East Pacific claim. Placer deposits were also found nearby, and a 5-mile-long ditch was built to bring water to work the claims. The camp was named for landowner Ruben Rader and its post office opened in 1868. Three stamp mills were erected about 1870 to process the ore. The Keating and East Pacific had yielded more than $3 million in gold and silver by 1904. During the late 1870s most of the easy gold had been exhausted and many of the mines shut down because of high freighting costs. But in 1883 a railroad was built, enabling cheaper transportation, and the properties were reactivated. Placer production was estimated between $500,000 and $1 million. Radersburg was the birthplace of famous actress Myrna Loy, who starred in many of the *Thin Man* movies during the 1930s.

RENOVA (Madison) Located in the Tobacco Mountain foothills, Renova was a station on the Northern Pacific Railroad. The site was a shipping point for gold, and also served as a trading and supply center. Most prominent of the lode mines was the Mayflower, discovered about 1892, but little mining activity occurred at the claim until 1896. After developing the property the American Development Company sold it to W.J. Clark for about $500,000. The mine operated until 1905, then reopened in 1935 and again became the chief producer. Renova's mines recovered about $3.5 million.

REYNOLDS CITY (Powell) The 1860s brought in the prospectors who found gold placers at Bear and Elk Creeks, and established the settlement. The first lode mine was found by a Mexican miner named Guayness. Reynolds City remained relatively small, but the region's mines did recover more than $1 million in gold within a two-year period.

ROCHESTER (Deer Lodge) During the 1860s rich gold outcroppings were discovered at a place called Watseca Hill near the Highland Mountains. When Rochester was established it became home to more than 800 miners, and acquired a post office in 1890. Since there was so little water, most of the placer gold was crushed using arrastres. Even the spring thaw didn't bring enough water. During 1868 a 10-stamp mill was built by two fellows named Hendrie and Woodsworth, followed later by a second mill. Regrettably, the mills cost more money to operate than the ore was worth. In 1871 an article appeared in *The Helena Daily*: "The mills have been idle for a time, and the town for let; Mr. Hendrie who recently visited the place, is soon to start up his mill, giving hopes for the repopulation of the town that is now deader than dead."[178] Roughly $2 million in gold was recovered.

ROCKER (Silver Bow) A small camp located along Silver Bow Creek, Rocker was named for a piece of equipment used to wash gold by rocking a steady stream of dirt and water to separate the gold. Ore was first discovered by Robert McMinn, who quickly sold his claim for $200. The next owner recovered more than $250,000 in gold. Nearby, another placer claim was producing nearly $1,000 every day for a period of time. But Rocker was short-lived and soon rocked itself into a permanent sleep.

SALTESE (Mineral) Located near the Idaho border, this camp was first named Silver City, then changed to Saltese about 1891 to honor a Nez Perce chief. The camp was originally a stopping place for freighters. It also accommodated travelers after an inn was built by a Colonel Mayers. Saltese later served as a supply point for the silver and gold mines in the vicinity.

SCRATCH GRAVEL HILLS (Lewis and Clark) A farmer named E.R. Tandy discovered a gold nugget while plowing his land. His find precipitated a rush of miners who raked the ground, finding more nuggets, thus the name Scratch Gravel. Placer deposits were also discovered in Iowa Gulch in 1874, which were soon abandoned because there was so little gold. However, other nearby creeks were found to have rich deposits, and by 1930 more than $1.2 million had been recovered. During the 1870s lode gold was discovered. Then in 1914 the Franklin Mine was located by Thomas Cruse, who also owned the Drumlummon Mine at Marysville. That property yielded Cruse about $500,000. Between 1916 and 1918 the region saw a time of great prosperity, but rising costs of supplies and labor eventually forced the mines to close.

SHERIDAN (Madison) This site was headquarters for silver and gold operations in the Tobacco Root Mountains, and lived its life as a supply and trading center. It was named for a Civil War leader, General Philip Sheridan, because many of the miners were former soldiers. The post office opened in 1866, with R.P. Batemen as postmaster.

SILVER CITY (Lewis and Clark) Gold was discovered in 1862 by William Mayer, whose claim was located along Silver Creek. Word got out, a rush began, and Silver City was founded. It was named for a Mr. Silver who, along with his wife, were traveling through the region, but she died shortly after their arrival and was buried near the creek. A big fight for county seat occurred between the camp and Helena. As with many struggles for the seat, Silver City's records were stolen in the middle of the night and taken to Helena. Before the site was deserted the placers yielded about $3 million in gold. Silver City later became a station on the Great Northern Railroad.

SILVER STAR (Madison) Silver Star had its beginnings in the 1870s, years after the first quartz lodes were discovered by Green Campbell, who realized about $270,000 in gold from his claim. The Broadway Gold Mining Company purchased the property and erected a 40-stamp mill, in addition to a special process used to extract gold from refractory ore. A tram was also built from the mine to the mill. Unfortunately the process did not succeed and the mill was closed. F.R. Merk later took over the property, and by the early 1900s had recovered more than $1 million. After the ore was depleted, Silver Star survived as a trading and supply center for the area.

STERLING (Madison) After gold and silver were discovered in 1867 the camp was established near Red Bluff and acquired a post office two years later. With the anticipation of riches, five stamp mills were erected. One of the most productive mines in the area was the Galena that realized more than $150,000 in gold and silver. Although other claims produced fairly well, their ore was of a lower grade. Investors spent more than $1 million developing the properties, but had disappointing results, and by 1880 there wasn't much left.

SUMMIT (Madison) Gold discovered in 1863 brought about the formation of this camp near Bachelder Creek. The first locations were mainly placers which yielded about $500,000 in a 20-year period. It was not uncommon at those diggings for miners to find nuggets that were valued between $100 and $600. In 1864 F.R. Steele began prospecting for lode gold and located the Oro Cache Mine, followed by the discovery of the Kearsarge Mine. Steele later sold the Oro Cache to Charles D. McLure. Much of the early lode production came from these two mines, which was estimated at $500,000 for the Oro cache, and $150,000 from the Kearsarge. To process the ore, a 60-stamp mill was brought in from Virginia City. Total estimated recoveries came to over $1 million.

SYLVANITE (Lincoln) Sylvanite was established about 1895 after Peter Berg and Bill Lemley discovered gold near Crawford Creek on Friday Mountain. This brought in a stampede of prospectors who eagerly staked their claims. The town grew with a hotel, saloon and feed store, in addition to a 10-stamp mill. Before a store was established the first winter, all supplies were brought in by pack train through snow drifts and rough wagon roads, with freight at 5¢ a pound. Total lode production was about $250,000. A forest fire in 1910 completely destroyed the town.

TROY (Lincoln) Placer mining was first conducted during 1884 along Callahan Creek, followed by more claims near Kootenai Falls about 1892. A year later Frank and James Stonechest and others discovered the Bangle Mine, as well as the Banner silver lode. These two properties were combined and called the B&B. Before Troy was established, it had been an Indian campsite. E.L. Preston, who was a surveyor for the Great Northern Railroad, purchased some mining claims in 1891 and laid out the town. He named the camp for Troy Morrow, son of a family he was staying with during his surveys. The railroad laid its line through the region about 1893. Large production of ore did not begin until the early 1900s. A concentrator was built to treat the ore, but it burned down in 1927. The B&B was sold a number of times; its name was later changed to the Snowstorm, and ultimately realized more than $4 million.

VIPOND (Beaverhead) John Vipond discovered the Mewonitoc Lode in 1868. He and his brothers located other claims, but transportation problems made development slow, so they had to build roads to Corinne, Utah, where the ore was to be shipped. Silver and gold were produced, but until the 1870s only a small amount of silver was recovered. The Lone Pine was discovered by the Partridge brothers, who later sold their claim to The Jay Hawk Company (an English syndicate) for $725,000. The properties were producing more than $30,000 a month in silver. By 1912 about $15 million had been recovered.

VIRGINIA CITY (Madison) In 1863 Henry Edgar and Bill Fairweather from Bannack discovered gold at Alder Gulch. After picking up supplies and equipment for their claim, they returned only to find they had been followed by about 2,000 prospectors, which marked the beginning of mining activity in the county. It wasn't long before the camp of Varina was founded, named for the wife of Confederate president Jefferson Davis. Within 18 months more than 10,000 people were living there. The name was later changed to Virginia City. It became the terminus on a branch of the Northern Pacific Railroad, which facilitated shipment of the ore from the region's mines.

The *Montana Post* wrote, "On arriving at this place what astonishes any stranger is the size, appearance and vast amount of business that is here beheld.... Indeed the whole appears to be the work of magic — the vision of a dream. But Virginia City is not a myth, a paper town.... The placer diggings will require years to work out.... Many persons are taking out $150 per day to the hand."[179] But someone else noted, "A street of straggling shanties, a bank, a blacksmith's shop, a few dry goods stores, and bar-rooms, constitute the main attractions of the 'city.' A gentleman had informed me that Virginia City contained brownstone front housed and paved streets, equal he guessed to any eastern town. How that man did lie in his Wellingtons! The whole place was a delusion and a snare."[180]

Within three years more than $30 million in gold had been taken out of the gulch, making this site the most productive in the state. In 1899 the Conrey Placer Mining Company brought in its dredging equipment which operated until about 1922 and recovered about $800,000 in gold.

WICKES (Jefferson) This site was named for mining engineer George T. Wickes. After placer gold was found along Prickly Pear Creek and Golconda Gulch in 1864, a town sprang up with a post office the same year. The district also became the largest producer of silver in the county. The Alta Mine discovered on North Hill yielded more than $32 million, although a relative of the mine owner believed it was only about $5.3 million. Other good producers were the Gregory Mine that recovered $9 million; Mount Washington at $1 million; and Ninah which yielded between $2 and $5 million. One of the largest smelters of its kind in the state was built at Wickes. By 1893 the boom began to

Blacksmith and trading company, Virginia City (courtesy Thomas Pritchard).

bust when the Alta Company's reductions works burned and the smelter was shut down. Demonetization of silver also lent to the camp's demise. The final blow came as a fire that burned the town to a crisp in 1900. Total production from placers came to about $25,000.

WINSTON (Broadwater) Lode mining began about 1866 with Billy Slater, who worked his claim between jobs to pay for his food and equipment. However, he never shipped any ore from his claim. Large-scale mining didn't get underway until the 1890s when both gold and silver were found in the nearby hills. Winston was then established and named for Northern Pacific Railroad contractor P.B. Winston. A road was built to some of the mines in order to haul the ore to the camp so it could be shipped out via the railroad. The East Pacific Mine, discovered by Robert A. Bell, produced more than $2 million. A total of $3 million was realized from all the combined mines.

WISDOM (Beaverhead) Located in the Big Hole Valley that is uphill from everywhere because it's higher than all the other communities, this small town was named after the Wisdom River. Originally called the Big Hole River, it was renamed by William Clark of the Lewis & Clark Expedition because he thought Thomas Jefferson showed a lot of wisdom when he bought the Louisiana Purchase. Settlers arrived about 1875, but most of them left because of the bitter winters, earning the valley the nickname "Land of the Big Snows." Although the town was mainly ranching, it became a supply and trading point for the region's mines. Most of the placer claims turned out to be not as rich as first believed.

Wisdom is famous for its hay and cattle; the hay was shipped to points east and used to feed race horses. Known as the land with 10,000 haystacks, hay was the only crop produced there. These haystacks weighed about 20 tons, measured more than 30 feet in diameter, and were produced using a mechanism called a beaverslide. This method is still used in the valley.

YOGO (Judith Basin) In a remote area east of Neihart, placer gold attracted a group of prospectors in 1865, but hostile Indians drove them off. During 1879 deposits of silver and gold were discovered, and the stampede was on. That same year the camp was established with tents strung out along a stream. The gravels gave out by 1883 and everyone left. About ten years later Jake Hoover and Frank Hobson were prospecting and got caught in a storm. Seeking refuge in an overhang, the men chipped away while waiting for the storm to pass and uncovered gold. They sought financing from S.S. Hobson to build a water ditch from the stream to their claim in return for a share in the profits. Unfortunately the gold didn't pan out as expected,

Ajax Mine south of Wisdom, 1850s or 1860s (courtesy Beaverhead County Museum Photo Archives, Dillon, MT).

and Frank went back home to Maine. While visiting a friend he showed her a box full of gold dust and some blue stones. They turned out to be sapphires. Thus, Yogo later became the center of a great sapphire mining industry.

Yogo was the home of a woman called Black Millie. She served as a nurse for a family there. Her fine cooking was known far and wide, and her table was always set to perfection. After the mines played out and the miners left, Millie was sent to a poor farm. But she refused to stay, and went back to Yogo, where she continued to set her table every day in expectation of the miners coming to her table, but they never returned.[181]

ZORTMAN (Phillips) Although gold was found by prospectors in 1868, there wasn't much interest in the area until about 1884 when ore was discovered in Alder Gulch and Beauchamp Creek. Located in the Little Rocky Mountains, this camp was named for mine operator Pete Zortman, and the post office was established in 1903. Although Pete is credited with the discovery of the Alabama Mine, many doubt that claim. The *Little Rocky Miner* wrote in 1907: "The Alabama was sold by the original locator for $40 to Spaulding, who later interested Pete Zortman and Putnam, both of Chinook. They did development work and got some ore yielding as high as $2,800 a ton."[182] Other lode mines were later developed, two of the most important being the Little Ben, producing $1.5 million, and the Ruby Gulch Mine, thought to have yielded about $5 million. Placer gravels discovered in 1884 yielded about $950,000.

NEVADA

ALDER (Elko) Miners from Montana came to Alder to seek their fortunes in 1869, but they didn't find enough gold to make it worth their while and left. Henry Freudenthal was more fortunate and struck gold along Young American Creek in 1885, naming his claim the Gold Bug Mine. He later sold the property to James Penrod and left the area to become the Lincoln County sheriff. A little more than five years later the Golden Gate Mine was located by Louis Burkhardt, followed by Emmanuel Penrod and his Oro Grande Mine. The place was platted about 1897 and named Delores, later changed to Alder, a transfer name from Montana. Speculators offered $20,000 for the Oro Grande Mine, but the owner turned them down. He should have taken the deal because investors realized only small returns, and Alder began to fade. During the early 1900s some mining was revived but offered only limited gold. In all the years of prospecting, a little more than 50 ounces of gold and 1,000 ounces of silver were recovered.

AMADOR (Lander) Amador is a transfer name from California, and honored a soldier of fortune with the Portola Expedition, Don Pedro, or his Indian fighter son, Jose Maria Amador. The site was situated about five miles north of the town of Austin. In 1863 silver was discovered at Amador Canyon, bringing in miners who combed the area for riches. The post office was established in 1864 with Isaac Sherman as postmaster. Amador once sought to get the county seat but lost the election. Everyone believed there were riches to be found there, but it was not to be. Property owners had minor success, and one smart owner sold his claim for about $32,000 in 1864. Two years later the town had faded into the sunset.

AMARGOSA (Nye) When the Las Vegas & Tonopah Railroad was being built in 1906, Amargosa was formed as its temporary construction camp and given the Spanish name for the bitter water from the nearby wells. It later served the nearby mines as a supply depot. When mining declined so did Amargosa.

AMERICAN CANYON (Pershing) Placer gold was discovered in the canyon southwest of Gold Mountain about 1871 by L.F. Dunn. He leased his claims to nearly 3,000 Chinese miners in the 1880s. These resourceful men were recovering between $5 and $10 a day in gold, and were thought to have ultimately recovered nearly $10 million in gold.

ARABIA (Pershing) In 1868 George Lovelock discovered the Montezuma Mine. The same year a county assessor's opinion was that the Montezuma and other claims "a mile square within which the Arabia mines are located would produce more bullion than any other ever known."[183] By 1875 the Montezuma had yielded more than $400,000. The nearby camp of Torreytown processed the ore with a 12-ton smelter, but the output was not as rich as initially thought, and within a year or so Arabia declined. The camp was located southwest of Lovelock and given an Arabic name meaning "country of plains," or "desert."

ARGENTA (Lander) Silver was first discovered about 1866, along with the accompanying camp of Argenta (Spanish for "silver") just east of Battle Mountain, The site became a stop along the Central Pacific Railroad two years later. Citizens thought Argenta would grow as a center of commerce when it became a shipping point for the mines in the

Bull Run Mine, Aura, 1906 (Northeastern Nevada Museum, Elko, Nevada).

Austin area. The post office was established in 1868, with William Westerfield serving as postmaster. But when Battle Mountain began to boom a year later, Argenta started its slow decline. The buildings were dismantled and moved to Battle Mountain in 1870. Argenta lay fallow until the 1930s when barite deposits were found that realized over $3 million.

ATHENS (Nye) This camp grew around the Warrior Mine that was discovered about 1912 in the Cedar Mountains. The property recovered about $20,000 in gold its first months of operation; between 1935 and 1939 it yielded about $29,000. The ore began to play out during the 1920s. Leasers later came in and worked the claims, but not a lot of gold was recovered and Athens disappeared.

ATWOOD (Nye) Located south of Gabbs, Atwood was part of the Fairplay Mining District created in 1903 after gold had been discovered a few years earlier in the Paradise Range. The camp's post office opened in 1906 and was named for miner Mr. Atwood. Not enough ore was found, and by 1907 the camp of more than 200 had disappeared.

AURA (Elko) Aura had its beginnings in the early 1900s, although Jesse Cope and his group found gold about 1869. Later gold discoveries gave rise to the settlement and Aura became a trading and supply center for the claims in the Centennial Range. It was also a stopping place for the Northern Stage Company. Aura is derived from Latin for "gold." Its post office opened in 1906 with Barney Horn as postmaster. There were some nearby lodes discovered, one of which was the Bull Run Mine. Ore recovered from all the properties came to nearly $1 million. Lack of investor capital and the fact that the mines played out saw the demise of Aura by 1917.

AURORA (Mineral) Aurora was so close to state borders that it was claimed by both Nevada and California, with Nevada winning the boundary decision. In 1860 prospectors James Braley, J.M. Corey and E.R. Hicks were riding one day and stopped to rest their horses. While the animals grazed the men took their picks and idly chipped away at the nearby rock outcroppings, hitting gold. They made more than 350 claims and two months later sold them for $30,000. The town emerged almost overnight. Because of its location on Last Chance Hill, every bit of food, explosives, and other items had to come via rough trails by mule to Aurora. A road that was later built from Bodie wasn't much better and it was still problematic to bring in provisions. The town had 17 stamp mills to process the silver and within ten years more than $30 million was recovered. By 1864 there was a population of about 10,000.

Aurora was not immune to the usual array of shooting scrapes. The Exchange Saloon was well known for its frequent fights: "At that time the crowd of miners and gamblers used to congregate at the Exchange Saloon, where frequent shooting scrapes would occur. Whenever trouble started every one would get out of the room. On one of these occasions a gun sent off in the crowd and Lee Vining went out the door ... and started up the street.... Shortly after, someone found him lying on the walk dead, and upon examination it was found that the pistol had gone off in his pocket, shooting him in the groin, from which he had bled to death."[184]

Aurora's lights began to dim when the mines played out. It lost its county seat status about 1869, but in 1880 there were about 500 people still living there. The *Esmeralda Herald* wrote in 1883, "There is no star that sets, but that will rise again to illuminate another sky.... A few more pushers on the wheel and Aurora will resume her place of twenty years ago." But two weeks later the paper said, "The few citizens of this place amuse themselves by going from store to store, saloon to saloon to tell each other how dull it is."[185] Mark Twain even tried his luck at the Aurora mines. He stayed for a few months, but had no success and abandoned his claim. He then worked at the stamp mill for a while, earning $10 a week which included room and board.

AURUM (White Pine) In 1869 two miners named Chisolm and Ramsdell found the first gold at nearby Silver Canyon. Further discoveries in the Schell Creek Range during 1871 brought in the miners and the camp of Aurum was established about 40 miles from Ely. Its post office was opened in 1881 with John Robertson as postmaster. A leading factor was Simon Davis, who discovered most of the mines in the region. A 10-stamp mill was erected at the camp by a Dr. Brooks which treated the gold from Ruby Hill. The Aurum Mine was located by Simon Davis and George Palmerton and its ore assayed at $300 a ton. They later sold the property for $35,000, and the new owners realized between $400 and $500 a ton. Unfortunately the lodes played out early and by the 1880s Aurum was history, until 1897 when there was a slight revival. Albert Erickson and Simon Davis discovered manganese with silver ore. Erickson mentioned, "Davis and I discovered some manganese outcroppings which, when opened up, produced some very high grade silver ore and for two years freight teams were hauling ore to Wells and to the railroad at Oasis, Utah. These mines were sold to a mining company in Salt Lake City ... but the vein pinched out after a year or two and the property was abandoned."[186]

AUSTIN (Lander) Austin was settled in 1862 after the discovery of silver by William M. Talcott. While working as a station agent for the Overland Stage, he came upon a rich silver float in Pony Canyon. After the word leaked out, more than 10,000 silver-crazed miners arrived. A newspaper journalist gave the new miners some advice: "He should purchase himself a plug [horse] for $40 to $75 at Sacramento, Carson City or Virginia City, array himself in homely apparel, hang a bag of lunch, a six-shooter and a canteen on the horn of the saddle, and tie carefully behind the same two substantial blankets. You may then establish yourself in the saddle and wheel off."[187]

The camp was named by David Buel for his partner Alva C. Austin. It may have also honored Leander K. Austin, an uncle of George Austin who developed the Jumbo Mine in Humboldt County. In 1863 Buel erected the first stamp mill and within four years there were eleven of them operating, serving more than 6,000 claims. A few of the properties were good producers: the Morgan and Muncy Mine realized more than $3,000, and another brought out $67,000. Production increased and a few years later the North Star recovered more than $250,000. Between 1866 and 1867 even larger lodes were found and went on to yield more than $50 million. By 1880 the ore was becoming marginal and Austin started to decline, even though a railroad had recently arrived. But the town prevailed when it became a distribution center for the isolated ranching and mining communities. Between 1862 and 1887 the production of silver was estimated as high as $20 million for the district. Gold recovery was about $58,000.

One of the exciting events of Austin's early years was the arrival of nine camels, including one that had been presented by the Sultan of Turkey to the U.S. government. Jefferson Davis came up with the idea of using camels as beasts of burden on the deserts of the Southwest. The experiment was unsuccessful and in 1864 the camels were sold at auction and used by mining companies to bring salt for the quartz mill from marshes 100 miles to the south. For years they were seen plodding patiently along the road.

BABYLON (White Pine) Babylon was a short-lived camp that arose in 1868 on Babylon Ridge. It was so cold because of its high altitude in the White

Pine Range, the men had to keep fires going even during summer. Only a few prospectors spent time there, living in either shanties or lean-tos, suffering through the high winds and deep snows during their first winter. There was no business district, but probably at least one saloon, a prerequisite for all mining camps. One of the residents told a new arrival that Babylon was as big as New York, it just hadn't been built up yet. The camp lasted less than a year.

BALD MOUNTAIN (White Pine) Initial silver discoveries occurred in 1869 when G.H. Foreman hit pay dirt, and a camp was established about 70 miles southeast of Elko. Mining was conducted for many years with minimal ore recovered, and later development was hindered because of high freight rates. The site was first named Joy, then changed to Bald Mountain (for its lack of vegetation) when the post office opened in 1897. The more prominent mines were the Genie, which yielded about $40 a ton, Bismark at $80 per ton, and the Nevada at $128 per ton, but it produced only a total of about $20,000. By 1877 little was going on. There was a slight revival about 1897, but not enough ore was found and the miners left.

BANNOCK (Lander) After Alex Walker and Sherman Wilhelm found a ledge of gold in Philadelphia Canyon, the camp was established on a steep hillside and soon populated by more than 200 gold-hungry men. Some of the better producing claims were the Pussin' Ken that recovered about $180 a ton and the Reno ($200 a ton in gold). The camp was very short-lived though and within a year it was all but abandoned. It was named for the Bannock Indians.

BARCELONA (Nye) Barcelona is a transfer name from Spain which honored Carthaginian general Hamilcar Barca. Silver ore was discovered in the Toquima Range by Mexicans about 1867. Not much development occurred until 1874 when the camp was established and it was added to the Austin-Belmont stage line route. There was some mining success, but by 1877 Barcelona had been abandoned. An attempt was made to extract ore in 1921 with a 10-stamp mill, but it operated less than a year.

BATTLE MOUNTAIN (Lander) This site took its name from the mountain, where in 1861 it was the scene of a skirmish between the Indians and whites. The town began as a station on the Central Pacific Railroad about 1868, then served the Battle Mining District when silver was discovered in the Reese River Valley. Its post office opened in 1870 with Thomas Reagan as postmaster. Many of the buildings from Argenta and Galena were moved here the same year. Adding to the camp's prosperity were the freighting businesses that hauled supplies back and forth to Austin.

One of the better paying mines was the Little Giant, which yielded about $1 million in silver. Gold lodes were also located, with more than $6 million recovered. By 1880 the mines began to play out and Battle Mountain went into a depression. Placers were not discovered until the early 1900s near Philadelphia Canyon, but the gold was deep in the gravels and lack of water hampered recovery. When the Natomas Company brought in its dredges in 1947, the placers realized about $2 million. Battle Mountain was revived in the 1930s after copper ore bodies were located. It later became a shipping point for ranching livestock.

BAXTER SPRINGS (Nye) Located north of Tonopah, this camp started out as a stagecoach station on the Belmont-San Antonio stage road during the 1870s. Baxter Springs was settled in 1906 after gold was found in a nearby gulch. Although it reached a population of more than 400 people, within three months the place was deserted after the placers were found to be insufficient.

BEATTY (Nye) Beatty became the southern terminus of the Tonopah & Tidewater Railroad and a supply center for the mining camps of Bullfrog and Rhyolite. It also handled most of the freight for the Bullfrog Mining District and called itself the "Chicago of Nevada." Beatty was settled in 1904 along the banks of the Amargosa River about 60 miles north of Goldfield, and flourished because of its almost unlimited supply of water from the river. Its post office was established a year later. The town was founded by Bob Montgomery, who built the first hotel and named the place for rancher Montillus M. Beatty. The population rose to about 1,000, but the number dropped when the mines played out about 1909. Beatty managed to survive as an entrance town into Death Valley National Monument, and catering to automobile travelers.

BELLEHELEN (Nye) Bellehelen came to the forefront during 1904 when gold and silver were discovered in the Kawich Range. It was located about 50 miles from the mining camp of Tonopah. When engineer George Wingfield and his party

Mine at Belmont (courtesy Jeff Lotze).

arrived in 1907 to check out the ore possibilities they were disappointed and left, as did others who were trying their luck. Bellehelen was brought back to life after the Nevada Bellehelen Mining Company came in and reopened some of the older mines. The ore taken out of the properties was yielding anywhere from $25 to $165 a ton. A good producer was the Cornforth & Sweet lode that realized nearly $30,000 in both silver and gold. Two others, the Never Sweat and Scotia, recovered more than $500,000.

BELLVILLE (Mineral) Taking its name from the Northern Belle Mining and Milling Company, the camp began about in 1873 and lived its life as a milling center for nearby Candelaria. After mill owners Samuel Youngs and A.J. Holmes at Columbus had an argument, Holmes left and came to Bellville where he built a new mill and took control of a mine at Candelaria. Out of spite to Youngs, he cut off the ore supply to Columbus. The Carson and Colorado Railroad also built its tracks through and Bellville became its terminus. When water was brought to the mines at Candelaria, Bellville's stamp mills were no longer needed and the town was abandoned.

BELMONT (Nye) This camp developed with a rush in 1865 and acquired a post office in 1867. Belmont later became the county seat until it was taken by Tonopah. The town bears a French expression for "beautiful mountain." Located on a slope of the Toquima Range, Belmont was a supply and trading point. The *Silver Bend Reporter* wrote in 1867, "Colonel Buel's mill has extracted $100,000 worth of silver bullion from 1000 tons of Highbridge ore, all taken within twenty-five feet of the surface; and there eighty stamps on the road ... which will make our bullion yield something over $200,000 per month before the summer is over. The hills are beginning to blacken with prospectors"[188] When the White Pine District began to boom the men left, and by 1900 not much was left, but not before more than $15 million in silver had been recovered. The discovery of turquoise in 1909 attracted renewed interest in the area.

BEOWAWE (Eureka) Prior to settlement, this site was a major Paiute Indian camp. Beowawe was established about 1850 when a trading post was built, and grew up around the Central Pacific Railroad. The post office was established in 1870 and closed its doors in 1995. Beowawe is a Shoshone expression for "gate." Because of the conformation of the hills on either side of the valley, the community looks like it stands in an open gateway. The town gained some prominence when it became a trading and shipping point for the Cortez Mining District. During 1910 George Wingfield had an electric power plant built here to run a cyanide plant used to process the ore from Buckhorn. Beowawe later profited from stock raising.

BERLIN (Nye) Located on a slope of the Shoshone Range, the Berlin property containing both silver and gold was discovered in 1895 by state senator T.J. Bell. He later sold it to John P. Stokes from New York. It operated until the early 1900s when the ore began to produce low values. The mine yielded about $362,000 between 1896 and 1909. Berlin was platted in 1897 and given a German name for "fishing village." In 1911 a cyanide plant was brought in to process the tailings which brought a little more than $2 a ton.

During the 1860s ichthyosaur (fish lizard) fossils were discovered, and later used by some of the prospectors as dinner plates. Excavation work began in the 1950s and the fossils were found to be more than 200 million years old. Berlin is now included in the Nevada State Park system which was established in 1957.

BERNICE (Churchill) Bernice was settled about 1882, although there was silver mining in the area before 1864. Located in the Clan Alpine Range, the camp served the nearby mines with its 10-stamp mill. The post office opened about 1882 as Casket, but the name was later changed. After yielding more than $300,000 in silver, the mines close down and Bernice folded in 1894.

BLACK HORSE (White Pine) Ore was discovered here quite by accident. Seeking shelter during a windy storm in 1905, prospector Tom Watkins chipped away at an overhanging ledge and found gold. He kept his find a secret until he couldn't contain himself any longer and told some friends at the camp of Osceola, who immediately staked their own claims, and were followed by a slew of others seeking their fortunes. The camp was formed with a post office in 1906, and took its name from a black horse owned by Watkins. The Black Horse Mine began operations the same year and some of its ore produced about $50,000 a ton. By 1911 most of the gold had been exhausted and the town declined. The surrounding mines yielded close to $1 million.

BLACKBURN (Eureka) Blackburn was established in 1874 as a railroad siding along the Eureka-Nevada Railroad until 1909 when the camp of Buckhorn began its rise. As a result, Blackburn became a shipping point for supplies headed there. It began its decline when Buckhorn's ore ran out about 1914. In later years the only thing left was a gas station that operated until about 1990.

BLAINE (White Pine) Located near the Egan Range, Blaine came into existence after W.D. Campbell found gold about 1894. He named his claim after presidential candidate James G. Blaine, and erected a 5-stamp mill. However, the camp did not last long because the ore was low grade, but Campbell continued to work his mine. He later sold it to Governor Denver Dickerson, who also experienced inadequate values that could not be worked profitably, so the boom busted. In all, the total yield from the mines came to about $200,000. The camp was abandoned by the early 1900s.

BLAIR (Esmeralda) Established in 1906 as a milling center, Blair was named for eastern banker John I. Blair. After land prices in Silver Peak soared, the Pittsburgh Silver Peak Gold Mining Company formed this camp for its facility. An aerial tramway was built to carry the ore from the Silver Peak mines to Blair's mills. Most of the ore was low grade and there were not enough profits, so the mining company sold its mills, and by 1918 the town was gone.

BONNIE CLAIRE (Nye) Ore found at Gold Mountain in the 1880s was processed with a 5-stamp mill which continued operations for about 20 years. A stagecoach station was also built to serve the miners between Bullfrog and Goldfield. The Bonnie Claire Bullfrog Mining Company purchased the mill, followed in the early 1900s by the establishment of the camp about 20 miles from Scotty's Castle in Death Valley, California. A few years later the Las Vegas & Tonopah Railroad laid its tracks through town. By 1909 the ore had run its course, and the place was abandoned. A year later the New Bonnie Clare Mining and Milling Company tried to rejuvenate the old mill, but it was a failure.

BOVARD (Mineral) Silver was discovered in the Gabbs Valley Range about 1908, bringing in a throng of prospectors. Bovard became a supply camp when it was founded by prospectors from the mining town of Rawhide, and a stamp mill was built by the Gold Pen Mines Company to crush the ore. Meanwhile, another camp nearby called Lorena was being formed. Rivalry occurred between the two when promoters tried to get Bovard people to move to Lorena. In turn, promoters from Bovard induced freighters hauling water to the camps not to deliver to Lorena, which forced the wayward miners to return. Both places did not survive long and were gone by about 1916.

BOWLERVILLE (Nye) Named for mine owner Fred Bowler, this camp was formed during the early 1900s and was very short-lived. His mine did not produce much and there were never more than 10–15 people living there at one time, and the place was abandoned within a year.

BROKEN HILLS (Mineral) This district did not produce a lot of ore. The only mine of any consequence was the Broken Hills, discovered by Joseph Arthur and James Stratford about 1913. They recovered more than $65,000 in silver before they sold the property to the Broken Hills Silver Corporation, backed by George G. Rice, a man with a checkered background. His real name was Jacob Herzig. He bribed a geologist to gloss over a prospectus of the area, publicizing rich ore and fresh water, which brought in eager prospectors. Unknown to them there was no water at Broken Hills; it had to be hauled in. An investigation was conducted on Rice, who ended up going to a penitentiary. By the 1920s Broken Hills ceased to exist.

BRUNO CITY (Elko) This camp began about 1869 after the Young America Mine was located, and named for hunter Baptiste Bruneau, or trapper Pierre Bruneau. Although gold and silver had been discovered about 1864, there was not much development because of the rush at Hamilton. Located near the Bruneau River, the camp served the miners who worked at a nearby hill called Silver Mountain. Although Bruno City was short-lived, the Young America yielded more than $100,000.

BUCK STATION (White Pine) This site was formed during the 1860s as a station for the Elko-Hamilton stage line. It was used mainly as a place for the freighters to change out their horses before heading to the outlying camps to deliver supplies. When Hamilton began its decline in 1870, Buck Station was no longer of any value.

During 1869 a Wells-Fargo stage was held up and relieved of the $40,000 it was hauling. The bandits were killed, but not before they buried the money. Tradition says it is still hidden somewhere near the station, but it's never been found.

BUCKHORN (Eureka) During 1908 a group of five prospectors discovered gold. A year later they sold their Buckhorn Mine to George Wingfield for $90,000. He formed the Buckhorn Mines Company and had an electric power plant built at Beowawe to run a cyanide plant. Several mines were located, but by 1914 the ore began to pinch out. The Buckhorn was the only property of consequence and yielded nearly $800,000. The site was situated about 35 miles from Eureka.

BUCKSKIN (Lyon) Gold was found nearby about 1906, followed by the establishment of the camp between Lincoln Flat and Smith Valley, and initially called Gold Pit. It was later changed to Buckskin, taking its name from a mine. By 1908 the gold was gone. Production estimates were between $1,500 and $9,000. During the 1930s Ambassador Gold Mines, Ltd., constructed a pipeline to facilitate hydraulic mining, but their venture was a complete failure. Buckskin gained renewed life after the discovery of copper in 1933.

BUELL (Elko) Named for Colonel David E. Buel, this camp was formed about 1869 and its post office opened in 1871. After silver was found in the Toana Range, Buel purchased a number of claims for about $125,000. A smelter was built to treat the ore, which was then shipped to San Francisco. The Black Warrior Mine produced about $16,000 in silver, but the smelter was shut down after processing only a few thousand tons of ore. By 1880 the town was on the decline. Mining activity ended for a time until copper was found and resurrected Buel.

BULLFROG (Nye) Frank "Shorty" Harris was responsible for the founding of Bullfrog. Located near the town of Rhyolite, this camp was laid out in 1904 and named for the Bullfrog Mine discovered by Harris and his partner, Eddie Cross. The gold was found in a green-stained rock that resembled a frog. On August 9, Harris struck a boulder of quartz with his pick and described his find: "Next morning we started west. Cross started down to the little hill to the south ... I found lots of quartz all over the hill and started to break it with my pick. Cross hadn't moved over 400 feet from me when I ran against a boulder, and I called out, 'Come back; we've got it.' The quartz was just full of free gold, and it was the original genuine green bullfrog rock. Talk about rich rock! Why, gee whiz, it was so great. We took the stuff back to the spring and panned it, and we certainly went straight up. The very first boulder was as rich in gold as anything I had ever seen." He later added, "No, I have never made a big stake. The biggest haul I ever made was $10,000. I let the Bullfrog strike get away from me because I had too much 'Oh, Be Joyful' on board, but I couldn't recover the property. I got only $1000 for the strike. But don't worry; the Short Man will never again sign anything when he is under the influence."[189] With constant freight tie-ups and heavy traffic, a railroad was finally built. Bullfrog couldn't compete with the town of Rhyolite, and foresaw its demise in 1906 when the *Rhyolite Herald* made its headline "Verily, The Bullfrog Croaketh."[190] More than $1.7 million in gold and silver was produced.

BULLION CITY (Elko) Located above a steep canyon (6,386 ft.) on the east slope of the Piñon Mountains, Bullion City came to life in 1870, and its post office opened the following year. Interest in the region began during 1869 when silver was found on Bunker Hill. The Palisade Smelting and Mining Company brought in the first smelter to process the ore; however, it was unsuccessful due to inferior construction and bad management. In 1872 the Empire City Mining Company built its smelter with better success, but since the ore had to be shipped at great expense to the town of Elko, no huge profits were made. The company closed its doors four years later. The Standing Elk Mine was discovered during the 1880s and produced about $111,000. In a span of about 15 years, the area mines recovered more than $3 million.

BULLIONVILLE (Lincoln) Another milling town about 10 miles from Pioche, Bullionville was founded in 1870 with the post office opening in 1874. The camp prospered processing ore from Pioche and other nearby mines. It was the ideal site because of the good water supply. A 5-stamp mill was brought in, followed by four more in 1872. This same year the Pioche and Bullionville Railroad was built by General A.L. Page, which transported the ore. After water was found near Pioche, Bullionville's mills were no longer needed and by 1877 the camp had faded away.

BURNER (Elko) Initial discoveries of gold and silver were made by Elijah and J.F. Burner in 1876, who named their first claim the Viola Virginia Mine. About 1879 the camp was established and named for them. Not much ore was recovered until Burner located the Mint Mine in 1884, which produced about $20,000. That same year another lode was found by miners from Tuscarora, Will and Gil Strickler, who recovered more than $15,000 in silver and gold.

Remnants of the Bullfrog Mine at the camp of Bullfrog near Rhyolite (courtesy Historic American Buildings Surveys, 12-RHYO.V, 1–27).

BURRO (Lander) After Anthony Dory discovered the Valley View Mine in 1906, a small camp sprang up named Burro. It died almost before it was born, since the gold proved to be low grade and was quickly exhausted. There were never more than 25 people living there at one time and the camp folded in less than a year. During 1907 there was some mining conducted by Gus Laurent, but he found very little and left for greener pastures.

BUZANES CAMP (Lander) A Greek prospector named John Buzanes came here from Cripple Creek, Colorado, in 1926 and found gold that was valued at about $17 a ton. Buzanes later sold his claim to the Magna Gold Mines Company for $200,000, which built a small stamp mill in 1928. But the ore values were low and the camp was abandoned.

CACTUS SPRINGS (Nye) During 1901 this region was first worked for turquoise until 1904 when the Cactus Nevada Mine was discovered. But the silver was so marginal there were no profits. During 1919 other claims were located, some averaging $30 a ton in silver. Capitalists invested more than $280,000 to develop the mines, but it was a lost cause. Other companies came in and tried to extract the silver, but again their efforts were futile. This region later became part of the Las Vegas Bombing and Gunnery Range.

CANDELARIA (Mineral) Silver veins were discovered in the Candelaria district during 1863 by a party of Spaniards. The following year a large deposit was found, but it wasn't developed. It would be relocated and become the great Northern Belle Mine. Candelaria was settled during the 1870s with a post office established in 1876, and became one of the leading silver camps in the state, thanks to the Northern Belle. Its ore was continually hauled by freighters: "Nearly every hour of the day and night found heavy freight wagons rumbling through the streets — all fetlock deep in dust."[191] The Carson & Colorado Railroad was built to town in 1882, which gave miners relief with lower freighting costs. That same year piped water was brought in from Tail Canyon in the White Mountains. A year later a fire almost destroyed the town. In all, about $20 million in silver was produced from the nearby mines, with almost three-fourths of the profits coming from the Northern Belle.

CANYON CITY (Lander) Located in the foothills of the Toiyabe Range, Canyon City was established in 1863 and had a population of about 200. Two stamp mills were erected to treat the ore, but the silver found at nearby ledges was in small amounts and the values were too low. By 1866 the camp was all but abandoned. Total production was about $500,000.

CENTRAL CITY (Nye) This town got its start in 1906 when gold discoveries were made and brought in more than 3,000 people. Central City grew with many businesses and saloons, but over the next few years fraudulent promotion schemes were hatched that gave the district notoriety and delayed its development. When big strikes were found closer to Manhattan, Central City declined. Placer mining reached its peak about 1912. The values recovered from both placer and lode mines came to more than $10 million in gold.

CHARLESTON (Elko) Charleston was established in 1876 as a supply center after placer gold was discovered on Seventy-Six Creek, and named for prospector Tom Charles. George "Alleghany" Mardis, who founded the camp, was quite a character and a Bible quoter. A self-proclaimed philosopher and preacher, he found his wealth on Seventy-Six Creek. He had a burro named Sampson to which he preached fiery sermons from the Old Testament. Mardis prospered and changed his old buckskins for fancy store clothes and paraded the streets as the first citizen of the town.[192] He was later killed by a Chinese man who learned he was carrying gold to Elko.

Charleston was probably the toughest town in the state. After a manhunt in this primitive area a sheriff said, "By darn, the trip was a success. I didn't get my man, but I rode plumb through Charleston without being shot at."[193] After the placers were depleted about 1883, Charleston ceased to exist. There are no production records for gold recovery at Seventy-Six Creek,[194] although it was estimated at $6,000.

CHERRY CREEK (White Pine) Gold was discovered nearby by a group of soldiers during 1861, but it was the 1870s before any development began. Named for the chokecherries in the canyon, this camp was established about 1872 and grew to a population of nearly 6,000. Many of the buildings from Schellbourne were hauled in to house the businesses and miners. Peter Corning and John Carpenter discovered what they called the Ticup Mine, which is Shoshone for "biscuit." The Exchequer Mine was also located and produced about $1 million in gold. The Mary Ann Mine yielded only about $35,000 in a 6-year period. By 1893 the decline began. Estimates of total production range between $6 and $20 million.

CLAN ALPINE (Churchill) Silver was found near the Clan Alpine Mountains about 1864 and a small camp was established along Cherry Creek. Its post office opened in 1866 and a 10-stamp mill was erected. But the ore was low-grade, and within two years the place folded.

CLAYTONS (White Pine) George Clayton found gold nearby about 1880. Because of the camp's isolation there were never many people, and supplies had to be hauled in over an old wagon road. In 1881 two small mines were located, but after less than $2,500 had been recovered the place was abandoned.

CLEVE CREEK (White Pine) After the Kolcheck Mine was located in 1923, a small camp was formed with a small population of perhaps 15. A stamp mill was built that processed about $15,000 of gold the first year, and was then abandoned. The property was relocated in 1915 and the new owners recovered about 235 tons of gold.

CLIFFORD (Nye) Located about 35 miles from Tonopah, Clifford was named for Ed and James Clifford, who found silver in 1905, causing the establishment of the camp, which was very short-lived. Prior to their arrival, the first discoveries were made by an Indian named Johnny Peavine. The Cliffords later sold their mine for about $250,000. When word got out about the sale, others came to try their luck. But by 1908 most of the ore was gone and Clifford was no more.

CLIFTON (Lander) Two miners named Marshall and Cole laid out this site about 1863 after silver was discovered in nearby Pony Canyon. It received a post office in 1867 and boasted a population of about 500. But the establishment of Austin spelled the demise of Clinton. Austin gained county seat status and Clinton was all but abandoned. It experienced a short revival after a 40-stamp mill was built by the Austin Silver Mining Company owned by J.G. Stokes, and the Nevada Central and Austin City railroads made Clinton its terminus.

CLINTON (Lyon) Although a few prospectors found silver here in 1863, active mining didn't begin until about 1880 when the camp was established at the mouth of Ferris Canyon. Prospects looked good after the Antiquarian Mine was located which was producing about $1,000 a ton of ore. As a result a stamp mill was erected. But the deposits began to pinch out, and by 1882 the place was abandoned.

COLUMBIA (Elko) Discovery of silver in Blue Jacket Canyon caused Columbia to be founded about 1868, which also became headquarters for those working the claims in the Centennial Mountains. The mines had produced about $25,000 in silver by 1870. A year later the Columbus Consolidated Mining Company had a 20-stamp mill erected at the camp. It lasted until only 1883, but not before producing more than $1.5 million in silver. Although the mines around Columbus made fortunes, the camp itself did not thrive. Supplies had to be brought in more than 75 miles away over rough roads. That and the later decline of ore spelled the camp's demise. Between 1875 to 1877 other properties were located, but only about $115,000 in ore was recovered.

COLUMBUS (Esmeralda) Columbus was created in 1865 with the discovery of gold and silver. A small stamp mill was brought in from the town of Aurora, but the ore was inadequate and Columbus began its swift decline. Adding to its demise was the fact that due to an argument between mill owners A.J. Holmes and Samuel Youngs, Holmes moved to Bellville and cut off the ore supply to Youngs' mill. A few years later borax was found in the region, bringing renewed life to Columbus.

COMO (Douglas) About ten miles southeast of Dayton, gold was discovered in the Pine Nut Mountains on the slope of Mount Como in 1860. The camp was settled in 1862 with a post office established the same year. This region was home to Paiute chief Numaga. Prospectors were cutting down the piñon trees to use in shoring up the mines and Numaga asked the men to stop, since the nuts from the trees were one of their essential foods. He was ignored, and later that day a group of Indians showed up a the camp, causing the miners to flee and request help from Fort Churchill. In 1864 a small stamp mill was erected by J.D. Winters, but it lasted less than a year because the ore was not profitable enough to even cover his expenses. By the end of the year Como was no more. During the 1930s some of the properties were relocated and the ore recovered was valued at about $200,000. The Hully and Logan Mine produced a total of $76,000.

CORNUCOPIA (Elko) Martin Durfee (sometimes referred to as Matt) discovered the first flakes of ore on Bull Mountain in 1872 and the camp was founded shortly thereafter. A year later more than 70 prospectors had established their claims. This was followed by a surge of excitement when someone discovered a vein of ore valued at more than $1,000 a ton. Beginning in 1874 the Leopard Mine owners constructed a 10-stamp mill to treat the ore, which was formed into bricks and hauled out by stagecoach. That same year Cornucopia experienced a severe winter with the result no supplies could reach the camp, causing most of the men to depart temporarily. By 1875 the mill was producing on a monthly basis more than $20,000 in gold and silver. When the wild camp of Tuscarora began to grow and Cornucopia's mill burned down, the decline began. Before the camp died the Leopard Mine had produced more than $1 million.

Elko County miner (Northeastern Nevada Museum, Elko, Nevada).

CORTEZ (Lander) Mexicans were early discoverers of silver near Mount Tenabo during 1863, and the first load of ore that was shipped to Austin aroused American interest. While following the Mexicans, the prospectors stumbled upon a number of silver ledges. By 1864 roads had been built to the camp and an 8-stamp mill was erected. Raymond wrote, "The mining operations in the Cortez District have not been carried on with much wisdom or economy. The Cortez company, owning some of the best deposits, had made thus far a partial failure.... There is a steady, though limited, transportation of rich ore from Cortez to Austin for reduction. As the hauling costs some $45 per ton and the milling about $45 more, it is evident that only rich ore can be profitably handled in this way."[195] Cortez was a busy camp until the price of silver dropped. One of the principal mines was the Garrison, which yielded over $2 million. The district's mines realized more than $14 million in silver. Gold recoveries amounted to about $170,000.

CORYVILLE (Mineral) Coryville was established near Mount Grant in 1883 with a post office the same year. Rich silver was supposedly to be found in the region, and as a result a road was built and mining equipment brought in. But the news turned out to be false, and Coryville was no more. The name commemorated J.M. Corey, one of the founders of Aurora.

CRESCENT (Clark) This camp was named after the nearby peak, which was crescent shaped. About 1863 Mexicans discovered gold and hauled the ore by burro to the town of Goodsprings where it was crushed with arrastres. In 1878 renegade Mormons who were traveling through the territory came upon the miners and killed them. Before their demise the Mexicans were thought to have recovered about $500,000. During the early 1900s the mine was relocated along with a few other small claims. The gold didn't last long and Crescent was gone by 1906.

CUCOMONGO (White Pine) During 1903 three miners found gold that assayed out at about $400 a ton. The news brought in the gold seekers and the camp was formed. The Joanna Mine was later located and a 40-ton mill was erected to treat its ore. A tramway was also constructed to bring the gold to the mill. Because of the camp's remoteness, water had to be hauled in from Egan Canyon. The following year a number of other properties were located, which were later taken over by the Hartford Nevada Mining Company. The camp was gone by 1909. Between 1889 and 1901 the Joanna Mine produced only about $20,000.

DANVILLE (Nye) While prospecting along the Monitor Range in 1866, a group of men found silver in Danville Canyon, but not much mining was conducted until about 1870. Most of the ore was hauled by wagon to Austin for treatment, but little values were found and the claims were abandoned.

DAYTON (Lyon) Dayton is one of the oldest camps in Nevada. It originally started out as a tent trading post until it was platted in 1861 by John Day. Dayton became a trading and milling center for the Comstock mines with a post office in 1862. The camp's earlier name was Ponderers' Rest because California-bound wagon trains halted here while deciding whether to continue westward or turn south and settle along the rivers. The first authenticated discovery of placer gold in Nevada was made in 1849 by Abner Blackburn, a member of an emigrant train to California, at the junction of Gold Canyon and the Carson River at the present site of Dayton.[196] Numerous stamp mills were erected to treat the ore, but by the late 1860s most of the people had moved on. Fires began Dayton's decline, one in 1866 and the second in 1870 which destroyed most of the town.

DEADHORSE WELL (Mineral) This was a little watering station established in the 1860s used by freighters between the Reese and Walker rivers. Ore discoveries in other regions such as Candelaria and Belleville created the need for more freight wagons to haul the ore. As a result, a stamp mill was built at Deadhorse to crush the gold before shipment.

DELAMAR (Lincoln) This site is located about 35 miles from Pioche in the Pahranaget Valley. During 1890 a mining district was created and Delamar became an important camp with more than half of the state's ore output up to 1895. Named for Captain John De Lamar, it was a principal gold producer up until the early 1900s. Delamar acquired the reputation of being a "widow maker" after miners suffered from constant coughs, which were later diagnosed as fatal silicosis. It came from the silica dust that drifted through the tunnels and mines with poor ventilation. Those more fortunate died quickly and didn't have to suffer the wracking, painful coughs that would take their lives. Delamar did not have a lot of water for mining, so a pipeline was built from Meadow Valley. In 1900 the camp was consumed by fire which spread so fast it destroyed most of the buildings. But the site was rebuilt and the mines continued production. The Delamar was yielding between $100,000 and $200,000 a month. Total production of the region was about $25 million.

DERBY (Washoe) Derby was a station on the Southern Pacific Railroad and named for one of its employees. Located east of Reno, this small camp was established in 1904 when gold was found nearby and the post office opened in 1906. The following year there was talk of the new booming camp where a mill would be built to process the riches. But not enough gold was found to make the venture worthwhile, and Derby lost its race for life.

DIAMONDFIELD (Esmeralda) "Diamondfield Jack" David heard of rich strikes near Goldfield, headed to the region, and located a gold ledge. When the news got out, the camp was established and filled up with eager prospectors. David came from Tonopah, where he had been convicted of murdering someone but was pardoned. By 1906 the town was nearly abandoned after news of richer strikes elsewhere. Two of the nearby mines, the Diamondfield Daisy and Great Gend, yielded only $30,000.

DIVIDE (Elko) Located along the divide between the Humboldt and Snake River basins, this camp was established in 1915. Miners rushed to the site after W.C. Davis and J.L. Workman discovered silver nearby. Unfortunately for the eager prospectors, there was not much gold. The April Fool Mine was sold for $65,000 to A. Backlund, but the ore was superficial and he abandoned the claim.

DOUGLASS (Mineral) During 1893 gold and silver ores were found in the Excelsior Mountains. The following year the camp was established, its post office opened, and a 5-stamp mill was built. When rich strikes were discovered near Tonopah, the camp quickly declined. During the first ten years of Douglass' life, the nearby mines produced gold and silver valued at about $500,000. Small-scale production continued to 1934.

DOWNIEVILLE (Nye) During 1877 rich silver was discovered in the vicinity. Since this camp was located not too far from Ellsworth, miners from there headed to Downieville because the Ellsworth area contained only low-grade ore. When the post office was established in 1879, it was named for Postmaster P. Downey. The best producing mine was the Downieville, which recovered between $7 million and $12 million in silver. By the early 1900s the camp was gone.

DUCK CREEK (White Pine) Duck Creek was settled in the 1860s and received a post office in 1872, but not much mining occurred until the turn of the century. D.C. McDonald discovered the

Success Mine and recovered more than $9,000 between 1909 and 1910. The McGill Mine yielded about $13,000 in a two-year period. The Lead King Mining and Milling Company leased numerous claims in 1913, and by 1921 more than $165,000 in silver and gold had been realized.

DUN GLEN (Pershing) Dun Glen was founded about 10 miles from Mill City and named for J. Angus Dun. After D.P. Crook found silver in 1862 he informed Dun, who laid out the town and encouraged people to purchase land, by advertising: "Great Auction Sale on Real Estate in the town of Dun Glen.... Persons wishing to purchase will do well to improve this opportunity, as these are the only remaining lots unsold."[197] Dun Glen grew to a population of more than 200, and for several years a small military contingent was stationed there to protect residents from Indian attacks. In addition to becoming a trading center, a 10-stamp mill was built to process the silver ore for the nearby mines.

The Telegraph Saloon was also established. Its owner put up a sign that read "Wines, Liquors and Cigars, such as the gods, looking down from their ethereal adobes must envy—and all for two bits a single chance! 'Man being reasonable, must get drunk; The best of life is but intoxication; Glory, the grape, love, gold—in these are sunk The hopes of all men and of every nation.'"[198]

Placer gold was also found in the nearby canyons, and it was estimated that nearly $4 million was recovered. Most of the placer miners during that time were Chinese. After the ore was depleted the region became known for its cattle raising.

DUTCH CREEK (Mineral) This site was named for a Dutchman who found some gold along Lake Walter about 1867. Tradition says that when the Indians found out about his discovery, they killed the man because he was on their land. The mine was located on the Walker River Indian Reservation, which was opened in 1906 and brought in the miners. But the gold was only surface, not much was recovered, and by 1907 Dutch Creek ceased to exist.

EASTGATE (Churchill) Located at the mouth of a narrow canyon, Eastgate was originally an Overland Stage station and stopping place for travelers. The site was named in 1859 by Captain James H. Simpson. Gold discovered nearby in 1906 caused a small camp to form. But none of the mines produced much, and Eastgate became a ghost town.

EBERHARDT (White Pine) Eberhardt sprang up with the discovery of a bonanza in 1869. Located in Applegarth Canyon, the camp took its name from the mine. A prospector wrote, "I wish I could send you some specimens as I have had given me. The ore from the stone wall is hard silver and you can cut it with a knife, just as bright as a dollar. The Eberhardt is black as ink and some of it so rich you can hammer it out ... but that pile of rock contains one million dollars in silver."[199] Another wrote to a newspaper, "The walls were of silver, the roof over our heads was silver, the very dust which filled our lungs and covered out boots and clothing with a gray coating was fine silver.... How much may be back of it Heaven only knows. Astounded, bewildered, and confounded, we picked up a handful of the precious metal and returned to the light of day."[200] The mine was worked as a glory hole that measured 70 feet long, 40 feet wide and less than 28 feet deep. One boulder full of the silver weighed more than 5 tons, and from one chimney nearly $80,000 was recovered in just a few days. The silver was so fine it could be hammered into sheets without processing. By 1885 the mine was history and so was the camp, but not before yielding more than $4 million between 1866 and 1878.

EDGEMONT (Elko) Small-scale mining was conducted after the discovery of gold during the 1870s, but operations were discontinued for about 20 years. The White Rock Gold Mining Company was created in 1897, and took over control of more than 15 placer mining claims. Then the Bull Run Mine was discovered near Aura, followed in 1898 by the Lucky Girl. The town of Edgemont was established two years later on a steep slope and a 20-stamp mill was built. Its post office opened in 1901. The mines yielded more than $1 million in gold.

Edgemont shared its disasters as did many mining camps that were built in the mountainous terrain. In 1904 some mining equipment and the mill were badly damaged by an avalanche. It was followed by another avalanche two years later that completely destroyed the stamp mill. When ore was found near the town of Cornucopia, many of the miners left for greener pastures. The final blow came in 1917 when yet another avalanche all but destroyed the town.

EGAN CANYON (White Pine) First discoveries of ore occurred about 1865 when men from a California contingent of volunteer soldiers were riding through the region. The site, which began as

Early Edgemont (Northeastern Nevada Museum, Elko, Nevada).

an isolated Pony Express station, was named for Howard E. Egan, a mail carrier for Chorpenning's "jackass" mail service between Salt Lake City and Sacramento during the 1850s. Since the place was so isolated, Indians were a constant threat. At one time stationmaster Mike Holden and one of the riders were surrounded by Indians who demanded bread. The men baked it as quickly as possible and the Indians ate until it was finally gone. The men were tied to a wagon and the Indians placed brush around them ready to set a fire when the U.S. Calvary came to their rescue.

During the 1860s the first ore mill in the territory was built by two men named Donohoe and O'Connor. In 1866 the Gilligan silver mine was discovered, followed by the Jenny Lind, which produced a little over $100,000. The place relapsed until the 1870s when some of the lodes were reopened. By 1883 about $400,000 in ore had been recovered before the mines and mills finally closed.

ELDORADO (Clark) Spaniards may have been the first to discover gold in Eldorado Canyon during the 1770s, but more than 100 years would pass before any serious mining took place. In 1857 the ore was relocated and Eldorado was formed, bearing an appropriate Spanish word for "the gilded one." A 10-stamp mill was built and the treated ore was shipped down the Colorado river on paddlewheel boats and barges, which carried the cargo more than 800 miles up the coast to San Francisco. Although Eldorado was a booming mining camp, its gold deposits never received much publicity, probably because of the Comstock boom. Early production of the area was estimated at between $2 and $5 million in gold. One of the major producers was the Techatticup Mine, which recovered more than $1 million.

Eldorado was a tough town. Its ore discoveries may not have been given much attention, but its lawlessness sure was. The road to Eldorado was paved with rough terrain, outlaws, and fierce Indians. Since the closest sheriff was more than 300 miles away, miners were forced to create their own vigilante group, since the law wouldn't bother to travel there to protect the men. Eldorado later disappeared under Lake Mojave.

ELLENDALE (Nye) Ed and Ellen Clifford were the first to locate a gold claim about 1909 near Salisbury Wash, precipitating a small stampede. The camp was located about 30 miles from Tonopah. A telegraph line was installed between Ellendale and Tonopah, and there were rumors of a railroad to connect the two towns. Only one ore shipment was hauled out for processing and little development occurred, so the town was abandoned after 1916. The only later activity was in 1938-39 when dump material valued at about $7,000 was shipped from the Ellendale Mine, the only claim of importance in the district. Total production of gold to 1948 was $166,015, but other estimates put them between $500,000 and $1 million.[201]

ELLSWORTH (Nye) Ellsworth was founded as a milling camp about 1871. Although silver had been discovered in 1863, the region was slow to develop. A 10-stamp mill was built and the post office opened in 1866. Because of difficulties in ore reduction, most of the workers left for better opportunities.

EMPIRE CITY (Carson City) This site began with Dutch Nick's Tavern and a station on the Overland Road about 1855. The town itself was laid out in 1865, and had a short but successful life with its mills strung out along the Carson River with a population of about 500. Until the Virginia & Truckee Railroad was built, the silver ore was hauled to the mills by wagons. Empire City also processed much of the Comstock's ore. Some of the residents called the place a seaport town because of spring floods from the Carson River. When the Comstock mines began to diminish, so did Empire City.

ETNA (Humboldt) Etna lived its life as a milling town. It was established in 1865 with a post office the following year. A mill built here served some of the mines at Arabia and Dun Glen. Etna is a Greek phrase meaning "to burn." By 1866 Etna was on a downward spiral after one of its mills was moved to the town of Oreana.

EUREKA (Eureka) Bearing a Greek name for "I have found it," Eureka began after the first locations were made in 1864 by five men from Austin, but little was produced until 1869 when large ore bodies were found on Ruby Hill. Growth was slow because the miners did not know how to smelt the gold and silver. Two of the major producers were the Eureka Consolidated and Richmond mines. By 1888 the ore was nearly depleted and the smelters closed, but not before the mines had produced about $40 million in silver and $20 million in gold. In 1905 the Richmond and Eureka properties were consolidated. They yielded only low-grade ore and could no longer be profitably mined, and increasing costs forced the workings to close.

At one time Eureka was known as the "Pittsburgh of the West." Since there were so many smelters, the heavy smoke created black clouds that left soot and dirt behind. During this period of time more than 100 saloons were supported by the population. Gambling halls were established, luxurious hotels were built, and a newspaper began. In 1879 a fire destroyed most of the *Eureka Sentinel*'s building, except for a portion that had been fire-proofed. Although the room was still very hot after the fire, the printers were determined to get their paper out. They covered themselves with wet blankets so they could get the next edition of the paper out on time. By 1890 the heyday of mining had dwindled and prospectors went to other boom towns. From a population of more than 10,000, Eureka dropped to about 1,500. In a hundred years this area produced more than $100 million worth of ore.

Eureka is located on Highway 50, which is called "The Loneliest Road in America." Coming into the valley of the Great Basin one can see two very thin lanes in the far horizon. Scarcely touched by civilization, people can drive without seeing another car for nearly 400 miles. Eureka holds a number of mining contests. Information for times and dates, (775) 237-6006.

FAIRVIEW (Churchill) Ore was discovered about 1905 and a mining district was organized the following year. Fairview became a booming mining camp after investors George Wingfield and George Nixon purchased a number of claims. Located about 10 miles from Middlegate, Fairview was quite isolated and supplies were expensive, such as water that sold for more than $2.00 a barrel. When the ore was exhausted the population drifted off to better diggings. Some mining continued until about 1917, but only high-grade ore was shipped out. Gold recovered was valued at about $800,000, and silver yielded a little over $3 million.

FALCON (Elko) Silver was discovered in the region during 1876 and the Rock Creek District was created. Falcon was established a few miles from Tuscarora the following year with a post office in 1879. The Falcon, Eagle, and Manhattan mines were operated by the Falcon Mining Company. A stamp mill was under construction but never completed, so the ore was shipped to the mill at Tuscarora. A group of men from San Francisco purchased interests in the Falcon Mine for $25,000. The following year water began filling the mine and attempts to stop the water proved futile. The property was later sold at a sheriff's auction for less than $4,000. By the time Falcon had begun its decline, about $400,000 in silver had been recovered.

FARRELL (Pershing) This short-lived camp was founded about 1906 and bit the dust by 1908. The site was named for a William Farrell, or mine owner Jack Farrell. Its post office opened for a very short time in 1907. Shallow veins of ore caused the miners to move elsewhere.

FRISBIE (Lander) Frisbie came on the scene after silver was found during 1883, and its post office opened the same year. The camp was located in the Shoshone Range about 25 miles from Beowawe. Silver that was recovered from the nearby claims was shipped to stamp mills at Austin. Only about $3,000 was recovered, and by the early 1890s the camp was through.

GALENA (Lander) Silver ore was discovered at the entrance of Galena Canyon in 1863, but the camp was not founded until three years later about 10 miles from Battle Mountain. The opening of numerous mines caused two smelters and a stamp mill to be constructed. Galena had daily stagecoach service and a small railroad was built. British interests came in and took over some of the mining operations. The White and Shiloh Consolidated Silver Mining Company operated until the 1880s after recovering about $450,000 of ore. Before it was depleted, nearly $5 million was realized. The buildings were later disassembled and moved to the town of Battle Mountain.

Fairview, ca. 1907–1911 (Churchill County Museum & Archives, Fallon, Nevada).

GENEVA (Lander) Charles Breyfogle led a group of prospectors to the Toiyabe Mountains in 1863 where they found gold and silver deposits. Breyfogle promoted the region and founded the camp the same year, which was located about 10 miles from Austin and had a population of about 500. Geneva is Celtic for "mouth," or "estuary." Its post office opened in 1867 with G.B. Moore serving as postmaster. A stamp mill was erected but operated for only a short time because the pickings proved not to be as valuable as believed. It wasn't long before the miners left for better opportunities.

GENOA (Douglas) Genoa is the oldest settlement in the state. It was originally a Mormon camp until John Reese arrived in 1850 and set up a trading post to serve the gold rush traffic. The *Daily Territorial Enterprise* was established there in 1858, and had the honor of being the first newspaper in Nevada. There were some small gold and silver deposits nearby, but they were insufficient and not worth the expense of recovery.

GETCHELL (Humboldt) Prospectors came to the region during the 1880s in their quest for riches. But gold mining was insignificant until Emmett Chase and Ed Knight's discovery of the Getchell mine in the Osgood Range about 1934. The following year mining magnate Nobel Getchell purchased the claim and a small camp was established. Large-scale, open-cut mining was started in 1938. During World War II the property was allowed to continue operations in order to produce tungsten and arsenic, which were in short supply. A scarcity of labor and materials forced the property to close in 1945, then the mill was dismantled in 1967. Total gold production was about $17 million.

GLENCOE (White Pine) During 1867 Overland Stage workers were the first to discover silver, but it would be ten years before any major development was initiated. Glencoe was established near the Kern Mountains and its post office opened in 1891. The Mammoth Mine was located, followed by the Glencoe, discovered by Frank Bassett and John Tippett. Investors offered the two about $100,000 for their claim, which they refused. They should have taken the offer because the mine failed to produce much. By the 1890s most of the ore had been depleted.

GOLCONDA (Humboldt) Named for a city in India known for its diamond cutting, Golconda was established along Gold Run Creek in 1863, with a post office in 1869. Located about 15 miles from the town of Winnemucca, Golconda prospered as a shipping point for ore when the Central Pacific Railroad laid its tracks to the area. To supply water for the mines, a 60-mile canal was built to Winnemucca. But engineering errors caused the canal to leak and the project was abandoned. During 1866 placer gold was mined sporadically along the creek, and a number of silver mines were opened. By 1871 the town had folded. Total production from both placer and lode gold was a little over $400,000.

GOLD ACRES (Lander) A late bloomer, Gold Acres was not established until about 1936 when the Gold Acres Mine was discovered. Owned by the Consolidated Mining Company, the property produced about $200,000 up until 1940. The company erected a mill, followed by a cyanide plant. Total production from the mine was about $10 million. The Gold Acres open pit was the largest operation that was mined solely for gold in the state. Dollar amounts before 1902 are unreliable, but it was estimated that gold taken from other lodes was a little over $3 million.

GOLD BANKS (Pershing) Gold discovered by A.P. Smith in 1907 sparked a small rush. But the deposits were quickly depleted and by 1909 the place was gone.

GOLD BAR (Nye) Gold Bar was established during the early 1900s near the original Bullfrog strike, and named for a mine owned by Benny Hazeltine. The Homestake Mine was later discovered and recovered ore that was valued between $30 and $150 a ton. The camp never had more than about 50 people, and by 1908 the ore had run its course.

GOLD CANYON (Storey) This is the place where the fabulous Comstock Lode would be discovered. Placers were mined in Gold Canyon as early as 1852, but it was not until 1859 that the rich silver lodes were discovered. Ethan Allan and Hosea Grosch arrived at the canyon in 1853 looking for ore, but abandoned their search. They returned in 1856 and discovered silver, but both died before anyone found out about the riches that lay below. In 1858 Henry Comstock came to the canyon and found prospectors working their claims. Somehow he finagled and convinced the men that he was owner of the property and took over what turned out to be the great Comstock Lode, running about 7 miles long, 400 feet deep, and more than 200 feet wide, which produced over $300 million in silver. The Comstock bonanza influenced politics and national interest was centered on its production when money was needed to conduct the Civil War. Total production of gold was $165,000.

As the silver was beginning to run its course, the Yellow Jacket Mine was located about 1869 with high hopes for riches in its veins. The Comstock was composed of different lode bodies, and the Yellow Jacket was located in the south portion of the vein. The *Enterprise* wrote: "This strike is one of the most important ever made in the mine or upon the great Comstock Lode.... Two big deposits have been found — one in the 900, and one in the 800-foot level, but the former is by far the richest and in every respect the most important. The vein in the lower level is fully nine feet in width, and is ... so rich that it is not rudely shoveled and tumbled about, as is the case with that of a lower grade, but is put up in sacks and tenderly cared for. It is ore that will yield some thousands of dollars per ton, (we don't exactly know how many, as we have seen no assays,) and is just the next thing to finding silver already moulded into bricks.... Let us hope that the barren belt of the Comstock has at length been passed.... This strike is certainly most encouraging to all mining companies holding ground on the Comstock, and we hope soon to hear of others striking down into the new White Pine of our own big lode."[202]

In April of the same year a disastrous fire occurred in the Yellow Jacket. The heavily timbered mine and natural gases provided the impetus for fire. Miners who were in the shaft at the time the fire broke out ran to the cages to get out. Two days later the shafts were opened, and the oxygen created a conflagration, causing the deaths of 41 men. Three years later a shaft was opened and, to the men's surprise, some of the stones were still hot.

The Gould & Curry Mine, located by Alvah Gould and Abram V. Curry, also contributed to the Comstock, since it was the first mine to uncover rich ore estimated at nearly $6 million. John Mackay and James Fair, mining magnates, got a few people interested in a scheme to run an underground exploration on the Comstock to see if they could find the real bonanza. They ended up taking over control of the Gould & Curry in addition to other properties and made their fortunes.

At one time Mark Twain wrote in his inimitable way that "Mr. Curry owned two-thirds of it — and he said that he sold it out for twenty-five hundred dollars in cash, and an old plug horse that ate up his market value in hay and barley in seventeen days by the watch. And he said that Gould sold out for a pair of second-hand government blankets and a bottle of whiskey that killed nine men in three hours, and that the unoffending stranger that smelt the cork was disabled for life. Fours years afterward the mine thus disposed of was worth in the San Francisco market seven million six hundred thousand dollars in gold coin."[203] Actually, Gould went on to continue mining in California, and Curry became warden of the territorial prison.

GOLD CENTER (Nye) This site was a trading center for the nearby mines, as well as the main water supply that was brought in from the Amargosa River. The camp was established in 1904. When the Las Vegas & Tonopah Railroad arrived in 1906, Gold Center briefly became a railhead for the mining camp of Rhyolite. A stamp mill was built, but not enough ore was found to make the project worthwhile and the place folded.

GOLD CREEK (Elko) This camp, located about 20 miles from Mountain City, sprang up about 1873 when gold was discovered near Hope Gulch. Rumor had it that "every sage hen had gold nuggets in its craw from feeding on gravel in the stream."[204] Some nuggets found in Gold Creek were worth $50. Emanual Penrod, C. Rouselle and W. Newton discovered the placers in 1873, and worked the diggings with primitive methods for about 20 years before any real interest was generated there. New lode mines were later found, but their veins contained very little gold. In 1897 the Gold Creek Mining Company came in and patented more than 4,000 acres of placer ground, but because there was not enough water, its efforts were hindered. An ambitious plan to build a ditch from a reservoir to the placers at a cost of $600,000 was abandoned. By 1900 Gold Creek ceased to exist. Early placers were thought to have been valued at about $800,000.

GOLD HILL (Storey) Surface ore was discovered in 1859 by prospectors from Johntown and the camp was established. A miner named "Big French" John Bishop later wrote "Where Gold Hill now stands I had noticed indications of a ledge and had got a little color."[207] Some of the men ignored Bishop's discovery because the gold was so fine they didn't think it was worth developing. But they changed their minds when the recovered ore began to pay between $15 and $25 a day per man. Word got out and Gold Hill blossomed to a population of about 8,000. By the early 1940s, the town was gone.

GOLD HITT (Esmeralda) This camp was located in the White Mountain Range about five miles from Basalt. Although gold was discovered in 1862, Gold Hitt was not established until 1905. During 1870 an Indian found a rich strike at what would become the Indian Queen Mine. A 4-stamp mill was erected to treat the ore, which was then hauled to the railroad via ox teams. The region's mines produced $1 million in gold and silver, the majority from the Indian Queen.

GOLD MOUNTAIN (Esmeralda) Thomas Shaw was the first to discover gold about 1864, but for some reason he soon abandoned his claim. About two years later other prospectors relocated his mine, but not much was accomplished because of hostile Indians and too little water. Gold Mountain was established about 1881. Shortly thereafter the Stateline Mining Company came in and built a water pipeline from Mount Magruder and Tule Canyon, in addition to a 10-stamp mill. The Stateline Mine and Oriental were the best producers, showing assays at about $170 per ton in silver and $13 in gold. During 1889 a particularly bad winter at Gold Mountain caused severe hardships on the residents since the camp was blocked by snow for almost four months. Pack trains and wagons could not reach the camp and the men were running out of food. Some of the local Indians brought flour over the mountains on sleds and shared it with the miners. By 1890 the town was gone because the ore didn't hold up to its assay figures.

GOLD PARK (Lander) The district dates back to the 1870s during which time the camp was established. In 1864 a miner discovered gold nearby but not much interest was generated. During 1880 his claim was relocated, but it would be about nine years before serious mining got underway. A stamp mill was built in 1893 by the Nevada Mines Company, and a post office was established the same year. Total production was about $500,000.

GOLD POINT (Esmeralda) This site was first

called Lime Point because of the lime deposits that were worked until the 1880s when silver was discovered. Since there was no mill in the area, the ore had to be freighted to the town of Lida at an enormous cost. As a result, mining ceased until the early 1900s. The Western Mine began operations in 1905, followed by the discovery of hornsilver which brought in a rash of prospectors in 1908. The *Goldfield Tribune* wrote, "New discoveries are reported daily. The town is growing faster than any new camp that has been discovered in this section.... Hornsilver is the latest wonder in Nevada mining districts! Hornsilver is a comer! Hornsilver is practically but a month old, and yet it is beginning to astonish the visitors who are flocking to it in an ever-increasing number."[208] The name of the town was then changed to Hornsilver. The Western Mine yielded about $500,000. After gold was discovered, the name was changed again to Gold Point. A resident named Martin Mitchell delivered supplies to the mines using his truck during the 1930s. He also hauled bullion to the Wells Fargo Office, and may have transported more than $1 million in gold. By 1942 the mines had run dry.

GOLDEN ARROW (Nye) Gold discoveries were made by Claudet and Marl Page in 1905, and the camp was formed the following year. It was named for the Golden Arrow Mining Company which laid out the townsite. The Pages later sold their claim to the company for $45,000. Other mining companies came in and recovered some gold and silver which lasted only a short time. The camp had a population of about 50 or so miners who left when the ore ran out.

GOLDFIELD (Esmeralda) Goldfield was born on the heels of Tonopah. A Shoshone Indian named Tom Fisherman found a gold specimen and showed it to Harry Stimler and Billy Marsh, then led them to his find. Stimler and Marsh convinced Tom Kendall of Tonopah to grubstake them, and in 1902 they staked their claim. A camp was established at the base of Columbia Mountain and initially named Grandpa because Marsh thought this area would be the grandfather of all the mining camps. Its post office opened in 1903. At first, low-grade ore was worked, but by 1904 high-grade silver was found and Goldfield began a time of high production. Charles Taylor located the Jumbo Mine that produced more than $1 million. Then the Mohawk was discovered, yielding more than $500,000. News of those riches brought in more than 15,000 men who stormed the hills. Mining stock was sold quickly,

The hard-working burro (courtesy Francis Nestor).

but most of the certificates were worthless. One of the investors who had been cheated wrote to the *Mining and Scientific Press* in 1907 on how to wildcat: "Sir — Get a claim near some mine, give it a name that can be mistaken for the adjacent property, leave a man there well supplied with telegraphic funds with instructions to use them prodigally; experience of mining is not necessary, provided he is prolific in mining lore, big imagination, and careless of telling the truth; form a company, get a lot of certificates printed."[205] During 1906 Federal troops had to be called in because of high-grading and management-labor problems in the mines. The district produced about 4,194,800 ounces of gold at a value of approximately $86 million. In 1913 the town was flooded due to a cloudburst, and then in 1923 a fire consumed Goldfield.

Life was not without its humorous moments in the mining camps, and the newspapers were known to play up stories. At one point in time, the *Bonanza* out of Tonopah printed the following about a young miner: "A well-known young man of Goldfield bought a pair of gloves for his girl and at the same time his sister bought a pair of hose for herself. The clerk in wrapping got the items mixed and the explosion came when the sweetheart opened her package and found that it contained a long pair of silk stockings. She blushed. Then she opened a dainty little sweet-scented note and read the following tender lines: 'I am sending you a little present. Oh, how I wish that no other hands than mine would ever be permitted to touch them after you put them on. But alas! Scores of fellows may touch them when I am not at your side ... I bought the longest pair I could get, and if they are too long you may let them wrinkle down. A great many girls I know wear them slipped down a little. I want to help you put them on and see how they fit your dear self when I call Tuesday night. You can clean them easily, dear, with benzene if you leave them on till they dry."[206] The newspapers were renowned for their stories, some not true, so it's not really known if this actually happened or if the *Bonanza* was just looking for a good laugh.

The town celebrates Goldfield Days in August, celebrating the gold rush era. Activities include live music, street dancing, historical tours and video presentations about Goldfield and the railroads. Contact the Goldfield Chamber of Commerce at (775) 485-3560, for dates and times.

GOLDVILLE (Eureka) During 1907 William Barney and Joseph Lynn located the Lynn Big Six Mine. It was a good producer, yielding more than $20,000 its first year of operation, news of which brought the inevitable stampede. Since the Lynn was the only big lode, prospectors made their money recovering gold from placers until they were nearly exhausted the following year. In 1912 the Lynn Big Six Mining Company reopened the mine and shipped the ore to Salt Lake City for processing, which yielded about $150 a ton. At that time Goldville was founded with a post office established in 1913. Between 1909 and 1927 about $225,000 was recovered from the nearby properties.

GOOD HOPE (Elko) The discovery of ore in 1873 brought in the Amazon Mining Company to start small-scale production. After silver was found along Chino Creek about 1879, the settlement was established with a post office in 1884. There were a number of active mines; the Buckeye and Ohio, two of the most important, yielded $100,000. Because transportation costs to a mill at the town of Tuscarora were so high, a 5-stamp mill was built at Good Hope. The Aurora was a fairly good producer, and the owner later sold it for about $30,000. New owners had bad luck though, as they struck water shortly after acquiring the claim. The camp lasted until the early 1920s when the mines played out.

GOODSPRINGS (Clark) The first discoverers of gold in the region were Mormon missionaries in 1856, who opened the Potosi Mine (Spanish for "great wealth"). They were unsuccessful in smelting the ore, and so abandoned the claim. After prospectors found silver in 1868, the camp was established at the base of the Spring Mountain Range and named for prospector Joseph Good. In 1892 the Keystone Mine was located, and went on to produce about $125,000. Completion of a railroad from Los Angeles and Salt Lake City in the early 1900s brought easier and more economical transportation for the ore, and as a result more miners. The Potosi was reclaimed and miners recovered more than $1 million in gold. Total production was between $15 million and $31 million in gold and silver.

GRANTSVILLE (Nye) Located on a slope of the Shoshone Mountains, this site was established after silver was found about 1863 by P.A. Haven, who also laid out the town. Reports of ore at White Pine caused Grantsville to be deserted about 1868. Ten years later mining resumed, a 20-stamp mill was erected, and the post office opened in 1879. Supplies were costly as they had to be freighted in from the town of Austin. Then in 1907 Bob Hanchett salted a tunnel with gold dust to lure in more prospectors, but the place was once again deserted soon after the scam was uncovered.

GROOM (Lincoln) Named for the mine, this camp was occupied during the 1850s, although little mining occurred until about 1869 when a British firm began developing the Groom Mine. Since the site was so isolated and the silver ore was low grade, the Groom Lead Mines Limited abandoned the property, which produced only about $300,000.

GUADALAJARA (Lander) Located at the head of Santa Fe Canyon, the camp was settled after silver was discovered by Mexican prospectors in 1863. Rumors of ore so rich one could build a town of silver bricks brought in a small stampede. A stamp mill was erected, but it ended up costing more to mill the ore than it was worth. Within a year Guadalajara was abandoned.

GWEENAH (Lander) Gweenah was a small camp established about 1908 after James Watt found gold and silver. The Gweenah Mine was the only producer, discovered by the Lemaire brothers. Total yield was about $15,000.

HAMILTON (White Pine) There was not much interest in this camp at first, although some silver was found and a stamp mill was in operation during 1867. It was named for mining promoter W.H. Hamilton. Albert S. Evans, a writer for a newspaper, once observed of Hamilton: "Long lines of mules and oxen, drawing heavy wagons lade with supplies of every kind ... 'honest' miners with salted claims, ready to sell to the newly-arrived greenhorns.... The stores and saloons were crowded with men in huge overcoats, the pockets of which were filled with big specimens, small silver bars, rolls of location notices and assay certificates."[209] A new arrival to town noted, "No one is favorably impressed with the appearance of the town when alighting from the stage. There are no adequate sidewalks, and you land in a sea of mud, which becomes worse as you go to the center of the street, until at times it is almost impassable," followed by another observer: "Buildings are going up everywhere; auction sales are taking place in a dozen localities; men are rushing about in every direction, corner loungers are in crowds, and all is life, bustle and excitement. You hear nothing but claims, real estate and rents; these are the all-absorbing topics of conversation, varied only by the saying 'Come take a drink'—an invitation not often declined."[210]

Thousands of miners flocked to the site, but to their disappointment most of the ore was superficial. The silver quickly diminished, and was followed by the Panic of 1893 which led to Hamilton's demise. Pushing it closer to its end was a fire in 1873 that was set by cigar store owner Alexander Cohen so he could collect the insurance money. He not only set fire to his store, but he also cut off the main water supply. By the time his deed was discovered, most of the town had gone up in smoke. Cohen spent seven years in prison for his offense. The buildings from Shermantown were moved here to replace those burned down. Hamilton's final blow came when the county seat was moved to Ely.

HANNAPAH (Nye) Gold and silver were first discovered in 1902, but little mining took place until the Hannapah Mine was located. The camp was established about 1906 and gone within two years. An enterprising individual hosted a barbeque to entice outside investors to purchase land and claims, but it failed and Hannapah never had a chance. However, in 1915 other silver locations were made in the area and the Hannapah reopened, recovering about $150,000. Mining continued sporadically until the 1920s.

HARDEN CITY (Humboldt) Founded about 1849 by James A. Harden, this mining camp was located in the Black Rock Range, but had a short life. Hardin was part of a 14-wagon emigrant train on its way to California. He was sent out with a group of men to hunt game for the travelers. While on their search the men found rocks that looked as if they might contain silver, but since the train was close to California and there was a threat of Indian raids, they continued on. Hardin returned in 1858 to relocate the rocks, but never found them. It was four years later that the site was rediscovered by other prospectors. Assayers believed the rocks didn't have enough silver to pay, but Charles Isenbeck removed the ore using flux and succeeded for a while. This brought in the prospectors, and a stamp mill was erected. But the flux method proved too expensive and the claims were deserted, as was Hardin.

HAWTHORNE (Mineral) Mining began in the 1880s with discovery of the Pamlico and LaPanta mines. The town was settled about the same time the Carson & Colorado Railroad built its line through the region. Folklore says a freighter's mules determined the town's site when they stood in a group where the wind was less severe. The post office opened in 1881 and was named for William A. Hawthorne, cattleman and contractor of a wagon road. This same year the Virginia City *Evening Chronicle* wrote, "The new town of Hawthorne, on the line of the Carson and Colorado Railroad at Walker Lake, will in the near future, give employment to a great many workingmen of all classes."[211] Between the two mines, more than $800,000 in silver was recovered. Then in 1906 the Lucky Boy Mine was located, yielding about $76,000.

By 1920 Hawthorne was in a decline, but was later rejuvenated due to an ammunition depot explosion in New Jersey. Political figures decided to move the plant to a less populated place. Two years later Hawthorne became home to the U.S. Naval Ammunition Depot.

HIKO (Lincoln) Hiko began its life in 1866 as a mining camp after Indians told the white men about silver nearby. After purchasing some of the miners' claims, William Raymond laid out the site with high expectations, even having a 5-stamp mill hauled in by oxen from 140 miles away. Hiko became the county seat the following year, but lost it to the town of Pioche in 1871. Transport of mining machinery and a stamp mill proved costly, and there were not enough workers for the mines. As a result, Hiko lost most of its residents and the town declined. Hiko is Shoshone for "white man," or "white man's town."

HILLTOP (Lander) It was 1907 before any important discoveries were made. First known as Marble City, the site was a stop for emigrants and a watering place for horses. The camp's name was changed to Hilltop for the district. Gold found nearby in 1907 gave impetus for the erection of a 10-stamp mill. The Maysville Mine was located and yielded about $33,000. Another lode, known as the Philadelphia West, recovered about $135,000, and the Kattenhorn brothers discovered a mine that produced about $95,000 in silver.

HOT CREEK (Nye) Indians lived here before the discovery of silver, and in 1866 they showed white prospectors numerous deposits of ore. The following year a post office was warranted with a population of about 300. Hot Creek was named for the nearby hot springs. Two stamp mills were brought in via oxen. By 1881 the ore was depleted, but the region did produce about $1 million.

HUMBOLDT CITY (Pershing) Silver was found by two Frenchmen along the Humboldt River about 1860, which generated a small rush. The Humboldt District was organized the same year, and the camp grew quickly to a population of about 50, with establishment of a post office two years later. The principal mine was the Imlay from which a considerable, but unknown, amount of silver was shipped. The Star Peak Mine yielded $130,000 in gold and silver up to 1935.

HUNTER (White Pine) Louis Barbeau may have been the first person to find ore, although it was William Armstrong and Chris Delker who found high-grade silver in 1871. That same year Hunter was established and named for Thomas and William Hunter who owned a ranch in the area. Its post office opened in 1877. Located near the Egan Range, Hunter was a prosperous camp after other mines were located. More than $200,000 was recovered in one year, but the profit was lowered because the district had to pay a bullion tax in the amount of $96,000. A man named George Kennedy purchased a number of mines, then built a smelter that produced more than $500,000 in bullion. Eventually the ore ran out and so did the miners. Operations finally ceased about 1884.

IONE (Nye) About five miles from the town of Berlin, this camp was established in 1863 after silver discoveries in the Ione Valley. A prospector known as Haven named the site for a mining camp in California. Ione's future looked bright with the anticipated wealth and it became the county seat. Newspapers wrote of the camp's first shipment of ore: "The first lot of bullion from the Union district amounting to 200 pounds was shipped from the Pioneer Mills. We hope next week to give a larger figure."[212] Another wrote, "Ione Lively—From the number of teams seen on our streets during the week several times we have blinked our eyes and imagined ourselves on C Street in Virginia.... Everybody coming, to look at the mines. So many make plans to remain permanently. Ione sees a brilliant future ahead."[213] The area recovered about $1 million in gold and silver. But with the boom at White Pine, the majority of men left. Then Ione lost its county seat status to Belmont in 1867. The *Belmont Weekly* wrote that county officials would head off to Belmont "as soon as the citizens of Ione have reconciled themselves to the fact that they have lost the county seat fight."[214]

IVADA (Elko) Bearing a name coined from Idaho and Nevada, this camp was established during the early 1900s. Silver was discovered along Taylor Creek by John Blosser and Matt Graham in 1907. Because of the camp's isolation, mule teams had to bring in supplies and equipment. Ivada died a quick death after discovery of ore at Jarbidge.

JACKRABBIT (Lincoln) Named for the abundant rodent, Jackrabbit was established about 20 miles from Pioche with the discovery of silver along the Bristol Range in 1876 by Isaac N. Garrison. First called Royal City, the name was changed after Garrison threw a rock at a jackrabbit and discovered the Jackrabbit Silver Mine. The ore soon ran low, but the camp revived after a small narrow-gauge railroad called the Jackrabbit Road opened. A smelter was constructed near the mine and operations continued for about two years. It is not known how much silver was recovered from the property.

JACOBSVILLE (Lander) Jacobsville was established as a way station for changing out mules and horses on the Daily Overland Mail, then later a Pony Express station. The site was named for

Washington Jacobs. After discovery of ore in the area, Jacobsville served as a supply center for the miners in the Reese River Mining District.

During the fight between Jacobsville and Austin for county seat, a letter was written by Shoshone chief Tortio to the local newspaper: "How do you do way up in Austin there? ... If you will come down and make us a visit here, I will take you around and show you some of the prettiest ladies on the Pacific Coast.... You fellows can brag of your rich mines ... but they are nothing compared to our pretty women at Jacobsville."[215] But Austin was growing too fast, claimed the county seat, and Jacobsville faded away.

JARBIDGE (Elko) In 1890 John Ross found a float (pieces of gold washed out from a gold vein), but was driven out by a snowstorm. The strike was later relocated by Russ Ishman, and as a result Jarbidge was founded with a post office in 1910. The camp was located 2,000 feet down along the steep confines of Jarbidge Canyon. The name is from the Shoshone word, jahabich, which means "devil," and comes from an Indian legend about a man-eating giant called Tswhawbitts, who lived in a crater. He captured Indians, put them into a basket he had strapped on his back, took them home and ate them.

The canyon divides the east and west forks of the Jarbidge River. When word got out that the Bourne Mine had recovered nearly $25 million in gold, more than 1,500 miners headed there and scoured the hillsides. To their disappointment the reports were severely exaggerated. The camp became a chief producer of gold in Nevada, but not without its disadvantages to the miners. Because of the location, supplies and equipment were difficult to acquire. Nearly $4 million in gold and silver was recovered.

The last stagecoach robbery in the west occurred at Jarbidge. An outlaw held up the stage in 1916, killed the driver and took more than $3,000.

JEFFERSON (Nye) Located in the Toquima Mountains near the camp of Round Mountain, silver was found in 1866, but not much work was conducted for about five years. More than 800 people moved to Jefferson and its post office was established in 1874. Two stamp mills were built after the Jackpot, Jefferson, and Prussian mines were located. The Prussian produced about $400,000, and the Jefferson close to $14,000. An ore specimen weighing 40 pounds from the Jefferson Mine was exhibited at Philadelphia during the Centennial Exposition. Operations continued until about 1878 when the mills closed. During 1908 attempts were made by New York investors to reopen the mines, but the venture failed and the town was abandoned shortly thereafter. Combined production for both gold and silver came to about $1 million.

JESSUP (Churchill) Gold and silver were found in 1908 on the east slope of the Trinity Range, followed shortly by the formation of Jessup. Its post office opened the same year. Two stage lines ran between Jessup and White Plains, and the town

Jarbidge deep in the steep confines of the mountains, 1911 (Northeastern Nevada Museum, Elko, Nevada).

grew to about 300 people. The ore that was shipped out yielded about $100 a ton. But it was short-lived, and by 1909 the camp was all but deserted.

JOHNNIE (Nye) Gold discovered early in 1869 caused no great furor, and the land lay silent until 1890 when five prospectors searching for the lost Breyfogle Mine accidentally came across an outcrop full of ore. The Jonnie and Congress mines were located and a 16-stamp mill erected, but by 1893 the veins had pinched out. During 1898 investors purchased the Johnnie and Congress, which were worked until about 1915, and the camp was settled between the two mines. The first recorded production of placer gold was in 1918. Gold recovery from the two mines was estimated at $4 million. The rest of the region had a recorded production of about $380,000, although it was also estimated at $1 million.

JOHNTOWN (Lyon) This camp was named for an old term used for Chinese miners who were called "John Chinaman." James Fennimore, nicknamed "Old Virginny," arrived in 1851 and also began placer mining. During 1855 the Chinese who had been digging a canal by the Carson River turned to placer mining and Johntown was settled at the mouth of Gold Canyon. The men experienced only fair returns and Johntown was abandoned.

JUMBO (Humboldt) Located in the Awakening District, the region's gold was discovered about 1910, and the camp of Jumbo was established on a slope of the Slumbering Hills. There was a small amount of production of gold and in silver while the Alabama Mine was in operation, yielding a little over $5,000. Then the Jumbo Mine was located in 1935 and a period of large-scale activity began. The mine, owned by Texas tycoon H.L. Hunt, recovered nearly $200,000. Other properties yielded between $5,000 and $140,000 in gold and silver.

KAWICH (Nye) Kawich honors a Shoshone Indian chief whose name means "mountain." The camp was formed along the Kawich Range about 1904 with the discovery of gold. One visitor to the camp commented, "Kawich is a hell of a place! No mines, no water, no feed, no women."[216] Water was scarce and had to be brought more than ten miles from Cliff Spring. Some of the claims were run by the Gold Reed Company, but not much was realized and the camp was defunct by 1907.

KEYSTONE (Eureka) This tiny mining camp began in 1898 with a post office the same year, and grew to a population of about 50. During the 1860s a prospector discovered the Keystone Mine, but not much mining occurred until the 1890s when R.D. Clark found both silver and gold. Although the ore assayed high, the end results were disappointing. Then John Turner struck his pick into what looked like putty, but turned out to be pure chloride of silver, worth $15,000–$25,000 per ton.

LA PLATA (Churchill) This former camp bears a Spanish phrase for "the silver." The discovery of ore near the Stillwater Range in 1862 led to its settlement the following year. A number of claims were purchased by investors from the East, but they realized very little and left. In 1864 the Silver Wave Mining Company had a mill built, but also failed to recover much of anything. When rich ore was discovered at White Pine, La Plata was abandoned.

LANCASTER (Pershing) This small camp was located at a stream crossing, and bore a Celtic name for "Roman town on the River Lune." It was hoped that Lancaster would become a big mill center for the surrounding mines. There were also plans to build a canal to bring water for the stamp mills, but it never occurred. By 1863 Lancaster ceased to exist. Some mining was conducted years later, producing about $200,000 in ore.

LANDER (Lander) Located near Battle Mountain, Lander was established ten years after gold and silver were discovered in the 1870s. It was named for Frederick W. Lander, an engineer for the Central Overland wagon route. During the 1870s three mines, the Campbell, Eagle, and Young, were located and yielded nearly $21,000 within a year. Two smelters and a 5-stamp mill were constructed. The mill processed ore from the surrounding mines, with the Silverside recovering a little more than $4,000 in one month. Combined revenues for the Campbell and Young mines were about $16,000, and the Eagle produced about $5,000. Between 1905 and 1921, more than $600,000 was recovered.

LEADVILLE (White Pine) Leadville was a very short-lived camp. When silver was found in 1887 about 25 miners came to the site, and a mine called the Winecup was opened. But the ore was located in very small pockets, and within a few months Leadville died a quick death.

LEDLIE (Lander) Named for a director of the Nevada Central Railroad, James H. Ledlie, this site was a railroad stop and shipping point for ore. Prior to the railroad's establishment in 1880, numerous freight teams and their mules were stationed at Ledlie. Because of the high demand for shipping ore and freight, it was decided to build another railroad the following year, the Ledlie and Cloverdale. But after only part of the road had been graded, the project was dropped. There was a lull in 1890 with a drop in freight, but when ore was discovered in 1906 near Skookum, business picked up again. Ledlie lasted until 1983 when the Nevada Central tore up its tracks.

LEWIS (Lander) Jonathan Green and E.J. George made the first silver discoveries in Lewis Canyon about 1867, but it was nearly seven years before any earnest development. The camp of Lewis was formed about 15 miles from Battle Mountain. When lode mines were discovered, a 10-stamp mill was built. Shortly after the discovery of silver, gold deposits were found at the Pittsburg and Morning Star mines, which produced about $1.2 million. The Defiance Mine was located by Tom Hildreth, who sold his interest in 1875 for $40,000 to two men named Fraser and Sweeney. The Betty O'Neal Mine was a big producer of gold and silver, but a boiler explosion and a fire that followed forced the mine to close. Noble Getchell purchased the property, rebuilt the mill in the 1920s, and the mine ultimately recovered more than $2 million. When it was speculated that the Nevada Central Railroad would build a spur line to the region, two more stamp mills were erected. But financial problems forced final railroad construction to cease in 1882.

LEXINGTON CANYON (White Pine) Silver was located in the canyon during 1870 and a number of mines opened. A small camp was formed and given a name commemorating the famous battle. Initial assays showed the ore was valued between $50 and $16,000 per ton in silver. Because most of the deposits were surface, the ore was quickly exhausted. The place lay dormant until tungsten was discovered in 1917.

LIBERTY MINE (Nye) Located about 25 miles from Tonopah, this mine was discovered in 1867. A stamp mill was built a year later to treat the silver ore which produced a little more than $110,000 in a seven-year period. When the Tonopah Liberty Mining Company purchased the property in 1910, it had a 125-ton mill built. As a result, the Liberty yielded another $500,000.

LIDA (Esmeralda) Gold was first discovered by Indian and Mexican miners. During 1867 the white man entered the region. About five years later the camp was formed and named for Alida, wife of David Buel from Austin. Because of high freighting fees, supplies and equipment costs were exorbitant. But Lida's mines were so rich that they paid a profit even after the freighting was deducted. By 1880 the camp was on a decline as the ore began running out. While the town of Goldfield was coming into its own, Lida rode on its shirttails and lived another three years. Although unsubstantiated, the region's mines were thought to have yielded nearly $1 million in gold and silver.

LIME MOUNTAIN (Elko) About 1870 silver and gold were found by prospectors from the camp of Cornucopia. A year later the Scottish Chief Mine was located by John Ford. Unfortunately, there was only a little ore, and it was about four years before other mining operations got started. When Cornucopia began to decline, so did Lime Mountain. Then in 1829 the Elko-Lime Mountain Gold Mine Company was established which made arrangements to have two mills built, but it never happened and the company folded. The following year the working mines were shipping the ore to a smelter in Utah. The Lime Mountain Consolidated Mining Company was formed, and between 1937 and 1940 a little more than $260,000 was realized.

LITTLE VALLEY (Washoe) Not much is known of this region located near the Carson Range. Placer mining was conducted sometime before 1900. It was believed that the placer diggings yielded between $60,000 and $100,000 in gold.

LOGAN CITY (Lincoln) During 1865 prospectors heard rumors of silver riches and came searching for the ore when they came across an Indian who took them to a silver-bearing ledge. While working the diggings in the springtime, the miners were attacked by Indians, and they didn't return until fall when Logan City was established. Although some rich ore was found, it quickly played out and by 1869 the place was dead. Miners then moved over to Pioche when richer strikes were found there.

LONE MOUNTAIN (Elko) Named for the mountain, this camp was settled in the 1860s when prospectors discovered gold. Henry Miller arrived in the 1870s and erected a station where the miners could bring their ore to be shipped out. The properties were fairly prosperous until water started to seep into one of the larger mines. Then during 1877 the King Mine was located by Edward Reilly, George Bliss and S.G. Weston. Its ore was valued at about $5 per ton in gold. Other lodes located were the Donna Roma, valued at more than $1,000 a ton in gold, and the Rip Van Winkle, which recovered more than $40,000 in both gold and silver.

LONGSTREET (Nye) Formation of this camp occurred during the early 1900s when Jack Longstreet discovered gold and silver. Because of his bad reputation, the camp's development was hindered. The ten or so miners who lived there left within a year. During the 1920s the Golden Lion Mining Company purchased Longstreet's claim, brought in a 100-ton mill, and recovered only about $100,000.

MCCOY (Lander) Located about 20 miles from Battle Mountain, gold discoveries were made in 1914 by James H. McCoy and before long the camp was formed. Activity was limited until about 1928 when high-grade ore was found, and McCoy grew to a population of about 75. In 1938 the Nevada United Gold Mining Company relocated the Dome mine which was worked until 1940. Total production of the region's mines was about $95,000.

MANHATTAN (Nye) Located about 10 miles from Round Mountain, Manhattan was established during the early 1900s, although George Nicholl had made an initial discovery in 1866. He was soon followed by others who located their claims, but the ore was superficial and soon played out. A cowboy named John C. Humphrey found gold and silver in sufficient quantities from a ledge in 1905, and when word got out more than 4,000 people moved to the new camp of Manhattan, a Lenape expression for "place that is an island." The next few years were marked by fraudulent promotion schemes that gave the region notoriety and seriously delayed its development.

Mining continued to about 1912, yielding a little over $4 million in lode and placer gold. When the San Francisco earthquake occurred in 1906, many of the investors withdrew their money from mine stocks, causing the mines to cease operations. But then rich ore was discovered at the White Caps Mine, creating another boom. It produced a little more than $2 million between 1918 and 1940. The Manhattan Gold Dredging Company began operations about 1938 and gleaned nearly $4.5 million in gold from the nearby gulch. Manhattan clung to life until 1946 when the dredge had completed its work.

MAZUMA (Pershing) Primarily a gold-producing region, there was also some silver recovered. Gold was found in Seven Troughs Canyon about 1905, but not much development occurred until 1907. The canyon derives its name from a series of seven troughs that had been placed below natural springs in the canyon for the purpose of watering stock. The canyon came to the forefront with discovery of the Mazuma Hills Mine which yielded more than $80,000. As a result the small camp was formed at the mouth of the canyon. A number of mining companies were established as were stamp mills. Although the area did not produce a bonanza, it did yield gold and silver valued at about $2.6 million. In 1912 the little camp of Mazuma died when a flash flood wiped out it out following a summer cloudburst. More than ten people died in the waters and the site was washed away.

MELLAN (Nye) This camp was established after Jess and Hazel Mellan found gold in 1930. With the ore assaying between $17 and $35 a ton, it brought in a few people. The place was deserted when only $5,000 in gold was realized.

MIDAS (Elko) First called Gold Circle, this camp was founded with the discovery of gold by James McDuffy in 1907, which brought in more than 2,000 miners. When the post office was established the same year the name was changed to Midas, for the fabled king and his golden touch. Golden it was not because the ore was low grade. Then about 1916 a cyanide mill began operating, processing a little more than $450,000 in gold. The Elko Prince Mine was later discovered as were several placer claims. Total recovered was roughly $2.4 million in lode and placer gold, most from the Elko Prince.

The arrival of Halley's Comet was an anticipated event until someone pulled a prank that drew attention away from the grand affair. It seems that when one of the residents saw a ball of fire on top of a hill, they cried out that the comet was coming. But the ball was actually a bundle of underwear some joker had attached to a long pole and set on fire.[217]

MILFORD (Elko) In 1869 Jesse Cope and his miner friends located a claim they called the Blue Jacket Mine. Before long a camp was formed as

Mill and Mine, Midas, ca. 1924 (Northeastern Nevada Museum, Elko, Nevada).

Blythe City for landowner Thomas Blythe. A stamp mill was also built at the site. The name was changed to Milford for mill operator William Milford. Since the region was low on water, two ditches were built but they didn't work. During 1885 James White purchased the Blue Jacket and stamp mill at a cost of $85,000. The ore was shipped out at a value of about $1,000 weekly. Milford was abandoned by 1917.

MILL CANYON (Eureka) Silver was discovered nearby about 1864 and Simeon Wenban, who purchased a number of mines there about 1867, built an 8-stamp mill to treat the ore, resulting in the establishment of the camp. Hauling the ore to the mill was difficult at best because pack mules had to haul it over a very narrow trail. The local mines produced more than $200,000.

MILLERS (Esmeralda) This camp was established as a milling town, named for Charles R. Miller, a director of the Tonopah & Goldfield Railroad. The site had the largest reduction plant built in the region by the Tonopah Mining Company. It also had a 100-stamp mill erected to serve the nearby mines.

MILLET (Nye) Millet was a former stage station, and later went on to become a freighting and milling center for the Toiyabe and Toquima mines. The camp was named for store owner Mrs. Albion Millet, and was settled after gold was found near the Manhattan and the Round Mountain region. Its post office was established in 1906. Millet lasted until about 1908.

MINERAL CITY (White Pine) While Thomas Robinson and a group of miners were prospecting nearby during 1867 they struck silver. The following year more than 1,000 claims had been recorded, and by 1873 Mineral City had a population of 600. The most productive years began about 1907 when more ore bodies were developed and stamp mills erected. Since no records were kept before 1902, there is only an estimate of about $40 million in recovered gold. The Chainman Mine was purchased by Charles D. Lane, who spent more than $150,000 to reopen its stamp mill. It ultimately produced about $325,000.

MINERAL HILL (Eureka) While prospectors John Spencer and Amos Plummer were searching for a stray horse in 1869 they found a rich silver float near the Piñon Range. As a result, Mineral Hill was established and a 10-stamp mill was built. A year later the men sold their property to George

Mine and workings at Mountain City (Northeastern Nevada Museum, Elko, Nevada).

Roberts and William Lent for $400,000, and the boom was on. A British mining company came in and purchased the mines and stamp mill for more than $1 million. The company turned around and resold it for $2.5 million. By 1872 most of the ore had been worked out. Total production came to about $7 million in silver.

MONTE CRISTO (Esmeralda) Settled in 1865, Monte Cristo was a supply and milling center for the nearby claims. A few years later a sawmill was built to provide timber for shoring the mines, and a 5-stamp mill was erected to crush the ore. The high-grade silver was then freighted to Austin where it was assayed up to $100 a ton. With the rush at White Pine and discovery of the Hidden Treasure Mine, Monte Cristo began its decline in 1868. The mill was reopened two years later and continued operations until 1872.

MOREY (Nye) Although gold and silver were discovered by prospectors from Austin in 1865, the camp wasn't established until 1867, the same year the post office opened. The Morey Mining Company that moved in during 1866 had a 10-stamp mill built two years later. To the company's dismay, the mill ran for only about a month before it was shut down. It was reopened in 1880 and processed about $80,000. By 1905 the place was just about deserted. Total production was approximately $460,000.

MOUNT HOPE (Eureka) This camp was established in 1872, a year after ore was found by a prospector named McGeary. The only mine of any substance was the Mexican, which had ore valued at about $100 a ton. George Bibbins was superintendent of a mining company and took an interest in the claims. But by 1878 all hope for Mount Hope was gone.

MOUNTAIN CITY (Elko) Placers were worked as early as 1870 by Chinese miners who reported they were recovering $2–$3 per day from the north side of the Owyhee River. Jesse Cope and others came to the region to work the diggings and a camp named Cope sprang up. During 1870 there were more than 1,000 miners living there. In order to provide water for the diggings, a ditch was built at a cost of about $10,000 to divert water from the nearby Owyhee River, and a 10-stamp mill was brought in. Gold placers were followed by the discovery of silver. Although transporting the ore was costly, the silver was so rich it more than made up for the deficit, with more than $1 million recovered before 1881. After the ore was exhausted Mountain City got a new breath of life with discovery of the immense Rio Tinto Copper deposits.

MUD SPRINGS (Lander) Stephen Bedell found gold about 1885 and located the Rattler Mine, which yielded about $500 a ton of gold, but it lasted only a few months. Gus Fowler from the mining camp of Beowawe discovered gold placers in 1907 which brought in a small flock of miners.

There were never more than 30 people living at Mud Springs. Since the majority of supplies and equipment were hauled in from Lander, there was only a saloon and one business in operation during this time. About 1907 the Grey Eagle Mine was located, and yielded about $25,000. By 1909 the ore had run out and the miners left for better diggings. Total production was only about $45,000.

NATIONAL (Humboldt) Although the region had been prospected during the 1860s, no development occurred until 1907. Rich deposits were discovered by two prospectors in the Santa Rosa Range. Contradictory to the typical prospectors with wagons or mules, these men arrived in a National car, for which the site was named. The ore was extremely rich, with some of it valued at $30 a pound. The National Mine alone recovered more than $8 million. It was later discovered that some of the workers at the mine had high-graded close to $1 million. Total production of gold from other properties came to a little over $3 million.

NELSON (Clark) Located in Eldorado Canyon, the camp was named for prospector Charles Nelson, who was murdered by Indians in 1897. The place served as a trading center for local mines and had a post office established in 1905. Soldiers from Fort Mohave discovered gold and silver in 1859, and by 1907 more than $4 million had been recovered. During 1909 a dredge was put into operation, but it was a total failure.

NEW PASS (Lander) The New Pass District was created in 1865 after the discovery of gold by an Indian who told prospectors of his riches. In 1868 a 5-stamp mill was brought in from the town of Austin, which treated more than 10,000 tons of ore. The camp received its name because one of the settlers thought he had discovered a new pass through the mountains. Some of the mines that were good producers included the Gold Belt, with its ore valued at more than $1,000 a ton, the Superior at $100 a ton and the Oriental that yielded about $62 per ton. By 1889 New Pass wasn't new any more; in fact, it was gone. That is until copper was discovered, giving it a temporary reprieve.

NEWARK (White Pine) Stephen and John Beard discovered silver nearby, and in 1866 the camp was established near the Diamond Mountains. The brothers didn't stay long and left after hearing news of richer strikes farther north. A Methodist Church invested money in mining ventures under the condition that some of the revenue be used to build a church at the town of Austin. With a population of about 200, Newark lived a successful life until about 1874 when the mines closed. One of the most productive was the Baystate Mine, which produced about $135,000. Newark was old hat by 1908.

NORTHUMBERLAND (Nye) Located near the Toquima Range, this camp was established in 1866 after silver was discovered on the mountain slope. The Blue Bell and Monitor mines were located and the ore had to be shipped to Austin until a 10-stamp mill was built at Northumberland. By 1891 all mining activity had ceased. There was a revival in 1936 after large deposits of low-grade gold were found. More than $1 million was realized from silver production, and the gold deposits amounted to about $730,000.

OLINGHOUSE (Washoe) Olinghouse was a small settlement formed in the Pah Rah Mountains about 1860 when the region was first prospected, but little work was done until about 1901. The site was named for rancher Elias Olinghouse. Placer deposits were found in the nearby ravines and produced considerable gold. The claims were worked until about 1902 when the decline began. It was estimated that about $500,000 was recovered. In 1906 another attempt was made to locate ore and a 50-stamp mill was erected. A few years later the Nevada Railroad built a spur line to the mines and mill, but there was not enough gold to sustain the mills so the camp and mines became history.

OMCO (Mineral) This camp's name is an acronym for the Olympic Mines Company from San Francisco. It was located near the Cedar Range and established during 1915 with a post office in 1917. A 70-ton cyanide mill was constructed, and after a fire in 1919 destroyed it the mill was rebuilt. Not much ore was recovered, and after producing $700 everything was shut down. A slight resurgence occurred in 1929, but after an earthquake caused the mines to cave in, it was all over.

OPHIR CANYON (Nye) After a Frenchman named Boulrand discovered a silver cache in 1863, he tried to keep it a secret. He would work his claim, then steal into Austin at night and get his supplies. Boulrand told a friend in confidence about his discovery, but of course the chap opened his big mouth and the rush was on. In 1863 a road was built into the canyon. Because of the site's location, nine wooden bridges had to be built to crisscross a creek that ran through the canyon. The following year R.B. Canfield bought the property he called the Murphy, and had a 20-stamp mill built. The Murphy Mine was predicted to be one of the best paying mines in the eastern part of the state. And good paying it was, producing nearly $3 million in silver, although some thought it only recovered about $54,000.

OREANA (Pershing) Oreana was a center of one of the first lead mining districts in the country, and was given the Latin name for "town of gold." The camp served as a milling and smelting site for the Arabia and Trinity Mining Districts. Its post office opened in 1867. The discovery of the Montezuma Mine brought in the gold seekers, and a 5-stamp mill was erected. Later, a larger mill was built near the railroad. Mine superintendent A.W. Nason then proceeded to build a huge smelter. Oreana blossomed, and in addition to the usual businesses boasted its own jockey club. Club owners were happy to take the miners' money from their regularly run horse races. The mills were processing a good $45,000 per month until 1869, when financial problems began to plague the mill owners. Oreana burned down in the 1870s.

ORIZABA (Nye) Gold and silver were discovered nearby in 1911, followed by the opening of the Orizaba Mine, from which the camp took its name. By 1918 much of the ore had been mined out. The area mines produced about $130,000.

OSCEOLA (White Pine) Gold lodes were discovered in 1872 by Joseph Watson and Frank Hicks. Placer deposits were also located in 1877 by John Versan. The site was located at Dry Gulch near the rugged Snake Range. Processing of the placers was first worked using primitive arrastres until a 5-stamp mill was built in 1878, the same year the post office opened. Miners had problems because of lack of water for their diggings until a water ditch was built by Ben Hampton. One of the largest nuggets in the state was discovered by a man named Darling, valued at about $6,000. This is the first place in Nevada where hydraulic operations began. The number of people in the camp demanded many supplies, so the White Pine Stage Company was established to serve them. Osceola experienced a bad fire in 1890 that just about totaled the place. By 1907 the placers had produced gold worth about $1.8 million, and lodes worth nearly $200,000.

PACKARD (Pershing) Two miles from the town of Rochester, this camp was settled by Henry Lund about 1912 after R. Ray discovered gold and the Packard Mine began operating. Located in a wash near the Humboldt Range, the camp prospered for a time after the Nevada Packard Mines Company built a 100-ton cyanide plant in 1915. Since there was little water in the immediate area, it had to be piped three miles from a spring. By 1923 the site ceased to exist, but not before the Packard recovered $5 million in gold.

PALISADE (Eureka) Palisade began its existence in 1868 as the northern terminal and headquarters of the Eureka-Palisade Railroad, which hauled silver bullion from the mines. Palisade was also a jumping-off place for the miners and a stop on the Central Pacific Railroad. It later served as a shipping point for supplies to the mining camps. In 1871 a fire burned down part of the town but it was rebuilt. The camp had shipped out more than 31 million pounds of bullion by 1878. But when the ore began to run out, the Eureka-Palisade Railroad quit making regular runs. Then in 1910 a huge flood almost destroyed the town and damaged the railroad tracks. As a result, freighters had to deliver and haul out provisions.

PALMETTO (Esmeralda) This camp bears a transfer name from South Carolina for the palm trees, so named by prospectors who hailed from there. Palmetto didn't last very long. Silver deposits were discovered nearby in 1866. A 12-stamp mill was built to accommodate the expected rich ore, but not enough was found to make a mill worthwhile, and Palmetto was no more. The discovery of slate brought some life back to the camp.

PARK CANYON (Nye) This silver camp was established about 1865 with the discovery of the Buckeye Mine. The ore was shipped to Austin for treatment until Park Canyon acquired its own stamp mill in 1867, but it closed down two years later. Owners of the Buckeye realized only about $12,000.

PHONOLITE (Nye) The Phonolite District was formed in 1906 after high values from the Paymaster Mine were reported. The camp was located near the Paradise Range and had a post office established in 1907. Other properties in operation were the Black Mule and Penelas, which was the main producer. The Penelas was located in 1913 by Severino Peneles, and was later purchased by the Penelas Mining Company that had a 50-ton cyanide mill

Freighters at Palisade after railroad tracks were washed out (Eureka Sentinel Museum, Eureka, Nevada).

built. It rewarded its owners with nearly $900,000 in silver by 1942. The Black Mule's ores initially assayed out at about $2,500 a ton in gold and silver. The Phonolite Townsite & Water Company laid out the town, which was incorporated at a cost of $1 million. By 1908 the camp was in a decline. Production of gold in the district was about $355,000.

PIERMONT (White Pine) During 1869 silver was discovered and developed the following year. The Piermont Mine, owned by Messrs. Godfrey, Laughlin and Wilson, was purchased by the Piermont Mining and Milling Company for about $60,000, which also brought in a 10-stamp mill. Total production for the region came to about $2.5 million.

PINE GROVE (Lyon) Gold was discovered in some nearby outcrops about 1866 by William Wilson after an Indian showed him the location. Soon afterward the Wilson and Wheeler mines were located. First called Wilsonville, the camp's name was changed to reflect its location near the abundant pine trees. The district produced steadily to 1893 with the Wilson Mine recovering about $5 million, and the Wheeler $3 million.

PINTO (Eureka) Moses Wilson found silver near the eastern foothills of the Diamond Range in 1865, but it would be four years later that any serious production was attempted. Pinto was settled about 1869 and lived its life as a milling and smelter camp. It was named for the pinto bean, Latin meaning "to paint." In 1871 a 20-stamp mill was erected which processed ore from the White Pine County districts. Its post office opened in 1870 with James Frank as postmaster. Although a mill town, Pinto did have a few claims working, one of which was the Champion that produced about $65,000 in 1870. Other nearby deposits proved to be shallow. After larger smelters were built at the town of Eureka, Pinto began its slow decline. There was some continued small-scale production backed by the Pinto Silver Mining Company, but during the 1880s the camp died. Recovery estimates came to about $226,500.

PIOCHE (Lincoln) Pioche was once the center of Nevada's most important mining districts located in a narrow canyon on the eastern side of the Highland Range. During 1863 miner William Hamblin was led to a silver ledge by a Paiute Indian. His discovery precipitated a boom, although little development occurred until a banker from San Francisco, F.L. Pioche, took an interest in the property.

During 1864 the camp was formed but because of its remoteness it became notorious for its lawlessness and was the wildest mining camp in the west. In a seven-year period more than 400 homicides occurred. One visitor wrote, "There is a fight every day and a man killed about every week. About half the town is whisky shops and houses of ill fame."[218] One of the bartenders learned his lesson too late. Mr. Faddiman had been warned often enough not to take a job at Pioche, that he was as good as dead. But he ignored their warnings, stating that he needed the work. After some drunk miner asked for another drink, Faddiman said, "You don't need another drink," and those were his last words.[219]

In 1871, during a celebration by Mexican miners to celebrate Mexico's independence, more than 300 kegs of powder exploded in a general store, causing a fire that destroyed many of the buildings. Then in 1873 another fire occurred followed by massive flooding. By 1876 the mining companies had pulled out. Talk of the Union Pacific Railroad extending its line to town was greatly anticipated, but with the demonetization of silver the tracks were never constructed. By the 1890s most of the ore had been recovered, roughly $40 million.

PIONEER (Nye) Named for the pioneering families who settled in the region, the site was founded about 1907 with the discovery of the Pioneer Mine. Its post office opened in 1909. During that time the Tonopah & Tidewater Railroad was planning to build its line to the camp, and ran advertisements touting the town and for all to come to Pioneer. But a fire destroyed the camp.

PLACERITOS (Churchill) Located on a slope of the Kamma Mountains, this camp was formed after gold was discovered in the 1850s. It was given the Spanish name for "little placers." Mahogany Jack and four other prospectors came to the region in the 1870s and found placers valued at about $30,000. Some mining continued up until the late 1930s.

PROSPECT (Eureka) In 1885 this camp was founded to serve the miners who worked at the nearby mines. Located near Prospect Mountain, it was named for the many prospect holes dug by miners. It had a post office in 1893 and a population of less than 50. A small smelter was constructed in the early 1900s, but not much ore was recovered and the town's prospects vanished.

QUEEN CITY (Humboldt) Queen City was located along Martin Creek and founded about 1874. It served as a milling center for the local mines. During 1879 no more than 100 people were living there. Because of its location in an isolated canyon, Queen City was unable to compete with other camps. The mills processed a little more than $200,000 of ore.

RABBITHOLE SPRINGS (Pershing) This site was founded on a slope of the Kamma Mountains by the Applegates in 1846. Before gold was discovered the springs were used by emigrants, their last water before crossing the Black Rock Desert. In 1916 about $3,000 worth of gold was recovered by two prospectors. Only small-scale placer operations were conducted that lasted only a short time.

RAVENSWOOD (Lander) Ravenswood was founded along the Shoshone Mountains after miners from Austin found silver in 1863. The site was so small it didn't even warrant a post office, and residents had to get their mail at Austin. Ravenswood was touted as being an important location in the region. Since there was plenty of water from a nearby spring, the U.S. Gold and Silver Mining Company purchased the claims. Unfortunately, the grandeur of the advertisements belied the poor ore in the area. Ravenswood was abandoned by 1870 after a paltry $10,000 was recovered. There was a small resurgence in 1906 when a Tonopah man named Taker Oddie purchased the mines, but he had no luck, and by 1908 the place was deserted.

RAWHIDE (Mineral) Rawhide was a spectacular camp that arose after the Goldfield boom. Located about 35 miles from Hawthorne, Rawhide began its rise in 1907 when placer deposits were found. There are a number of versions of the camp's name: A prospector may have used rawhide strips to mend his clothes, the mailbox was large piece of rawhide, or rawhide baskets were used to carry out the ore. But prospector Charles A. McLeod begged to differ: "I went into what is now Rawhide on Feb. 12, 1907, with Albert J. Bovard and prospected around that section ... Charley Holman was very bitter about his recent experience at Buckskin, a new boom camp not far away. They had told him they didn't want any more prospectors making claims around here. He was fuming about all this all through supper and finally he exploded. 'By God!' he said, 'They think they're so darn smart getting that name of Buckskin on the Map! We'll go them one better and get a Rawhide on it!'"[220]

With the discovery of gold, Rawhide acquired a population of more than 5,000 that may have risen to 10,000. The camp boasted more than 100 businesses and homes, one church, and about 90 saloons, which brought in the gamblers to lighten the prospectors' pockets, one of whom was Riley Grannan. After he died the Reverend W.H. Knickerbocker held services for the gent, and in his grand eloquence, spoke: "God flings the auroral beauties 'round the cold shoulders of the north; hangs the quivering picture of the mirage above the palpitating heart of the desert; scatters the sunbeams like lamellated gold upon the bosoms of myriad lakes that gem the verdant robe of nature; spangles the canopy of night with star-jewels and silvers the world with the reflected beams from Cynthia's mellow face; hangs the gorgeous crimson curtain of the occident across the sleeping room of the sun; wakes the Coy Maid of Dawn to step timidly from her boudoir to climb the steps of the orient and fling wide-open the gates of the morning." [221] His dialogue brought tears to even the most hard-hearted miner, even though his sermon had nothing to do with the man in the ground.

Most of the productive placers were located on Hooligan Hill, with more than $250,000 recovered. From 1908 through 1935 a little over $1 million in placer and lode gold and was produced. One windy day in 1908 a curtain blew over a gas stove and within two hours Rawhide was gone. Some thought the camp could be rebuilt and rise from the ashes, but it never did.

RAY (Nye) This site was named for a judge named L. Ray, who discovered gold in 1901 and had previously prospected with Jim Butler of Tonopah fame. Miners came to the region when they heard the ore was paying about $240 a ton. Working the claims and stockpiling their riches, the men were in anticipation of a new wagon road to be built so they could have the ore hauled out. It took so long that only a few wagonloads were shipped out two years later, and Ray just shriveled away.

RHYOLITE (Nye) While out prospecting one day in 1904, Frank "Shorty" Harris and Eddie Cross stumbled upon a quartz gold outcropping which became their Bullfrog Mine at the nearby camp of Bullfrog. Of course, the news spread like wildfire and by 1905 Rhyolite had been established. The ore assays were good and reports were glowing. Investors and mining speculators came to the camp, followed by more exuberant miners. Residents believed Rhyolite would become the center of all southern Nevada, and the population soared to nearly 10,000. Property values also climbed. Three water systems and three railroads were built to Rhyolite. During the camp's early years water had to be hauled in from the town of Beatty about five miles away. A report was written that said, "The monstrous mineral ledges of the Bullfrog District are quite without parallel in this country. The discovery of veins measuring from 20 to 30 feet in width are much too common to excite any unusual interest or attention."[222]

In 1906 the first freight and passenger train arrived, and the *Herald* wrote, "The first week of railroading into Rhyolite has been rather a strenuous one for the Las Vegas and Tonopah people. By the end of the week 100 carloads of freight will have been landed here. This freight has been accumulating on the sidings from Las Vegas to Gold Center, waiting to be moved to the metropolis of the Bullfrog District."[223] Other mines were located and everyone believed they would find magnificent riches. But most of the veins were superficial and by the end of 1908 Rhyolite turned to dust. About $3 million in gold and silver was recovered.

ROBERTS CREEK STATION (Eureka) This site was a station on the Pony Express and Overland Stage lines, named for Pony Express superintendent Boliver Roberts. There was some small scale mining, and in 1870 the O'Dair Mine was discovered. There were only small values of gold and silver, and by 1877 the place was forgotten.

ROCHESTER (Pershing) During the 1860s prospectors from Rochester, New York, found gold in Rochester Canyon. But it would be 1912 when the silver bonanzas were located by Joseph Nenzel that the big rush began. More than 1,500 men comprised the camp of Rochester. Two years later the Nevada Short Line Railroad was built, the Rochester Mining Company was established, and a 100-ton mill was constructed. The company closed down in 1929. The placers yielded $11 million in gold.

ROCKLAND (Lyon) Rockland was settled about 1868 after the Rockland Mine was located, and a 10-stamp mill was built by a Mr. Keene. He experienced financial problems after the output began to outweigh his intake. One of the miners, a Mr. Rhodes, did not get paid and burned down the mill, causing activities to be suspended. Former governor Henry Blasdel took over the operations, but had no better luck. When a chunk of rock that was later found assayed at $500 to a ton of ore, mining resumed. Not much was recovered and by the early 1930s Rockland became a pile of dust.

ROSEBUD (Pershing) During the 1870s placer deposits were worked by the Chinese, but they realized only about $2,000. In 1906 silver was discovered and Rosebud was founded. Its post office opened the same year and was named for the nearby

Present open-pit mine at Round Mountain (courtesy Nevada Bureau of Mines and Geology, Jon Price, photographer).

rose-colored hills near the Kamma Mountains. Miners flocked to the region, but the ore proved to be nearly worthless, and like the last rose of summer, Rosebud had wilted within a year.

ROUND MOUNTAIN (Nye) In 1906 gold was discovered by Slim Morgan, although some believe it was Louis D. Gordon who made the first location. The claim was located on the south slope of Round Mountain and a camp was formed at its base. First called Gordon, the site was renamed for the mountain because of its shape. That same year placer gold deposits were discovered below town by Thomas Wilson, who went on to become known for inventing a dry-washing apparatus for placers. The area was very productive and recovered more than $3 million in less than ten years. Hydraulics came into play when the Round Mountain Hydraulic Mining Company arrived and built a pipeline more than five miles long to bring in water for its operations. Total gold production of the district was about $8 million. Today there is still open-pit mining being conducted by the Round Mountain Gold Corporation.

RUBY HILL (Eureka) Owen Farrel, Alonzo Monroe and M.G. Clough were the first white men to discover gold and silver on Ruby Hill in 1869 after an Indian led them to the ore. Their mines, the Buckeye and Champion, yielded more than $50,000 in one year. Further discoveries brought in the Eureka Consolidated and the Richmond Consolidated mining companies. A narrow-gauge railroad was also built to the mines, which hauled ore to the smelters at Eureka. Ownership and litigations ensued between the two companies with Richmond having to pay Eureka Consolidated more than $2 million. By 1878 Ruby Hill's days were numbered. Total production came to more than $19 million from the Eureka Consolidated mines, and $4 million from the Richmond-Eureka property.

RYE PATCH (Pershing) Located on the west flank of the Humboldt Range, Rye Patch Canyon was the scene of silver discoveries in 1863. The following year the Rye Patch Mine was located and a camp was established shortly thereafter, named for the wild rye grass that grew there. The site later became a station on the Southern Pacific Railroad.

During the 1860s a British interest purchased the mines and erected a 10-stamp mill. Previous to this, the ore had to be hauled to the town of Auburn. The Rye Patch produced more than $1 million through 1881. Although predominantly a silver mining area, some gold was recovered, which amounted to about $195,000.

SAFFORD (Eureka) Ben C. Safford was the first to discover silver in 1881, and named his claim the Onondaga Silver Mine. News of his find brought in more miners and the camp was formed in 1881, with a post office the following year. Gabriel Zenoli and Francisco Thomas later located the Zenoli Mine. Safford sold his property to the Onondaga Gold and Silver Mining Company, but to the stockholders' distress the ore quickly ran out. The camp lasted only a short time. Between 1882 and 1886 the Onondaga produced a little more than $30,000, and the Zenolie about $67,000. The best producer was the West Mine, which recovered more than $1 million.

SAN ANTONIO (Nye) Gold was discovered near the San Antonio Mountains in 1863 and two years later the camp was established as a stage station, which also had a 10-stamp mill built. Its post office opened in 1873. The place had a minor boom when the silver ore was assayed out at $600 a ton and there was an ample water supply to run the mills. But the mines failed to produce high-grade ore, so the mill was dismantled and hauled to the booming town of Belmont.

SAND SPRINGS (Churchill) This site was an early stopping place for travelers. Although there was prospecting for gold during 1905, major production didn't commence until about 1919. Only about $30,000 in gold was found, with the majority of it from the Dan Tucker Mine.

Sand Springs was the place where the tall tale of the Chicken Craw gold rush began in 1907. Larry Hunt, who owned an inn there, was dressing some chickens when he found gold nuggets in the craws of two birds. He proceeded to kill all the chickens, and found a few more nuggets of gold. Alas, he never found the wealth of gold since the chickens he purchased came from a farmer more than 70 miles away, and they in turn had been sold four times over.[224]

SANTA CLARA (Pershing) This little camp arose about 1862 when silver was found at Santa Clara Canyon in the Humboldt Mountains. It was a busy town, housing more than 200 people. Hopes were high as the *Register* wrote in 1864: "This is the high old place to live in; richest town in the country, for items; too rich in fact; can't pile em up rightly."[225] This was an obvious come-on to get people to move there. The same year a notice of sale of stocks for delinquent payments of assessments by the Union Tunneling Company spelled the death sentence for Santa Clara, and by 1868 it had passed into oblivion.

SCHELLBOURNE (White Pine) During 1871 James McMahon was the first to find silver. In prior years, Fort Schellbourne had been built to protect a mail station from Indian attacks after a stationmaster had been killed. It was named for Army major A.J. Schell. Along with his partners, McMahon established the McMahon Gold and Silver Mining Company after discovering the Woodburn Mine, which assayed at about $800 a ton in silver. The camp was founded by R.D. Ferguson and a 5-stamp mill was erected. But development of the ore showed poor results, and with the rich strikes at Cherry Creek the miners left. Many of Schellbourne's buildings were hauled to the new site at Cherry Creek.

SCOSSA (Pershing) The discovery of jewelry rock in 1907 brought about the establishment of the camp. Located about 40 miles from Lovelock, the place lasted only a few months. It wasn't until 1930 that the camp came back to life. Charles and James Scossa discovered gold, but by 1937 the settlement was gone once again. Total production of gold came to a little more than $17,000.

SEARCHLIGHT (Clark) Searchlight's name origin has two interpretations. A miner named G.F. Colton was lighting his pipe with matches that carried the Searchlight brand name when he noticed some rocks with what looked like gold in them. It may be that Colton and his partner were checking out the area when they found a rock that looked like it had gold in it. One of the men told the other that if there was gold in the rock, one would need a searchlight to find it. More than likely the first theory is correct, because the Searchlight matches were often used here, and people have found many of their wooden box crates.

Searchlight was a busy mining town in the early 1900s. The biggest production of gold and silver came from the Duplex (owned by Colton) and Quartette Mines, two of over 300 properties in the area. The Quartette was originally discovered by

John Swickhart, who sold it to Ben McCready for a team of mules, some plug tobacco and $5. The Quartette yielded more than $2 million in gold, and the Duplex about $650,000. A narrow-gauge railroad was built from the Quartette to a mill along the Colorado River. In 1907 the Barnwell and Searchlight Railroad connected with the Santa Fe Line that ran from Needles and Mojave in California. But after a massive washout, the trains were stopped and never restored. Searchlight overlooks Joshua tree forests that are the densest and largest in the world.

SHERMANTOWN (White Pine) This camp sprang up about 1868 as a result of the Treasure Hill bonanza, and served as a milling center for their mines. Located about five miles from Hamilton, the site was founded by Joseph Carothers and called Silver Springs. The name was later changed to commemorate General William Tecumseh Sherman. The camp had an abundant water supply and prospered with its 10-stamp mill, which was later expanded to 15 stamps, then was followed by the addition of six more mills. Shermantown had a population of more than 3,000. A local newspaper wrote, "Shermantown can boast of the coolest and best drinking water in the State. Whilst all the water consumed in Treasure City is carried there by wagons from Rock Spring and Hamilton, which becomes warm by the time it reaches its destination, necessarily causing the use of ice, having to pay good figures for both water and the ice, we have an abundance of water in our wells at a temperature that will make the teeth ache to drink it."[226] The mills processed more than $200,000 of silver in a two-month period during 1868. Since the camp's economy was tied to the Treasure Hill riches, by the 1870s the ore was beginning to run out and so was Shermantown's life. After it was abandoned, many of the buildings were torn down and taken to Hamilton to replace those burned in several fires.

SIEGEL (White Pine) Prior to 1872, reports of rich silver ore spread and before long a camp sprang up known as Centerville along Siegel Creek near the Schell Creek range. The site was renamed Siegel for a Utah mining financier. Initial discoveries were made by Samuel Jameson and E.G. DeMill. They later sold their claim for about $28,000 to the Tehama Consolidated Silver Mining Company, which built a 20-stamp mill. But the ore was disappointing and the mill ceased operations. Some of the smaller mines produced moderately, but by 1874 there were only a few people left at the camp.

SILVER CITY (Lyon) Although gold had been found in 1849 by a party of California-bound Mormons, silver ore didn't come to the forefront until the discovery of the Comstock Lode. The ore was found while placer miners were searching the hills for gold-bearing veins. This camp was located about three miles from Virginia City and grew rapidly, first serving as a place where the animals that hauled the ore wagons were housed. It later became a mill center with eight stamp mills. Someone described the camp as "The deep pits on the hillsides; the blasted and barren appearance of the whole country; the unsightly hodgepodge of a town; the horrible, confusion of tongues; the roaring, raving drunkards at the bar-rooms, swigging fiery liquids from morning till night; the flaring and flaunting gambling-saloons, filled with desperadoes of the vilest sort; the ceaseless torrent of imprecations that shocked the ear on every side; the mad speculations and feverish thirst for gain."[227]

The placers were worked for about ten years and yielded nearly $1 million. Silver recovery values were indeterminable because the ore was included in the Comstock values, but they were estimated at about $12.7 million.

SILVER GLANCE (Nye) Located near Hannapah, this camp was formed after the Silver Glance Mine was located in 1905. The following year *The Standard* "printed the front page of one entire edition in ink mixed with gold dust assaying $80,000 to the ton."[228] The mine was worked until 1909, then reopened in 1919 by the Silverzone Mines Company and yielded about $300,000 in silver. The camp was gone by the 1920s.

SILVER HILL (Churchill) Silver Hill should never have been founded. Some silver was discovered by prospectors near the Stillwater Range, and the ore supposedly assayed up to $1,800 a ton. Promoters thought it would become another Comstock. The site was established in 1861 in early spring and by June it was gone.

SILVER PEAK (Esmeralda) Silver Peak was settled between the mining camps of Tonopah and Hawthorne during the 1860s when gold-bearing ledges were discovered. As a result, the Silver Peak & Red Mountain Company was established and a 10-stamp mill was brought in from the camp of Jacobsville. Two years later a 30-stamp mill was erected, but when the ore values declined the mills ceased working. Then in 1870 the Pittsburgh Silver Peak Gold Mining Company purchased the properties and gleaned what was left of the ore. Total production was between $8.5 million and $11 million. In 1948 most of the town was destroyed by fire.

SIMON (Mineral) Silver deposits were found in 1879 when the Simon Mine was located, followed a few years later by gold discoveries. The camp was named for a landowner named "Pop" Simon, who also organized the Simon Silver-Lead Mines Company. Its post office was established in 1919, and about two years later a 100-ton mill was erected to treat the ore. Gold was discovered at the Copper Contact Mine in 1902, followed by the Olympic Mine in 1915. Before 1919 the Olympic mine yielded more than $700,000 in gold. The Simon produced about $200,000 between 1926-1927.

SKOOKUM (Lander) Bearing a Chinook name meaning "strong" or "powerful," Skookum was formed in the early 1900s when an Indian discovered gold and silver float and sold his claim to the Lemaire brothers. Then in 1908 rich hornsilver was found, bringing in hordes of miners. That, plus the fact there was more than sufficient water available made the people believe that the camp would greatly prosper. The railroad at Ledlie also fared well with Skookum's mines. Unfortunately, the boom lasted only a few years and the camp was history. Total production was under $15,000.

SPRING CITY (Humboldt) Spring City sprang up about 1873 after the discovery of silver and gold a few years earlier and a 10-stamp mill was erected. The post office was established in 1878. During the 1880s more than $1 million in gold and silver had been recovered from the nearby mines.

This camp was a wild one, with a spot nearby called Gouge Eye, and for good reason, as related about one miner who made the following request in a saloon: "I'm a roarin' ripsnorter from a hurrah camp, an' I can't be stepped on. I'm an angel from Paradise Valley and when I flap my wings there's a tornado loose.... Give me some of your meanest whiskey, a whole lot of it, that tastes like bumblebee stings pickled with vitriol. I swallered a cyclone for breakfast, a powder mill for lunch, and I haven't begin to caught yet. Don't crowd me."[229] The bartender obliged the man by mixing up some fusel oil and rubbing the glass with red pepper.

SPRING VALLEY (Pershing) Placer gold deposits found in this region were thought to have been the most productive in the state. The site was located about 12 miles from the town of Oreana. During the 1880s hundreds of Chinese miners dug shafts and systematically drifted along pay streaks to recover the ore. Total production was estimated at more than $10 million.

SPRUCEMONT (Elko) Named for its location in a pine forest, Sprucemont began after W.B. Latham located the Latham Mine in 1869. He later sold his property for $30,000. There was enough silver there to warrant a small rush and the following year a number of lodes were discovered, followed by the construction of a 50-ton smelter. When the facility could no longer handle the ore it closed down. Total production came to a little less than $3 million.

STAR CITY (Pershing) Star City experienced a wild boom with discovery of the Sheba Mine. Silver was found in a nearby canyon, and by 1865 there were two hotels, a telegraph office, a Wells Fargo office, and a population of more than 1,000.

Gold Bars, Elko County (Northeastern Nevada Museum, Elko, Nevada).

The local saloon was making money hand over fist with the eager miners: "It assured the thirsty that its liquors were as good as those in Frisco and invited smokers to enjoy prime Havana—'"so soothing when it has been bought with the last quarter, and a fellow's busted.' Its marble-bed billiard tables promises [*sic*] improvement for the mind and body; the game 'employs the mind, while interest in your loaned money keeps counting up. Exercises your legs, by sending you around the table to get your ball out of the pocket. Gives grace to the body, by introducing that elegant posture, one leg in the air, hat for a bridge, and cue harpoon style. Exercises the lungs, by the large amounts of whiskey after bad shots.'" [230]

A 10-stamp mill was built by the Sheba Company, which operated until the boom went bust in 1868, as a local newspaper wrote: "So sudden has been its decline that the daily mail, the express office, and telegraph office are all in operation yet, though the entire population consists of a single family, the head of which is mayor, constable, postmaster, express agent, telegraph operator, and, I believe, sole and unanimous voter!"[231]

STOFIEL (Elko) Named for settler Walter Stofiel who moved there in the 1870s, the community became a shipping point for supplies headed to Gold Creek after ore discoveries there. It also had a stage station that was run by the Butler family, and a post office was established in 1891. There is still a gas station and store still in business.

TAYLOR (White Pine) Two prospectors named Taylor and Pratt were led to a silver strike by an Indian in 1872. They paid him $500 for the Argus Mine, which was renamed the Taylor. Three years later they sold their claim to the Martin White Company for $14,000. In 1880 the camp of Taylor was formed, followed a year later by the establishment of the Monitor Mill and Mining Company with its 10-stamp mill. New discoveries in 1883 brought in eager prospectors, and the population rose to about 1,500. The Taylor Mine brought out more than $250,000 in silver until disaster struck. One of the drills hit some unexploded dynamite, causing the death of one miner and blinding another. The mill closed down in 1886 after making its last shipment of ore valued at $36,000. A drop in the price of ore forced the camp to fold.

TEMPIUTE (Lincoln) Tempiute was formed about eight miles from Rachel a few years after silver was found near Grants Peak in 1865. It bears a Southern Paiute expression meaning "rock water people" Since water had to be hauled in from more than ten miles away and the camp was located in a remote area, it was difficult to work the mines. A nearby stamp mill was dismantled in 1871 and miners were forced to take their ore to the camp of Tybo. The silver was not as rich as first believed and Tempiute folded. One of the mines, the Wyandotte, produced only $720. The camp's life was renewed for a time when tungsten was discovered in 1916.

TENABO (Lander) Silver ore was first located in the 1870s, and gold placer discoveries followed in the early 1900s. The camp was established in Mill Gulch on the east side of the Shoshone Range, with a hotel, assay office, school, stamp mill, and a population of more than 1,000. By 1909 the Tenabo Mining & Smelting Company had purchased many of the properties. But three years later operations ceased because the cost of hauling in water from Indian Springs and the expense of milling was higher than the ore was worth. Total gold production came to about $3.2 million.

TONOPAH (Nye) One of the mothers of all mining camps, Tonopah was established during the early 1900s. The discovery of silver by Jim Butler gave rise to this great mining camp, a Shoshone or Paiute word for "small spring." In 1900 Butler was camping in the area when he accidentally came across a silver outcrop while hunting for his loose burro. He took samples and had them assayed, which were found to yield about $600 a ton. Butler went back to stake his claim, then later sold it to the Tonopah Mining Company, which was organized in 1902.

The camp boomed with more than 2,000 residents. Supplies and water had to be brought in from the town of Belmont, more than 50 miles away. Other properties were located and by the end of the year more than $4 million in silver had been recovered. Tonopah's riches ultimately exceeded $30 million in gold and silver. It also became the county seat until wrested away from Belmont.

Tonopah features a tour to an underground mine at Mining Park. It's open from April 1 to September 30, seven days a week, 9–5, and October 1 to March 30, Wednesday through Sunday, 10–4. It also hosts the Nevada State Mining Championships held during Jim Butler Days on Memorial Day weekend. Information: (775) 482-0274.

TRANSVAAL (Nye) Settled in March of 1906,

the camp was born with the discovery of gold near Beatty and Rhyolite. Transvaal is part Latin and South African meaning "across the murky river," and referred to the camp's location near the Amargosa River. Not a lot of gold was found, as noted by one of the miners: "I carried away from Transvaal enough gold in my boots to buy a dinner in Beatty."[232] Within four months Transvaal was gone.

TREASURE CITY (White Pine) This camp got its start after a starving Shoshone Indian named Napias Jim stole some food from prospector A.J. Leathers, then gave him some silver ore in exchange for the vittles. Excited, Leathers and his partners got the Indian to show them where he got the ore. Jim took them more than 9,000 feet above sea level to Treasure Hill. In 1868 they discovered the Hidden Treasure Mine, which was sold two years later for $200,000. The camp was established in 1867 and major development began. Although the altitude made life bitter because of the cold, the rich silver still drew in the miners who suffered through icy winds and deep snowdrifts.

A tongue-in-cheek account of an incident in 1869 noted, "A singular case of frost bite occurred on the Divide between Treasure and Hamilton during the extreme cold of Sunday night. A jolly fellow, much elevated — we are elevated about nine thousand feet, but he was spiritually elevated — making the journey on that night between two places, after answering an imperative call, carelessly exposed himself, and upon reaching Hamilton discovered that he was frozen — well, where women could not be. Suddenly sobered by the appalling calamity, he frantically rushed for professional aid, and the medicine man at once applied an abundant poultice of snow. Partial relief was thus given, but it is expected that early amputation will be necessary in order to enable a respectable — chignon. We have heard of weather causing such disasters to brass monkeys, but not to the genus homo.[233]

More than 200 mines were located and ten stamp mills were erected. Raymond wrote that "The ore deposits of Treasure Hill are richer than any that have been discovered during the present century.... Not long before my visit, the miners of the district held a meeting, at which they were strongly urged to adopt at once the system of 'square locations,' and abandon the farce of staking out claims on ledges which do not exist. This proposition was defeated; and every man on Treasure Hill now claims so many feet of a vein, running, he does not specify in what direction, and dipping, he cannot tell at what angle, from a hole which he has made at random in the neighborhood of some already exposed body of ore. If he gets down to the ore, all the better; he can then work night and day, extract a large quantity of rich chloride, and send it away, before the neighbor, who has a prior location, can prove the identity of the deposit."[234] By the end of 1870 it was just about over, but not before more than $20 million in ore had been realized. The Hidden Treasure itself recovered more than $2 million. In 1874 a fire destroyed what was left of the camp.

TULE CANYON (Esmeralda) Mexicans were believed to have mined in the canyon as early as 1848. About 20 years later the Americans found gold placers. Chinese miners were also working the diggings, and may have recovered more than $1 million in gold. A post office was opened for a short time in 1905.

TUSCARORA (Elko) During 1864 gold placers were discovered along McCann Creek by the Beard brothers, and a ditch was built to bring water to the claims. It was not uncommon for the men to find nuggets from their diggings that weighed more than one ounce. Many thought the wealth was in placer mining, but it would be lode gold that provided Tuscarora's riches. The year 1869 brought the Central Pacific Railroad which employed more than 2000 Chinese who stayed after the tracks were completed, and gleaned what was left out of the abandoned mines. Prospectors moved their camp here in 1871 when silver was discovered, and the post office opened on July 18 the same year. It was named for the Tuscarora Indians, interpreted as "hemp gatherers."

There was no intense mining activity until 1876 when J. Woods found some of the first silver that would become the Grand Prize Mine. This prompted the *Engineering and Mining Journal* to write: "A rich strike is reported in this mine [the Grand Prize].... Some eighteen or twenty tons of rich ore from the new bonanza have been forwarded to the Leopard mill at Cornucopia for reduction."[235] More properties were located and the camp boomed. The Grand Prize recovered $152,000 in just one month. Its total production before closing was more than $2.5 million in silver. Another good producer was the Independence that recovered $600,000. By 1886 the bonanza ores were exhausted. It was estimated that about $10 million was realized in gold and silver.

Folklore says that a group of men skimmed off more than $100,000 in gold from the mines, and

Commonwealth Mine, Tuscarora, 1893 (Northeastern Nevada Museum, Elko, Nevada).

put the gold dust and nuggets in an oak water barrel and buried it along McCann Creek. A band of Shoshone Indians were watching them, and finally attacked, killing some of the miners. Those who lived went back later for their stash, but it was gone, probably taken by the Indians. Investigations over the years found that the Indians did not take the gold, so it may still be buried along McCann Creek.

TYBO (Nye) Tybo is about 10 miles from Warm Springs and bears a Shoshone word for "white man." During 1870 silver was found by an Indian who showed it to a white prospector. He took a sample and had it assayed, which was valued at about $2,000 a ton. Discovery of the Two G Mine in 1871 brought on rapid development of other lodes and a smelter was built. Gold was also mined as a byproduct of silver. Freight wagons hauled out a good 250 tons of silver bullion each month, and returned with supplies from Eureka. A stage line was also established and advertised that it would go "Through to Eureka in 18 hours. Leave Tybo on Tuesday, Thursday and Saturday's at 8 P.M. for Eureka. Leaves Tybo on Arrival of Eureka stage for Belmont. Experienced Drivers and Excellent Accommodations."[236] The Tybo Consolidated Mining Company built a mill in 1879, but because of reduction problems it was discontinued the next year and Tybo began its decline. But in the early 1900s the Louisiana Consolidated Mining Company brought in a mill which operated for about three years before closing down. The mines yielded more than $9 million in silver. Gold production was about $500,000.

UNION (Eureka) In 1879 prospectors found a rich lead-silver deposit, but serious mining did not begin until about 1886. A smelter was built by James Lindsay, who also took control of most of the mining claims. They produced about $100,000 and then the site lay idle until about 1915 when the Union Mines Company was created by William Fairman. A post office was established the following year. The ore from the mines was shipped out to Midvale, Utah, for processing. But the company paid almost $243,000 for ore extraction, and yields were only $225. The company closed its doors in 1918.

UNIONVILLE (Humboldt) Unionville was situated on the east side of the Humboldt Range. Silver ore was found in 1861 by Indians who took it to Virginia City and had it assayed. The camp was quickly formed and a rush was on. Claims were staked and then the Arizona Mine was located. Mine speculators were advertising the riches to be found, but the editor of the *Humboldt Register* wrote that miners had been in the canyon for at least two years but had not yet driven "their tunnels in far enough to protect themselves from the rain." The editor also proposed a State of Buena Vista, whose seal should show "a mountain with good croppings and float; on the mountain side, untaxable hole in the ground, six inches deep: near by, cedar tree; under its shade, two miners with pack of cards, working assessments."[237] The *Territorial Enterprise* called Unionville "the richest mineral region upon 'God's footstool.'"[238] Production of silver was on the decline about 1870, and when the Central Railroad bypassed the site the end was near. More than $3 million in silver was recovered from the mines. Those people who stayed after the mines closed went into the fruit-growing industry.

VANDERBILT (Eureka) This camp began in 1870 when silver was discovered in Secret Canyon about seven miles from Eureka. The Vanderbilt Mine was located the same year and a post office opened in 1871. A stamp mill was also erected. This was a short-lived town because Eureka was beginning to boom, and then a fire destroyed the mill. Total production from the mines came to about $700,000.

VARYVILLE (Humboldt) Varyville was established shortly after a mining district was formed about 1875. It was located in the Pine Forest Range, about 65 miles from the town of Winnemuca. Gold was discovered at nearby Bartlett Creek and first worked using an arrastre until two stamp mills were erected. Gold production was much lower than expected, and by the 1880s Varyville ceased to exist.

VERNON (Pershing) Vernon served as a supply and trading center for the mines in the Seven Troughs Range and was settled about 1905. Drinking water had to be hauled in from a nearby canyon which turned out to be unclean, but it still sold for more than $5 a barrel. The camp lasted only a year.

VICKSBURG (Humboldt) Also known as Ashdown, this camp came to life about 1904. Some mining was conducted during the 1860s, but Indian attacks kept most of the miners away. Although there was some placering, most of the ore came from lode mines, notably the Ashdown, developed by the Ashdown Gold Mining Company, which produced about $400,000. There was a small resurgence in the 1930s, with nearly $100,000 realized.

VIRGINIA CITY (Storey) The wild, lusty silver town was named about 1869 for an old prospector named James Fennimore, who called himself "Old Virginny." Tradition says he named the site after he broke his whiskey bottle, and said, "I battize this spot Virginia Town."[239] Bancroft described Fennimore as "an intemperate Virginian, without either brains or education, who for some breach of lawful etiquette committed elsewhere had found it convenient to remove to Carson Valley in 1851, where he had remained ever since, digging his season's wages out of the earth to pour it down his throat in bad whiskey during his leisure months."[240]

Virginia City was a product of the famous Comstock Lode discovered in Gold Canyon, and lived high off the hog until the Comstock was depleted. James Fair, William O'Brien, John Mackay and James Flood secretly purchased stock in the area at low prices, then announced the bonanza, which brought them in more than $160 million. Ore extracted from the Consolidated Virginia Mine has been estimated at $62 million. The Arizona Comstock Mining Company recovered low-grade ore from a claim previously owned by William Chollar that was valued at more than $1 million. Other nearby mines containing high yields were The Belcher ($36 million), California ($64 million), and the Savage ($15 million). Tailings from the mines were averaging between $7 to $19 a ton. The district recovered close to $1 billion in gold and silver, which also included revenues from the Comstock.

The town experienced a fire in 1863 that ending up costing $7 million in losses. Then in 1875 another fire started after an oil lamp tipped over, setting off a firestorm at an estimated loss of $12 million. "John Mackay fell to his knees and cried out to God Almighty to save his mine. As the million-dollar hoisting works blazed on the surface, Mackay made his ultimate deal with the Lord. If his shaft was spared, he would build 20 churches."[241] But God did not help him, rather his workers dynamited the nearby homes to form a rocky fireproof circle around the costly underground workings. Another report on the fire stated, "The streets were filled with flying people, furniture brought out of the houses was burning on the sidewalks ... loose horses from the stables were dashing madly to and fro seeking to escape, with the hair all burned from their backs.... The flames came tearing on with a front 200 yds. wide and 200 ft. high.[242] Virginia City did prevail, and is a thriving town today.

Virginia City hosts its Annual International Camel Races in September, and many events in relation to the Comstock. Call the Chamber of Commerce at (775) 847-0133 for more information.

WARD (White Pine) Located about 20 miles from Ely, Ward was established about 1876 and named for miner T.F. Ward. Silver was first discovered in 1872 while freighters William Ballinger and John Henry were searching for their stray bulls and accidentally came across rich outcrops. Their claim would become the Paymaster, which was later sold to a Judge Frizell and yielded more than $300,000 between 1875 and 1878. As other mines were located, two smelters and a 20-stamp mill were built to accommodate them. The camp grew to a population of about 1,500 and its post office opened in 1877. As the lead content of the ore declined, one of the smelters was converted into a mill, which was a failure. Ward's demise began in 1883 after a fire destroyed much of the town, followed by a decline in ore and a new mining boom at Cherry Creek.

WARDELL (Elko) Although L.H. Wardell located the Wardell Mine and recovered small amounts of silver in 1879, the camp was doomed from the start. A fire in 1880 at the claim killed two men, and the silver panic caused the mine to close. It wasn't until 1905 that the Union Mine was found by Richard O'Neil and George Vardy, which was valued at about $450 a ton. Another property was later opened, but its production was low, and it wasn't long before Wardell bit the dust.

WASHOE CITY (Washoe) This camp sprang up with the discovery of the Comstock Lode and served as a smelting and milling center. Since other mills were too far away from the Comstock, Washoe was an ideal site. Water for the workings was piped in from the Sierras. Some of the ore from Virginia City was also processed at Washoe, and a sawmill was cutting timber for shoring up the mines. By 1862 the camp had a population of more than 2,500. After the Virginia & Truckee Railroad was built, the camp began its decline. Once the county seat, Washoe saw its end when the city of Reno grew rapidly and took the honors.

WEDEKIND (Washoe) George Wedekind was a piano tuner by trade who found gold about 1896 in the hills near Reno. His discovery brought other gold seekers who established the camp in 1901. Wedekind's claim yielded him about $100,000. He later sold his mine to Governor John Sparks, but after erecting a stamp mill on the property Sparks shipped out very little gold, and the place folded.

WEEPAH (Esmeralda) Bearing a Shoshone word for "rain water," Weepah was founded about 1902 after the Indians found gold. Word of the ore brought an influx of miners, but within a few weeks they were gone because so little was found. It was 1927 when Frank Horton and Leonard Traynor relocated the previously worked claims. When news got out that their diggings assayed at about $70,000 a ton, the rush was on with more than 1,500 prospectors. By 1938 the ore had been depleted after about $1 million was realized, and Weepah was gone.

WHITE CAPS (Nye) Located about three miles from Manhattan, this camp came to life about 1915 when the White Caps Mining Company purchased a number of claims. A 10-stamp mill was built, in addition to a cyanide plant that produced only about $120,000 although the company had initially invested $200,000. The ore continued to diminish and by 1918, only $87,000 had been realized Then in 1936 the mine and mill burned to a crisp.

WHITE PLAINES (Churchill) This site was a milling camp that served the Desert Queen and other silver lodes in the region. It was located about 30 miles from the town of Fallon. In 1864 a 5-stamp mill was built to treat the ore. Between 1883 and 1931 the Desert Queen produced only about $25,000. The mill closed down because it was too difficult to get water and fuel.

WILLARD (Pershing) Named for a boxer, this small camp sprang up in 1915 about five miles from the town of Lovelock. A few hundred miners took up residence, but production was so low that the camp folded within a few months.

WONDER (Churchill) Two prospectors came to Wonder about 1906 and told their friends about the rich silver they had discovered, bringing others to stake their claims. For a few years the area was an important producer of gold and silver, and at one time single lots sold for as much as $8,000. It received its name because it brought wonder to the

Camp of Wonder, ca. 1910 (Churchill County Museum & Archives, Fallon, Nevada).

eyes for its desolation: "No vegetation meets the eye when gazing on the vast expanse of dirty white alkali."[243] Both gold and silver were found near Chalk Mountain, the leading producer being the Nevada Wonder Mine, which recovered more than $6 million between 1911 and 1920.

YANKEE BLADE (Lander) Yankee Blade was founded about 1863, four miles from the mining town of Austin. Named for a newspaper in New England, the camp lasted about ten years. Silver discovered in Yankee Blade Canyon brought in the miners, and a stage company began running its line between Yankee Blade and Austin. The first paying mine was the Whitlatch Yankee Blade, which brought about construction of a number of stamp mills. Other mines were opened that also produced well, but in 1867 water struck the claims and they were closed down. The following year a fire destroyed most of the buildings and mills. The Whitlatch recovered about $131,000 in gold and silver, and the others combined produced about $170,000.

NEW MEXICO

ALBEMARLE (Sandoval) Although prospectors found gold during the 1870s, boundary disputes with Mexico dampened early enthusiasm. Albemarle was settled in the Jemez Mountains about 1894. During this time the Albermarle deposit was located which consisted of a consolidation of four lode mines. The camp's name was transferred from North Carolina, where it honored George Monck, duke of Albemarle. The Cochiti Gold Mining Company later purchased the property, developed the mines, and the camp began a period of prosperity. It was established in Colla Canyon where rock had to be blasted out in order to accommodate the mill. As the mines got deeper the value of ore declined, and by 1905 operations came to a standstill. The district produced more than $1 million in gold and silver.

ALMA (Catron) Located about 10 miles from Mogollon, Alma lived part of its life besieged by Indian attacks. Apache chief Victorio and his men ravaged communities including Alma in 1880. Then about 1885 Geronimo began raiding and killing the residents in other nearby communities. In spite of that, Alma continued to prosper as a trading and supply point for the gold and silver mines. It was named for the mother of townsite owner Captain J.G. Birney, with a post office in 1882.

AMIZETTE (Taos) This boom camp was thought by some mining entrepreneurs to be the next Cripple Creek (as in Cripple Creek, Colorado). It was settled about 15 miles from Taos, where gold and silver were discovered during 1893. Al Helphenstine and William Fraser formed the camp and named it after Helphenstine's wife. The ore was difficult to extract and expensive to haul out of the canyon to a mill, plus there simply wasn't much to be found. Gerson Gusdorf and his brother refused an offer of $20,000 for their claim, believing they had a bonanza, but they got very little out of it. Amizette began to decline within two years of its founding.

ANCHOR (Taos) Anchor came to life in the Cimarron Mountains about 1884 when prospectors created a small mining district. Since the men had little equipment, getting the gold out was laborious, and after about three years of hard work with little yields and litigation over land rights, Anchor disappeared.

BLACK HAWK (Grant) This short-lived mining camp was located at the foot of Bullard Peak in the Burro Mountains. About 1881 a prospector discovered a rich silver float nearby, which later became the Black Hawk Mine. A small camp sprang up in 1885. A few other silver lodes were found, but by 1887 the camp was already on the decline.

BLAND (Sandoval) Bland was built in a narrow canyon that was only 60 feet wide. When more room was needed for a building, the rock was simply blasted out. It must have been pretty tight quarters, since there were more than 50 structures and 3,000 people living there. The site was named for Richard P. Bland, who waged a fight against the demonetization of silver. A post office was opened in 1894. By 1904 the boom had busted, but not before the mines brought out close to $1 million.

BONANZA CITY (Santa Fe) Once a rendezvous for thieves and outlaws, this camp was established in 1879 with the discovery of silver by Frank Dimmitt and Robert Hart. It was located about 10 miles from Santa Fe and given the Spanish name for "prosperity," which turned out to be a pipe dream. Its post office opened in 1880. There were high hopes of riches, so a smelter and reduction works were built to accommodate the mines. The men were receiving only about $3–$4 a day, and by 1885 the town faded away after disappointing results. Only small amounts of gold and silver were recovered.

BONITO CITY (Lincoln) Miners discovered silver and gold about 1880 in Bonito Canyon, where the camp was established because of its good water supply. The post office opened in 1882, and given the Spanish name for "pretty." It is not known how much ore was recovered from the claims. The town was later submerged when a dam was built by the Southern Pacific Railroad.

BRICE (Otero) Before the discovery of gold, turquoise was mined near Brice by the Indians. During 1879 an old miner named S.M. Perkins was prospecting in the Jarilla Mountains when he was approached by hostile Apache Indians. They spared his life when they saw he had a hunched back, which to them probably had some spiritual significance. Continuing his search, Perkins made his strike, then later sold his claim for two barrels of water after his own well had gone dry. The prospect hole would later become the Nannie Baird Mine. Brice was settled in the early 1900s as Jarilla, but when the El Paso & Northeastern Railroad ran a line through the region, it was changed to commemorate a mining superintendent. The post office was established in 1904. Gold lodes were mined in addition to dry placer operations, but little ore was recovered. By 1920 Brice ceased to exist as a gold camp.

CARBONATEVILLE (Santa Fe) Founded in 1879, Carbonateville was named for the Carbonate Mine owned by two men named Dimmitt and Hart. Businesses were established in the following year. Since the town had no water, it had to be hauled in from either Alamo Creek or the Galisteo River. After the Autocrat Mine was located, it began yielding gold between $50 and $200 a ton. But for the most part the mines were not very successful, and within a few years the town folded.

CERRILLOS (Santa Fe) Bearing a Spanish name for "little hills," Cerrillos was founded about 1879 and became a supply and shipping point for a number of mining camps. Frank Dimmitt and a partner came to the region and found some gold and silver. They went to Leadville, Colorado, to have their samples assayed, which brought in a horde of more than 300 miners. The following year the Atchison, Topeka & Santa Fe Railroad laid its tracks through town. When the gold ran out, Cerrillos managed to survive after coal deposits were located and used by the railroad for fuel. Gold values are included with Dolores since they were both in the same mining district.

CHANCE CITY (Grant) Named for the mine, this town was settled about 1880 with a post office in 1882. Three prospectors, William Kent, William Hyters and J.L. Dougherty, found silver in the Victoria Mountains during 1880, one of which became the Last Chance Mine. The ore was hauled to the Southern Pacific Railroad about four miles away. William Randolph Hearst later purchased the property. The lodes were not rich and produced only about $500,000.

CHLORIDE (Sierra) Located in the foothills of the Black Range, Chloride was founded after silver was discovered in Chloride Gulch about 1879 by Harry Pye, a mule skinner and prospector. He was hauling freight at the time, and when his contract was over a few years later, he and two partners returned to the site and staked a claim. Pye was later killed by Apache Indians. Miners who followed also left because of Indian troubles, but returned armed to the teeth to protect their claims.

In order to entice people to live there, the first lady who would move to the camp was given a free lot. Another incentive was a seat on the city council to the father of the first new-born baby "if it is known who he is. Fortunately, virtue was not a deciding factor for the recipient."[244] The town declined with the drop in silver in 1893. A total of about $500,000 was produced from the properties.

CLERMONT (Catron) James C. Cooney found silver along Mineral Creek about 1875, followed by the discovery of gold outcroppings near Copper Creek. Clermont was established as a supply and trading camp for the mining population. It was located about five miles from Mogollon with a post office established in 1881. The Galveston Company came in and purchased a number of claims with high aspirations, but did not have much success. After a road was built to the town of Mogollon, Clermont faded into the sunset.

COOKES (Luna) During 1876 Edward Orr discovered lead-silver ore on Cookes Peak. Not much interest was generated until the arrival of two prospectors named Taylor and Wheeler in 1880. Two years later the camp was established by George L. Brooks while he was having a road graded to the site, which was originally an Apache Indian stronghold. Since the workers feared the Indians would return, a small unit of soldiers was sent there to guard them. The camp was named for Captain Philip St. George Cooke, leader of a Mormon battalion. Total production of silver is unknown.

COUNCIL ROCK (Socorro) Billy Hill discovered silver nearby about 1881, which brought in a group of prospectors who discovered the Old Boss Mine. The camp was established the same year and named for a huge boulder that was earlier used by Indians to hold their councils. Hopes were high for riches in the Iron Mountains, with the result that land speculators took over and sold lots for about $250. But the ore was disappointing and the miners left.

DOLORES (Santa Fe) Dolores was born about 1828 with the discovery of placer gold in the Ortiz Mountains, and may have been the location of one of the first gold discoveries in the state. Tradition says the ore was found by a Mexican herder while looking for his stray cows. About five years later gold quartz veins were discovered by a Spaniard named Don Cano a half mile from Dolores. Water for the placers had to be carried in by barrels, or the miners had to haul the dirt to a stream more than two miles away, but they managed to recover between $300,000 and $500,000. The richest placers were at the mouth of Cunningham Gulch.

At one time Dolores had a population of about 4,000. During 1846 Lt. James W. Abert visited the town and wrote, "The houses were the most miserable we had yet seen, and the inhabitants the most abject picture of squalid poverty, and yet the streets of the village are indeed paved with gold."[245] Total output from the district was a little over $2 million, mainly from placers before 1900. In 1900 Thomas Edison brought an invention to Dolores to test it. He tried to extract gold using static electricity, but after a few unsuccessful attempts it was abandoned.

ELIZABETHTOWN (Colfax) During 1866 a group of men were conducting assessment work and did some panning along Willow Creek, discovering rich placer deposits. John W. Moore aided in establishing the camp and named it for his daughter. Prospectors discovered gold on Baldy Mountain and the camp soon burgeoned with a population of more than 7,000. Apache Indians

Elizabethtown (New Mexico State University Library, Archives and Special Collections, 02860006).

plagued the miners because the area was considered their hunting ground, but they eventually drifted away. Water had to be brought to the placers, so the Moreno Water and Mining Company began work on a 40-mile-long ditch from the Red River, which was completed in 1868. Building the aqueduct took considerable effort since it had to cross arroyos, and then a trestle had to be raised nearly 80 feet over the valley floor. But the ditch did not function as expected because of water seepage. Placer deposits yielded more than $5 million.

About ten years after the town's establishment, an observer noted, "It makes one lonesome to walk the streets of Elizabethtown. Although not an old place, it is deserted, and instead of the crowded streets ... a sort of grave yard stillness ... the vitality of the village life had departed."[246]

During Elizabethtown's prosperity the deluge of miners kept the hotel and boarding-house rooms full, so for many there was no room at the inn. One proprietor, Charles Kennedy, found a ghastly way of handling the situation. A new boarder would arrive, rent a room, pay his money, then suddenly disappear, and the cycle continued. No one noticed what was happening because the boarders were strangers, until one of the new boarders happened to be a friend of a "former" lodger. Through a chain of events Kennedy was discovered by a group of vigilantes tossing the missing person's dismembered body parts into the fireplace. Kennedy never made it to trial, because residents were so outraged they put a rope around his neck, whereupon he was dragged up and down the street behind a horse until he croaked.[247]

FLEMING (Grant) Fleming was established about 1882 when a prospector was grubstaked by a mining engineer named John (or Jack) Fleming. The town was located about 10 miles from Silver City with a population of 200–300 people, and became a busy trading center after a road was built to Silver City. Fleming experienced numerous attacks by the Apache Indians, so the townsmen formed a company to protect residents. The women also organized their own company named the Calico Guards. The men joked that some of the female members could shoot as well as the men with both eyes closed.[248] By 1889 the town was gone.

GAGE (Luna) This site served primarily as a shipping point for the mines in the Victorio Mountains. It was established about 20 miles from Deming, and may have been named for William Gage who, along with others, were responsible for development of the mines. The name was also thought to have honored a General Thomas Gage. The mining camp of Chance City lacked water and had to have it hauled in from Gage at a cost of more than 2¢ a gallon.

GEORGETOWN (Grant) Named for a mining company executive, George Magruder, this camp sprang up in the 1870s and served as a milling center for the nearby silver mines. During 1866 prospectors discovered ore near the Mimbres River, and within a few years the Mimbres Mining and Reduction Company was established. Ox teams were used to haul the silver from the mines to the mills. During the 1880s Georgetown boasted a population of more than 1,000. But a smallpox epidemic forced many to leave, then the demonetization of silver drove the rest away. After a fire broke out, most of the place went up in a cloud of smoke.

GLENWOOD (Catron) When gold and silver were discovered in the mountains about 1893, the town of Graham was established at Whitewater Canyon. It was named for John T. Graham who built a mill the same year. He was also superintendent of a mine at Mogollon. The name was later changed to Glenwood with a post office in 1895. A water shortage problem at the mill was solved after a pipe was built in the mountains that had to run down three miles to the mill site. Unfortunately, mining and milling were not very successful and the camp declined. During Glenwood's earlier days the canyon was a haven for Apache chief Geronimo and his band. It was also a hideout at one time for outlaw Butch Cassidy and his gang.

GOLD DUST (Sierra) Not much is known of this site. It began in 1880 when a mining boom brought in the prospectors, and within a year it was a busy tent town. Prospectors had problems with Apache chief Victorio and his band, who once attacked the camp in a surprise raid. Luckily, some 30 men were working in the canyons at their claims when this occurred. When the ore was exhausted, Gold Dust bit the dust.

GOLD HILL (Hidalgo) In 1884 gold was discovered by David Egelston, Robert Black and Tom Parke, who named their claim the Gold Chief Mine. That same year the camp was established and continued to prosper for about 20 years. Although there was a moderate amount of gold, little water was available to work the diggings. The miners were also plagued by threats from the Apache

Indians. During a Fourth of July celebration in 1885, two American flags were erected over the hill. Unwittingly, this action saved the lives of the residents, since Geronimo and his men were planning on attacking the camp. When they saw the flags waving in the breeze, Geronimo thought the cavalry had come to protect the camp and they left without further incident. Gold values are included with those of Lordsburg.

GOLDEN (Santa Fe) Initial placer deposits were discovered in 1839 near the future camp in Lazarus Gulch near a branch of Tuerto Creek. Miners made a mad dash to Golden and worked their claims on the west side of the San Pedro Mountains, most of which were mined out by 1880. That year a number of mining companies were formed and the camp of San Pedro was established after the San Pedro Milling Company began large-scale development of the region. The name was changed to Golden to reflect its anticipated riches. During 1880 hydraulic mining came into play, but water shortages and various litigations halted operations. Years of drought caused the men to abandoned their claims by 1884. Even with all the problems, it was estimated that the district produced about $2 million to 1904.

GRAFTON (Sierra) Located in the Black Mountain Range, Grafton was established during the 1880s with the silver rush. In 1883 it had more than 75 residences and a school. Serving as a trading and supply center, Grafton was short-lived. By 1892 the mines had been depleted and only 20 people were left in town.

HANOVER (Grant) Hanover was established in 1841 when prospectors from Germany tried their luck at mining. The camp was given a German name for "high bank," and its post office opened in 1892. Prospectors attempted to work their claims, but the Apache Indians attacked and drove them away. In 1891 the Santa Fe Northern Railroad laid its tracks to town. But with the silver demonetization in 1893 not much happened, and the place was abandoned.

HEMATITE (Colfax) Gold was discovered during the 1890s along Moreno Creek, and shortly thereafter the camp was established about five miles from Elizabethtown, given a name for the red iron ore nearby. Most of the ore was recovered from placer diggings, and the Athens Mining & Milling Company built a stamp mill to crush the ore. Unfortunately, the diggings were superficial and before long Hematite ceased to exist.

HERMANAS (Luna) Bearing a Spanish name for "sisters," this camp was established about 1879. Some silver ore was found in the Tres Hermanas Mountains, which brought in a few mining companies. But there wasn't much production, so they left. Hermanas was sustained for a short time by the ranching industry and the Southern Pacific Railroad when its line was constructed, but eventually it, too, was abandoned.

HILLSBORO (Sierra) Hillsboro was established in 1877 after Daniel Dugan, Dave Stitzel and Joe Yankie located gold deposits in the Black Range. They had the ore assayed which was found to be valued at about $160 a ton. The rush was on and more than 300 miners came to Hillsboro, located roughly 35 miles from Truth or Consequences. Because the Apaches posed a threat to residents, a small contingent of soldiers was stationed at the camp. Gold lodes and placers were worked extensively from 1884 to 1905, and recovered more than $2 million. Silver production was about $5 million. Fire consumed much of the town in 1904, followed by a flood in 1914. Four years later the killer influenza visited Hillsboro.

JICARILLA (Lincoln) Although placer mining began about 1850 with Mexican miners, it was the 1880s before white men entered the scene and the camp was formed. Its post office opened in 1892 as Jicarilla, Spanish meaning "little cup." Miners experienced moderate success and the camp lasted until the 1930s. During the Depression Jicarilla was a haven for poor people who conducted mining and earned about $7 a week. When the Depression was over, most of the people moved on as jobs became available.

KINGSTON (Sierra) During 1880 a group of prospectors discovered silver nearby, and were soon followed by other money-hungry fellows. An enterprising man named Barnaby arrived with supplies and opened up the first store in a tent. Other mines were discovered, including the Iron King, and the Solitaire, located by Jack Sheddon in 1882. Situated in the foothills of the Black Range, the town was formed by the Kingston Townsite Company the same year. Miners had to contend with Apache chief Victorio and his band, who killed many of the men. But it continued to grow with more than 7,000 people, boasting nearly 20 saloons, and a

One of the last buildings at Lake Valley (courtesy Thomas Pritchard).

theater where the famous Lillian Russell and her group once played. In order to get supplies, a wagon road had to be cut through to the town. By the early 1900s more than $7 million in silver had been recovered.

LA BELLE (Taos) During 1866 gold was discovered, but it wasn't until the 1890s that enthusiasm was spurred in the Red River region. This gold mining camp was established with a post office in 1895, and named for Belle Dixon, wife of a prospector. The ore proved disappointing and La Belle declined. Famous outlaw "Black Jack" Tom Ketchem came to La Belle on occasion, staying at its hotel. He finally got nabbed and was hanged at Clayton.

LAKE VALLEY (Sierra) This region was one of most prosperous silver-producing districts in the state. The first person to find ore was George W. Lufkin in 1878, and the news brought in a swarm of prospectors. By 1888 more than $4 million in silver had been recovered. The camp was established in 1882 as was the post office, and named for the nearby lake beds. A stage line was also established, and more often than not driven by Sadie Orchard, a British madam. She also opened brothels at Kingston and Hillsboro. Lufkin later sold his claim for about $100,000 and moved back east, a big mistake.

George Daly took over managing the property and opened up one of the biggest hornsilver pockets, which was called the Bridal Chamber Mine. Unfortunately, Daly was killed by Apache Indians the day after the discovery. The camp literally bulged to a population of more than 4,000 avid miners. The silver was so pure it was shipped to the mint in Denver without being smelted. The walls just sparkled with silver crystal: "In the 'Bridal chamber' pure silver may be melted off with a candle, and Gov. Safford offered $50,000 for the ore that he could extract unaided in 10 hours."[249] The mine produced more than 2.5 million ounces of hornsilver, worth more than $3 million. One of the pieces was exhibited at the Denver Exposition, valued at $7,000. Total production of the region came to nearly $25 million. The town saw its demise after the drop in silver prices, and then in 1895 a fire that was started by an angry drunk destroyed most of the buildings. Lake Valley experienced a small resurgence when manganese was found.

LORDSBURG (Hidalgo) Lordsburg was settled while the Southern Pacific Railroad was building its line through the region in 1880, and named for one of the construction engineers. The town acquired a post office the following year. Prospecting began about 1870 after someone salted the area with diamonds, but the ensuing stampede did result in the discovery of a few silver deposits in spite of some joker. Recorded production of the region came to more than 220,000 ounces of silver. When the ore was depleted, copper was discovered, thus assuring Lordsburg a long life.

MALONE (Grant) Malone was a very small, flash-in-the-pan camp. It was named for John B. Malone, who found gold there in 1884. After his ore ran out, Malone disappeared. The place never had more than 50 people.

MIDNIGHT (Taos) This short-lived town located along Bitter Creek was sort of a sister camp to Anchor, and named for the Midnight Mine, although folklore says it was because that was the wildest time of the day. Located in the Cimarron Mountain Range, Midnight was settled in 1895 with a post office the same year. The camp only survived for a few years. Land litigation and little free-milling gold ore caused the place to pass away.

MINERO (Grant) Minero was founded about 1875 a few miles from Hachita after ore was discovered in the Little Hatchet Mountains. First called Eureka, it was changed and given a Spanish name for "mine," or "mine operator." Minero served as a water stop and loading point for ore that was shipped out when the El Paso & Southwestern Railroad was built in 1902.

MOGOLLON (Catron) Mogollon was named for the state's provincial governor, Don Juan Ignacio Flores Mogollon, who may have been the first to discover ore there during the 1700s. Sergeant James C. Cooney, a scout and guide, located a rich ledge in the Mogollon Mountains in 1870, but kept it a secret until he was discharged from the cavalry. He and his companions staked the first claims in the Mogollon District during 1875, but hostile Apache Indians prevented any shipment of ore until 1879. Intermittent Indian raids continued to harass the miners until the defeat of Geronimo in 1885, whereupon the place was besieged with eager men who rapidly developed the mines. Cooney was later killed by the Apaches.

Mogollon was established at the bottom of Silver Creek Canyon. By 1889 three stamp mills had

Mogollon (courtesy William Dilworth).

American Mine at Nogal (Johnson Sterns Collection, Lincoln County Historical Society, Carrizozo, NM).

been built and the camp boasted a population of more than 600. Up to 1905 the mines had produced more than $5 million in silver and gold. Mogollon began to fade when a flood almost destroyed the camp, followed by the depletion of ore.

NOGAL (Lincoln) Located in the foothills of the Sacramento Mountains, this camp was formed in 1882 with a post office the same year. First called Dry Gulch, the name was changed to Nogal, Spanish for "walnut." Gold placers were first discovered in the gulch about 1865, followed by lode deposits three years later when the American Mine was located. Major development did not begin until the land was withdrawn from the Mescalero Indian Reservation. Good producing mines were the Helen Rae and the Nogal Peak. The Helen Rae was discovered by Henry George and Charles Epps. They later sold their property to John Rae for $100, who named it for his daughter. Initially recovering about $1,000, Rae sank a shaft and brought out more than $14,000. He sold his claim to Rolla Wells for $15,000, but the mine yielded only about $5,000. The Nogal Peak Mine produced about $500,000.

NUTT (Luna) This town began as a terminal for the Atchison, Topeka and Santa Fe Railroad, and was a point of departure for the miners heading out to the silver and gold fields. The railroad station had a sign posted which read: "Ho! For the Gold and Silver Mines of New Mexico. Fortune hunters, capitalists, poor men, sickly folks, all whose hearts are bowed down and yet who would live long, be rich, healthy, and happy: come to our sunny climate and see for yourselves!"[250] Nutt's access to the railroad made it an ideal place to ship the silver ore on to refineries, in addition to serving as a trading point for miners in the Black Range. When the railroad extended its line to Lake Valley 1884, Nutt's population declined. The camp was named for a stockholder of the railroad, Colonel Nutt.

ORGAN (Dona Ana) This name was taken from the Organ Mountains, so called because the peaks resembled old organ pipes. Mining was king here at one time and Organ saw a tremendous increase in population after silver, copper and lead were discovered near the mountains and the railroad came

Louis Bentley in his assay office at Organ, ca. 1910 (New Mexico State University Library, Archives and Special Collections, 00140189).

through in 1881. The post office was established this same year and is still in operation. After the mines played themselves out, Organ reverted to a bedroom community. Silver and gold yields were about $2.5 million. Today, many of the residents work at the White Sands missile facility.

At one point in time, store owner and assayer Louis B. Bentley found the town drunk passed out on the street. He carried him off to the jail to sleep it off, leaving the door open and returned to his store. When he returned the next morning he saw the fellow was dead. On his way out of the cell, Bentley found an empty bottle of Yellowstone Whiskey. With a twinkle in his eye, and to honor the drunk, he placed the bottle in the man's hand. Bentley told some of the miners of his death, who then went to the jail and saw the bottle of whiskey. Saloon owner Billy Viscarra couldn't figure out why his bottles of Yellowstone Whiskey sales had suddenly dropped.[251]

Billy the Kid hung out in the area and was shot about six miles from Organ. The town's cemetery dates back to the 16th century and holds the remains of one of Quantrill's Raiders, plus some of the men who fought in the Civil War. Famous frontier sheriff Pat Garrett was shot near the camp in 1908.

OROGRANDE (Otero) Orogrande was settled near the Jarilla Mountains with the gold rush in 1905. The camp was initially called Jarilla Junction after the El Paso & Northeastern Railroad erected its station there in 1897. It was renamed Orogrande when the post office opened in 1905, a Spanish expression for "big gold." Water for the site had to be piped in from the Sacramento River, 55 miles away. The Southwest Smelter and Refining Company built its plant in 1906 to process the gold, but so little was found the facility was closed. Nervous promoters started to plant placer gold to lure in the unwary, but the diggings that actually did produce were shallow and yielded only about $40 a ton. There was no more interest in the region until turquoise was later discovered.

PARSONS (Lincoln) Parsons was established along a fork of the Bonito River in 1892 and named for miner R.C. Parsons who discovered silver there in 1886. About 100 people made the camp their home. A mill was built to treat the ore, followed by a post office in 1888. During the early 1900s a huge mansion was built (later called the Parson Hotel) and entertained such notables as Mr. Studebaker, and writer Mary R. Rinehart. Up until the 1970s the hotel was still open for business, but it later burned to the ground.

PIÑOS ALTOS (Grant) Spaniards were the first to discover ore about 1800, but they were killed by hostile Apache Indians. A group of prospectors named Colonel Snively, Hicks, and Birch came to the region in 1860 seeking their fortunes. Birch saw gold flakes while he was getting water from Bear Creek; within one month the place was full of miners and Birchville was established. But Apache Indian raids during this time drove many of the men away. After 1864 the men returned and renamed the site Piños Altos, Spanish for "tall pines," since the place was located among the majestic ponderosas. There were still problems with the Indians, so the military built Fort Bayard for their protection. The *Mesilla Times* noted, "Thomas J. Marston is pushing ahead his work of grinding quartz, although constantly annoyed by Indians."[252] The Pacific Mine was located in 1860 by Thomas Marston, followed by other lodes that proved to be very productive until the 1920s. Silver was also discovered nearby in 1891 by the Dimmick brothers, who were ranchers. While rounding up a stray cow, one of them prepared to throw a rock at the animal when he noticed it contained silver. The region was estimated to have recovered between $1.5 and $8 million in gold and silver.

Parson and mill, 1800s (Johnson Sterns Collection, Lincoln County Historical Society, Carrizozo, NM).

PITTSBURG (Sierra) Pittsburg was a short-lived mining camp that was established about 1903 when placer gold was discovered by Bernardo Silva. After he went to the town of Hillsboro and got drunk, his loose lips excited more than 500 prospectors who stampeded to the site and staked their claims. The post office opened the following year. But the deposits were shallow, miners left for better diggings, and the town was abandoned.

PYRAMID (Hidalgo) This 1880s gold and silver camp was located at the north end of the Pyramid Mountains, hence its name for one of the peaks that resembled an ancient pyramid. It was originally established as a water station on the overland route to California and owned by settler Eugene Leitendorf. The discovery of ore turned it into a mining town. A 20-stamp mill was erected and operated until the Panic of 1893. There were only a few silver deposits, but the gold recovered from the nearby mines came to about $4.6 million

RED RIVER (Taos) Miners from Elizabethtown were prospecting in this vicinity during 1869, but there was no real interest until a smelter was erected by the Waterbury Company ten years later. The Mallette brothers may have been the first settlers and mine owners during the 1880s. Mining promoter E.I. Jones purchased the Mallette claim and formed the camp of Red River. Mining was moderately successful, but hauling the ore down the steep mountains was hazardous at best, then the stamp mill burned down in 1889. A few of the residents had earlier raised a flag on top of a nearby mountain to show Red River was a prosperous town, but it came down when the ore ran out.

Although the men were hard-working, they did have time for practical jokes. In one instance, some of the miners took time off from their diggings and hit the local saloon, which was built on piles over Bitter Creek. The men told the saloonkeeper that a wall of water was headed their way and most likely would destroy the place. The man made a hasty exit while the miners drank to their hearts' content, then made off with as much of the liquor as they could carry.[253]

RILEY (Socorro) This site began as a Hispanic ranching settlement along the Rio Salado and was named for a sheep rancher named Riley. It became a temporary trading center for the miners in 1897 when gold and silver were discovered. After the mines closed, Riley reverted to a ranching and farming community. Over time the water table dropped, making it difficult to irrigate, so most of the farmers left for greener pastures.

ROBINSON (Sierra) With the prosperous mines in the Black Range, Robinson popped up in 1882 with hopes the camp would become a shipping point for the mine companies. It was also anticipated that the Santa Fe Railroad was going to extend its line through Robinson. The site was named for M.L. Robinson and had a population of more than 300. The railroad never came through and Robinson's residents scurried out of town.

ROSEDALE (Socorro) During 1882 Jack Richardson found gold near the San Mateo Mountains. He named the infant camp for his wife, Rose, but problems with the Apache Indians kept many of the gold seekers away until the 1890s. Richardson's claim was later developed and sold to the Rosedale Mining and Milling Company. Operations continued until about 1937 with the district recovering about $500,000.

SAN AUGUSTINE (Doña Ana) Named for a fourth-century priest, Saint Augustine of Hippo, the camp began in 1849 after a Mexican miner found silver in the Organ Mountains. Prospector Hugh Stephenson teamed up with the man, and later purchased the Mexican's interest. A small camp sprang up with a stockade built for protection against continuous raids by the Apache Indians. Its post office was established in 1876. During the 1850s Stephenson sold his mine to a man named Bennett. By 1889 most of the silver had been taken from the mine and the camp was abandoned. The Bennett property recovered about $1.2 million in silver.

SHAKESPEARE (Hidalgo) Shakespeare grew up with the discovery of silver ore about 1870, and more than 3,000 miners settled there. The silver lodes proved rich for a short time, but by 1893 the place was fading away. William C. Ralston, who founded the Bank of California, got involved in a fraudulent diamond scheme to try and keep the town alive. Rumors were circulated that there were diamonds at nearby Lee's Peak. Actually, two rogue miners named Arnold and Slack salted the site with diamonds and showed them to Ralston. He had Tiffanys check out the gems and found they were genuine. Ralston then bought out the miners' claim for $600,000 and began selling stock. It wasn't long before the hoax was discovered, but by then Arnold and Slack were long gone. During 1997 what was left of Shakespeare burned to a crisp.

SILVER CITY (Grant) Silver City actually produced much more gold than silver. Originally it was a camping place used by the Apache Indians. Placer diggings and about 30 mines were in operation during the 1860s. The region was soon abandoned due to the Civil War in addition to the depradations of the Apaches. Residents were angry because the Army wouldn't do anything to protect them, so they started a fund to pay for a scout to patrol the region, and remunerated those who brought in Indian scalps. It wasn't until the 1870s that mining picked up again. Located in the foothills of the Piños Altos Range, the site became a shipping point for the mines when it was founded in 1870 with a post office the following year. After the ore ran out Silver City became a center for the ranching industry. The most productive placer deposits were found along Bear Creek, Rich, and Whisky gulches. Total gold recovery was about $3.1 million.

SOCORRO (Socorro) Nearby silver discoveries led to the establishment of Socorro during the 1860s. Located in a rich mining district, the town was a major supply and shipping point for more than 200 wagons that served the mines. Its post office opened in 1875. Socorro was a wild, bustling community that boasted more than 44 saloons. Its prominent mines were the Socorro Tunnel, Torrence, and Merritt. Up to 1882 the Socorro had produced $430,000 in silver and $10,000 in gold. Total gold shipped out that same year came to a little over $1 million. During 1893 the drop in silver prices caused the mines to close.

Russian Bill was a resident who was well liked by the townspeople, but he had a penchant for practical jokes. The end came one Christmas Eve after he became unruly during a poker game and mischievously shot off the finger of one of the players. That did it. As well as the men liked him, they'd had enough and immediately hanged him in the dining room of the hotel. The charge against him read: "Hanged for being a damn nuisance."[254]

Today, Socorro is home to the New Mexico Bureau of Mines and Mineral Resources, and the New Mexico Institute of Mining and Technology.

STEEPLE ROCK (Grant) First called Carlisle, this camp was established in the 1880s and named for Claude Carlisle Fuller, the first child born there. It was later changed to Steeple Rock for the shape of a nearby formation. Gold was found in 1881, and a few years later the Carlisle Mine began operations and a 5-stamp mill was erected. This mine produced about $3 million in gold and silver. As a young man, President Herbert Hoover acquired a job here in 1898 as assistant superintendent at the Steeple Rock Mines.

SYLVANITE (Hidalgo) Silvanite began about 1908 when "Doc" Clark found gold near the Little Hatchet Mountains. In a few months there were more than 500 gold-hungry prospectors conducting placer mining. By 1909 most of the gold had been worked out and the camp folded. Roughly $1 million was recovered.

TELEGRAPH (Grant) After A.J. Kirby hit silver pay dirt in 1884, the camp was formed. Kirby named his claim the Tecumseh Lode. The camp took its name from the mountain, for the Army heliograph station located there in the 1880s. About 100 people made Telegraph their home. A road was built over the mountains that connected the camp with the Gila River, followed by a reduction works erected at the camp to treat the ore. But when the Tecumseh Lode closed, the site was soon abandoned.

VALEDON (Hidalgo) During 1855 Valedon was established about three miles from Lordsburg after a blacksmith named Sam Ransom and his friends discovered silver in the Pyramid Mountains. They named their claim the '85 Mine. Most of the men who worked at the property lived in Lordsburg and walked to work every day, a distance of five miles. A mill was erected built to crush the ore and structures were built to house the miners. Valedon never had a business district, since most of the men spent their leisure time at Shakespeare. In 1913 a spur line was extended from Lordsburg by the Southern Pacific Railroad which by then had a population of about 2,000. The camp lasted until about 1931 when the Phelps Dodge Company purchased the property, and within a year closed the mine and ordered the people to leave. What buildings remained at Valedon were moved to other locations.

VERA CRUZ (Lincoln) Located on the west side of the Tucson Mountains, this small camp was established about 1880 and served the mines in the region. It bears a Spanish name meaning "true cross." Most of the people living there worked in the Vera Cruz Mine, but the ore proved to be low grade and it was shut down. About 10 years later capitalists from Denver intended to reopen the mine and install a stamp mill. But it never happened and Vera Cruz just idled away.

VIRGINIA CITY (Colfax) Three men were sent to the region in 1866 to do assessment work at Baldy Mountain. While camping for the night at Willow Creek they discovered gold placers, which precipitated the rush to Elizabethtown, whereupon Virginia City got on the bandwagon and tried to ride on that camp's coat tails. It was named for landowner Lucien B. Maxwell's daughter. The camp never stood a chance because most of the residents moved to the more active Elizabethtown.

WHITE OAKS (Lincoln) This site was named for the white oak trees prevalent in the area and was settled about 10 miles from Carrizozo after discoveries of gold and silver. During 1879 the Homestake Mine was located in the Jicarilla Mountains by George Baxter, George Wilson, and Jack Winters. After the news got out, the region teemed with miners. A 20 stamp mill was erected by a man named Glass. As a bonus, some coal was found nearby which provided steam power to run the mill.

White Oaks (courtesy Thomas Pritchard).

Baxter sold his interest in the Homestake for $300,000 that went on to yield more than $3 million in gold.

During 1887 word of a railroad being built created immense interest. The *Interpreter* printed, "White Oaks is bound to be one of the best towns in the Territory. It has the precious metals ... right at its own doors and when the railroad reaches that portion of the country, White Oaks will grow rapidly."[255] But it was not to be. Greedy businessmen overpriced their property that was to become the right-of-way, so the railroad went to the town of Carrizozo instead, and White Oaks lost the rest of its golden acorns. Total production of the district was nearly $6 million.

OREGON

ALLENTOWN (Josephine) Gold was discovered in Sailor Gulch as early as the fall of 1853. With winter coming on, the miners realized they could not all stay because their provisions were low, so some of the men left until spring. But one fared better than the rest after finding plenty of meat in addition to a small gold claim. When he returned to the gulch, he found the men who stayed behind nearly starving and kindly shared his food with them. Not wanting anyone to know of his claim, he drank with the miners, waited for them to get drunk and pass out, then took off to work his claim.[256] Placer mining was conducted until the 1870s when hydraulics were put into operation and continued until about 1900. Gold recovery values are included with the camp of Waldo, since the claims were all in Sailor Gulch.

At one time a priest named Father Croke came to the diggings soliciting money for his church at Jacksonville. He wrote a letter to an archbishop of his success: "I arrived at Sailor Diggings ... and commenced collecting at a place called Allen Gulch, where the greater part of the miners are working. In two days I collected there four hundred dollars in cash, a sum far above my most sanguine expectations for a place where there are not more than seventy men in all."[257]

APPLEGATE (Jackson) This site lived a double life as a mining town and lumber camp, named for settlers Jesse and Lindsay Applegate. It was established about 1852 with a post office opening in 1858. Gold was discovered near town and the miners realized moderate results. Between 1852 and 1900 they were recovering an annual production close to $450,000. Local folklore says that a Chinese miner named Chiny Linn discovered rich diggings that yielded him nearly $2 million in gold, but that has never been substantiated.

ASHWOOD (Jefferson) Ashwood was settled in the 1870s along Trout Creek near the community of Madras. It was named for the volcanic ash deposits left from ancient lava flows on Ash Butte. A sheepherder named Thomas J. Brown accidentally found a piece of quartz containing gold and silver in 1897, and before long the camp was alive with miners. Its post office was established in 1894 with James Wood as postmaster. One of the more promising mines was the Oregon King. Plans were made to build a stamp mill, but the mine proved to be low in ore and Ashwood began its decline. The discovery of cinnabar brought renewed prosperity.

AUBURN (Baker) While loading up with supplies at Portland in 1861, Henry Griffin and David Littlefield overheard someone talking about a fabled gold mine. They managed to get the stranger to lead them to his find, but when they arrived he told them he lied. However, they did find a small gold prospect hole near Blue Canyon Creek, and shortly thereafter a camp was established with a post office opening in 1863. The ore was so rich that the two men in one season panned out nearly $100,000. Supplies were brought to the miners via pack trains. Auburn mushroomed with a population of nearly 5,000 and more than 1,000 claims were recorded. Two Frenchmen discovered a placer bonanza and recovered more than $100,000. After word got out of the riches in Idaho during 1867, the men left for better diggings and Auburn declined. It also lost the county seat. Perpetrators from Baker stole into Auburn early one morning in 1868 and took the county records before Auburn residents knew what was happening. Production of gold from 1906 was a little more than 19,000 ounces, worth roughly $392,000.

Auburn had a few problems with shootings and outlaws as many a mining camp did. At one time two outlaws were caught attempting a robbery. While sitting in jail, they wrote a note to the sheriff: "On the trail we met two fools, and took from their horses and mules, we left with them their pocket knives, which served perhaps to save their lives." In turn, the sheriff answered with his own note: "When bad boys steal horses and mules Auburn lets them hang around for a while."[258] They were swiftly hung.

AUSTIN (Grant) Named for settler Minot Astin, who built the first general store, Austin was established as a trading and supply center for the nearby mines and its post office opened in 1888. Located in the Blue Mountain region, Austin was also the terminus of a small narrow-gauge railroad from the town of Sumpter. It was expected the busy camp would have a bright future, but when the mines rapidly declined so did Austin.

BAKEOVEN (Wasco) This former site received its name because of the baking ovens built by a trader. Working his way from The Dalles, the man stopped here overnight. When he woke the next morning, he discovered the Indians had stolen his horses, but left his flour. So the enterprising baker constructed clay ovens on the spot, and sold his bread to miners heading to the Baker mining district.

BAKER CITY (Baker) This town was named for state Senator Edward Baker, with a post office established in 1866. During 1864 J.S. Ruckel built a 10-stamp mill along the Powder River, followed by P. Basche, who set up shop in 1865 and prospered by selling equipment for the stamp mills and mines. Ruckel's mill operated until 1898. Production in the Baker area was mainly placer mining at Griffin Gulch until discovery of a gold-bearing quartz mine called the Virtue, located in 1862 by James W. Virtue. Six years later George W. Grayson purchased the property and operated it until 1884, which then lay idle until 1893. Ore from the Virtue was transported by horseback about 200 miles to a landing point at The Dalles. Supplies were hauled in from Umatilla Landing until a railroad was built to the region in the 1880s. Total production was about $2.5 million in gold, most of it recovered from the Virtue Mine.

BLUE RIVER (Lane) Located near the Blue River, this camp was created in the 1860s and the post office opened in 1886. After discovery of the mines about 1863, arrastres were built to crush the ore. Stamp mills were erected in 1898 to replace the arrastres, but they ceased to operate because the ore proved to be low grade. The only major producing mine was the Lucky Boy, discovered in 1887 by Nate Standish and others. Other mines in the region recovered between $50,000 and $100,000 before 1896. By 1924 a little more than $150,000 was realized.

BOHEMIA (Lane) A prospector named James "Bohemia" Johnson found gold about 1863 in the Calapooya Mountains while hiding there after he had killed an Indian. The camp was situated between the Willamette and Umpqua rivers. Placer mining was active at nearly 2,000 claims until 1877, but only a little gold was recovered. There was a period of inactivity until it was discovered that the vicinity was full of quartz ledges, and lode gold came into play. In 1891 the Musick Mine was located by James C. Musick. That same year the Noonday Mine began operating. Dr. W.W. Oglesby commenced to erect the Champion and Noonday mills near the Musick Mine. Early yields of gold were estimated between $500,000 and $1 million. After 1902 more than $400,000 had been recovered.

BONANZA (Baker) This camp was established along a fork of the Burnt River after the Bonanza Mine was discovered in 1872. About 50 men worked the property, and the ore was hauled to a 20-stamp mill via an aerial tram. The mill was later refitted with 40 more stamps to handle the huge amount of gold. At one time a group of men stole some of the ore (called high-grading), and it was estimated that more than $1 million was lost. The Bonanza produced nearly $2 million in gold before 1904. Placer claims recovered more than $200,000.

BOURNE (Baker) First called Cracker City, this camp was established in the 1880s inside of a canyon, so it was restricted to one long street. The name was changed to honor U.S. senator Jonathon Bourne when the post office opened in 1895. Gold placers were found along Cracker Creek, but most of the gold values were located in lode mines. In 1887 the North Pole was discovered by Tom McEwen, who sold it to an English interest for about $10,000. It would go on to yield nearly $9 million in gold. Bourne saw its end when the post office

Restored post office at Buncom (courtesy Reeve Hennion).

closed, followed by a cloudburst that sent a thundering wall of water down Cracker Creek into the canyon. Total gold produced in the Blue Mountain region was about $75 million, with $40 million of it attributed to placers.

BRIDGEPORT (Baker) Placers were discovered in Clark Creek about 1861, followed by the establishment of Bridgeport along the Burnt River. Miners received their supplies from the town of Baker which had to be hauled across the river. As a result, a bridge was built over Clark Creek to shorten the route, hence the town's name. Placer production amounts are included with the Baker City.

BROWNTOWN (Josephine) Established at the mouth of Walker Gulch (a tributary of Althouse Creek) in the 1850s, this small camp was named for Henry H. Brown, operator of the general store. He also hauled supplies to nearby claims with his pack train. Brown later went on to become a newspaper publisher in California. Frank and Phil Althouse were the first to discover gold. Although not much was found, it didn't stop the thousands of miners from coming here seeking their wealth. One of the prospectors did find a nugget valued at about $1,200, and another discovered one worth more than $3,000. The forest later reclaimed the camp.

BUNCOM (Jackson) This once thriving mining camp (and one of the last standing ones in Southern Oregon) was established in 1854 when gold was found along Sterling Creek, bringing in thousands of miners who conducted placer mining. Prospectors brought their gold to Buncom to be weighed and stored until it was transported to Jacksonville. Hydraulics later went into operation. When the gold ran out, Buncom kept its economy moving as a supply and trading center for the outlying camps. The post office was established in 1896 as was the first general store. Buncom's death was a result of the automobile. What would have taken miners a half day to ride into Jacksonville for supplies changed when the cars were able to make the trip in less than half an hour.

Fowler and Roberts wrote: "It is as if the little 'almost a town' remained for its own reasons, a carefully kept secret until now [1995] when we who long for a peek at the past shake its rickety foundations for the answers."[259] There are still a few building still standing: the post office, cook house, and bunkhouse. In 1990 the property was purchased by Reeve and Lyn Hennion and its three buildings were repaired.

At one point in time Buncom was taken off the maps, but four years ago Reeve Hennion lobbied to have it put back, which recently occurred. "It needed to be used as a point of reference," he explains, then adds, "I knew the county uses it, that you guys use it [*Mail Tribune*]."[260]

Buncom has a historical society that sponsors Buncom Day, held on the Saturday of Memorial Day weekend, an all-day affair with live music, craft and food booths, and various events, including a "chicken splat" contest. The money received from this event goes towards preservation of the buildings. Located about 20 miles southwest of Medford on State Highway 238, the drive to Buncom is all paved. For information, e-mail the Buncom Historical Society at buncom@buncom.org.

CANYON CITY (Grant) This city sprang up in Canyon Creek after the discovery of gold in 1861 along Whiskey Flat. William A. Allard panned for gold and took out about $20 on his first attempt. It was first called Whiskey Gulch because the liquor really flowed, and a post office opened in 1864. Joaquin Miller wrote a few tidbits about the camp in the *Mountaineer*: "I saw three live merchants, all busy, of course. Bruner was lying on the counter, tickling his cat under the tail with a straw. McNamara was sitting at the breakfast table, picking his nose with a fork. Felsheimer had finished his breakfast and sat at the table quietly paring his toenails with the butter knife. A merchant sold a pair of shoestrings last week and went to his drawer for the change, but the thing had rusted in its place from idleness, and he had to use a crowbar to open it."[261]

Miners received supplies by pack trains, which were later supplanted by a regular freight service that stopped there about three times a week. In 1897 Ike Guker located the Great Northern Mine and recovered about $65,000. During the 1880s dredges operated in the canyon. The camp suffered a severe setback when most of the buildings burned in a disastrous fire during 1937. The region produced about $15 million from the placers, and more than $1 million in lode gold.

Canyon City holds its '62 Days in June, a celebration of the town's legacy, hosted by the Whiskey Gulch Gang. For exact dates, call the Grant County Chamber of Commerce at 1-800-769-5664.

Canyon City, 1880 (Grant County Historical Museum, Canyon City, OR, #80-366).

CONNOR CREEK (Baker) Placer mining began on Conner Creek near the Snake River during the 1860s, and about ten years later the Connor Creek Mine was found, later to become a heavy producer. A 20-stamp mill was erected to process the ore, but the chunks of rock were so hard the mill couldn't crush it. Then someone cracked open one of the huge boulders only to find it loaded with gold. Out of four tons of the rock, more than $30,000 was recovered. The mine closed about 1910. Total production between lode gold and placers came to a little over $2 million.

CORNUCOPIA (Baker) This camp was named by miners from Cornucopia, Nevada. One of the earliest prospectors was Lon Simmons, who found gold on the slopes of the Wallowa Mountains during the 1880s. Before long the region was teeming with more than 1,000 men, and a 20-stamp mill was built to accommodate the claims. The camp was very remote, and as a result transportation problems and old equipment slowed operations. Horses and mules were later used to transport the ore until arrival of a railroad that greatly alleviated shipping costs. During the 1880s the Union Companion Mine began operating and became one of the area's big producers. It changed hands a number of times. The first owner sold it for about $800. Then in 1895 the new owners paid $60,000 for it. Two years later it yielded $200,000, followed by another $200,000. More than $2 million was recovered from the area.

Cornucopia had an ordinance that allowed no hard liquor in town, only beer. But that didn't stop the resident gentlemen. One Labor Day the place was crowded with miners, ranchers and residents. Some of the prospectors from the hills arrived with a bottle of elixir of pure Pine Creek water mixed with corn squeezins and passed it around. Known locally as "tangle leg," it was said that a few drops would turn a pine squirrel into a screaming panther.[262]

ELDORADO (Malheur) After the discovery of gold in 1863, Eldorado was established near the mining camp of Malheur and served as a supply center. It was expected the camp would be very prosperous as noted in 1873 by the *Bedrock Democrat*: "During the day the streets of Eldorado are deserted, all men at work in the tunnels back of town. Great hopes are placed on the benefits to be derived from the 'Big Ditch' under construction by Packwood and Carter; water will be in the ditch in less than ten days. The future of Eldorado looks very bright."[263] The ditch project was 120 miles long at a cost of $250,000. In addition to bringing in water, it also floated logs that were used to timber the mines.

GALENA (Grant) Named for the lead ore containing silver, Galena was founded during the 1860s about 20 miles north of Prairie City near a middle fork of the John Day River. The post office was established about 1901 when miners "took" the post office from Susanville. A fair amount of gold was found, but the rush didn't last long and Galena began to decline. It saw a slight resurgence during the 1930s when dredges were brought in to get the last of the gold from the creek beds.

GALICE (Josephine) Placer mining began in the district along Galice Creek during 1854. The Old Channel and Benton mines were among the important gold producers in southwest Oregon. Located about a mile from camp, the Old Channel was one of the largest hydraulic mines in the state and began operating in 1860. The Benton, a lode mine in the Mount Reuben area, was discovered in 1893 and worked until 1905. A Mr. Hutchins discovered the Golden Wedge Mine in 1893, which was taken over by the Gold Road Mining & Milling Company about 15 years later. This property recovered about $50,000 in gold. By the 1880s the richer deposits were mined out and the Chinese took over many of the operations. The region recovered about $5 million in gold. Galice was named for a French doctor named Louis Galice, who came here about 1852 to search for gold. But folklore says it received its name after an Indian asked for food from members of a pack train, specifically "gleece," which supposedly means "bacon."

GOLD BEACH (Curry) Shortly after the gold rush in California, news of rich ore on the beaches of Oregon in 1852 brought more than 1,000 men to wash the sands filled with gold that was run-off from the nearby mountains to the sea. Initial placer diggings were located along the mouth of the Rogue River near Gold Beach. Until a road was built to the area in 1890, supply teams had a hard time reaching the beaches. But the beach sands didn't hold up to their potential and the miners left.

GOLD HILL (Jackson) This town owes its name to gold found in the hills during 1853. Placer mining was not in paying quantities so Gold Hill grew slowly until a big strike in 1859 with the discovery of lode gold. A miner who found the quartz went to Yreka, California, to show a specimen to his friend, George Ish. George hurried back to Gold Hill and with another prospector friend located the ledge, which turned out to be 20 feet wide and 20 feet long. The Gold Hill pocket in its first year yielded about $400,000. Just a few miles away the Revenue recovered nearly $100,000. The first quartz mill in the county was erected at Gold Hill in 1860. Another 5-stamp mill was built in 1892 by the Union Iron Works at San Francisco. It was shipped in by railroad, then hauled seven miles to Sardine Creek using mules and sleds. Josiah H. Beeman purchased a mine, and with his riches he built his home at Gold Hill, which today houses its historical society. Total production from the Gold Hill District was about $700,000.

Tales abound of the gold rush days, and Gold Hill had its moments. The Rhoten brothers struck it rich, and wanting to celebrate they went into Medford and hosted a huge party at the local saloon. When the owner wanted to close the bar, the brothers bought the saloon outright, partied all night, then gave the saloon back to its original owner.[264]

Gold Hill is very much alive and thriving. Nearby is the famous House of Mystery at the "Oregon Vortex." It is a naturally occurring visual and perceptual phenomena that can be captured on camera. For more information and directions, call (541) 855-1543. On display at the Gold Hill Historical Society and Museum is the stamp mill once used on Sardine Creek. It is located at the corner of Fifth Street and First Avenue. For hours, call (541) 855-1182.

GRANITE (Grant) Located northwest of Sumpter, this camp was established with the discovery of gold in 1862. First called Independence, the name was changed to Granite when the post office was established in 1878. Close to town a prospector found a nugget worth about $500, then another with a value of more than $1,500, bringing interest to the region. Mining was not very profitable until new equipment was brought in to work the lodes. During 1886, E.J. Godfrey and Clark Tabor located the Red Boy Mine, then erected a 20-stamp mill and concentrator. By 1896 more than 7,000 feet of mine shafts had been dug. The Red Boy closed in 1916. In later years it was discovered that water from the mine carried arsenic and other metals to a number of creeks, threatening the habitat and different species of fish.[265] Total production of the region was about $1.8 million.

GRANTS PASS (Josephine) This town and county seat came into existence as a stopping place on the California Stage route, then became a trading and banking center for the local mines. It was named for General Grant when residents heard the news that he had captured Vicksburg. The Granite

Red Boy Mine located near the camp of Granite, 1898 (Grant County Historical Museum, Canyon City, OR, #77–76).

Hill Mine was discovered about 1890 and developed by the American Goldfields Company during the early 1900s. A 20-stamp mill was also erected. About $65,000 in gold was recovered from the property. Another major lode was the Daisy, which produced about $200,000. Placer claims in the Grants Pass area were the Rocky Gulch, discovered in 1856, which yielded more than $150,000, and the Red Jacket in which more than $65,000 was recovered. Between 1932 and 1957 the mines yielded a total of $150,000. Earlier production records are not known. After the mining era, the Grants Pass economy was stimulated by the timber industry.

GREENHORN (Baker) Located in the Blue Mountains, Greenhorn received its name after two novice miners came to the camp about 1890. They asked the more experienced men where there might be a good place to find their riches, and were directed to a specific site where it was known no gold existed. Yet gold they did find, in the form of blossom rock which was a very rich ore. They named their claim the Greenhorn Mine. The town took the name and its post office opened in 1902. Richard Baird purchased the property in 1914, but with the increased cost of mining it became unprofitable. At nearby McNamee Gulch someone found a nugget that was worth about $14,000, and of course this news brought in more seeking their fortunes. The placer diggings yielded about $1.5 million in gold. By 1925 the ore had run out and so had Greenhorn.

JACKSONVILLE (Jackson) About five miles from Medford, Jacksonville was established in the 1850s after gold was discovered along Jackson Creek and Rich Gulch. It may have been a Mr. Sykes who was the first to find a placer claim. Originally called Table Rock City, the name was changed about 1860. During the camp's first winter it was impossible to get supplies and prices soared. An important commodity was salt, of which there was none to be had. An enterprising merchant just happened to have a keg of butter that was old and over-salted, so he rendered it to settle the butter, then sold it for about $3.00 an ounce to the prospectors.

Mining ceased temporarily in 1855 when the Rogue River Indian wars started. Activity later continued and Jacksonville built up with an express company and a business district. Since there were no roads at that time, the gold was shipped out by pack trains to a mint in San Francisco. The Sterling Mine was full of gold, but lack of water slowed its development, so in 1877 a 23-mile-long ditch was built to the claim, with the result that the Sterling brought out nearly $3 million. Two important mines were the Town (produced about $100,000 in gold) and Opp (nearly $100,000). Chinese miners later moved in and gleaned the placer diggings. They also dug a labyrinth of tunnels under the streets for gold.

During one of Jacksonville's early celebrations, the *Jacksonville Miner* insisted the streets be thoroughly swept to remove excess gold nuggets left by miners as the visitors strolled through town. The mayor admonished: "If the dogs in town would do their digging on the flats or in the back yards instead of in the city streets ... there wouldn't be all this gold lying around for people to stumble over. Yesterday the Marshall reported that Emil Britt's Cocker Spaniel got out of the yard and started to bury a bone in front of Charlie Chitwood's place, and rolled a big hunk of gold down the hill as big as your hat. Roy Smith came driving along and nearly busted a wheel on it. Such business has to stop or people will be afraid to come to the celebration."[266]

Jacksonville experienced a series of disasters. Smallpox reared its ugly head in 1868, then a flood

Today's Jacksonville (author's photograph).

the following year caused Jackson Creek to rush into town and flood the business section. A fire later destroyed many buildings, but the people rebuilt. After the ore was fairly well depleted, residents were still washing out some gold from their backyards, enough to pay for their groceries. Aside from the rich Sterling Mine, production after 1904 was about $500,000.

Jacksonville today is still a lively little town with more than 90 original buildings that date back to the 1850s. They were restored to their original condition using old photographs as guides. Located in the foothills of the Siskiyou Mountains, it is only five miles from the city of Medford on Jacksonville Highway. Call the Chamber of Commerce at 541-899-8118 for the many events held throughout the year.

KERBY (Josephine) James Kerby settled on a donation claim in 1854, established a general store, and Kerby grew up as a trading center for the surrounding mines. Its post office opened in 1856. The claims in the Illinois area were located a few miles from Kerby, and more than 500 miners supported the town. A pocket lode known as the Chatty was valued at about $8,000. P.W. McGuire found placer diggings at the mouth of Red Dog Creek and recovered nearly $100,000.

At one time, a nearby camp had ordered a pool table for the miners' recreation. While being hauled by mules on an arduous wagon road from Crescent City, California, one of the animals disappeared during an overnight stop near Kerby. The next morning he was found dead. The packer decided he'd had enough and set the pool table up at Kerby.

LELAND (Josephine) Leland was settled about 1852 and became a trading and supply center for the Grave Creek mines. Before the town was founded, Bates and James H. Twogood supplied the miners by hauling in provisions with their pack train. The post office opened in 1855, and was named for settler Martha Leland Crowley.

MALHEUR CITY (Malheur) Prospectors were about ready to give up their search for gold here in 1863 when they found ore along Willow Creek near Eldorado. The placers were fairly rich, but the water supply was sporadic, so the Eldorado Ditch was built to bring more water for sluicing and was completed in 1875. Nearly $150,000 in gold was realized that year. As the placers pinched out, some lodes were found, but they were not very productive. Malheur is French for "misfortune," a most appropriate name for the district, since total production came to only about $200,000.

PRAIRIE CITY (Grant) Miners from the south discovered placer gold along Dixie Creek in the John Day Valley during the 1860s, bringing a swarm of more than 400 more prospectors. The creek was located near Prairie City which served as a trading point, named for its location in a lowland. The placers were thought to have realized about $1 million. Prairie City went on to support farming and livestock after the gold was depleted.

RANDOLPH (Coos) This community was settled about 1853 and served as a trading and supply center after gold was found along the nearby beaches. But since the ore was so fine, it was hard to separate from the sand. Then a storm in 1854 that buried the miners' equip-

Dixie Meadows Mill near Prairie City, 1903 (Grant County Historical Museum, Canyon City, OR, #81-8-10).

ment caused the place to be abandoned. The camp was located along the Coquille River a few miles from the ocean. An observer once wrote of early Randolph: "Storekeepers were weighing out flour, sugar, and coffee. Miners, up from the sluices on the beach, warmed chilled bones in improvised saloons, the steam from their sodden garments mingling with the reek of whiskey, wood smoke, and kerosene. Along the beach, honeycombed by the workings of the miners, at the mouth of the creek that is still called Whiskey Run, sprawled the town, a jumble of board and log houses with shake roofs, stone and mud chimneys, oiled-paper windows, and puncheon floors. Pack mules floundered deep in the mud of the streets."[267] The prospectors who lingered a while longer high-tailed it to another tributary down the Coquille River when "Coarse Gold" Johnson found gold there.

ROBBINSVILLE (Grant) In 1864 William Daggett and a group of men found gold near the headwaters of Olive Creek located in the Greenhorn Mountains. The camp was named for William Robinson, and had a population of about 1,000. The Virginian was one of the most productive lodes, yielding about $70,000 in gold. As the veins ran out the place declined. Then in 1884 a fire all but destroyed Robbinsville.

SANGER (Baker) This site was established as a placer camp during the 1860s along Eagle Creek near the Wallowa Mountains. The gravels what were washed out yielded almost $500,000. A miner named Sanger discovered the Sanger Mine about 1870, and the town was established just below the property with a post office in 1877. It was originally called Hogum because some of the prospectors were hogs (greedy). The Sanger was the largest producer of gold, recovering about $1.5 million.

SCOTTSBURG (Douglas) Levi Scott founded this town in 1850 and opened a post office the following year. Prospectors hurried to the vicinity when ore was discovered on the north slope of the Siskiyou Mountains. In 1852 the gold excitement spurred the U.S. government to appropriate about $120,000 to build a wagon road from town to Stuart Creek. Scottsburg was located along the Umpqua River where steamers offloaded supplies and equipment for the mines, which made it an ideal trading and supply point.

James Cluggage and James Poole were packers who carried the provisions between Scottsburg and Yreka, California. Others joined them and by 1852 freighters with more than 500 pack mules were in business. The wagons were loaded with everything from Bibles to whiskey. Frequently a barrel of whiskey was lashed close to the driver's seat and a straw inserted through a small hole, where one could imbibe during a long journey. Near the point of delivery, the barrel was refilled with water from a nearby creek and the hole replugged. In 1861 a flood destroyed the town.

SPARTA (Baker) The first early prospectors to the region may have been the Leiken brothers, who found a surface mine they called Old Dry. Since there was no water nearby, they had to haul the ore by oxen to Eagle Creek for processing, and recovered close to $30,000 in gold. Later, Squire Morris and Neales Donnelly found gold nuggets near the lower Powder River and named their claim the Shanghai-Moultrie. This was one of the richer placers that yielded about $500,000. Prospector Tom Koster was right behind them and staked his claim at Maiden Gulch. Sparta was established with a post office in 1872. The placers were hard to work with so little water, so in 1873 William H. Packwood had a 32-mile-long ditch built, which was completed within two years, but by that time the gold was running out. After they were abandoned the Chinese who had worked on the Transcontinental Railroad set up camp and gleaned what was left of the diggings. The gulches produced about $157,000 in gold before 1900. Some quartz veins were later found and realized a little more than $650,000.

SUMPTER (Baker) Sumpter was established in 1862 before any discoveries occurred, and its post office opened in 1874 with Joseph D. Young as postmaster. The first claim was located by prospector Jack Hazzard about 1877 who worked it until he sold it to the Geiser brothers. Ore was also discovered along Cracker Creek. Until a narrow-gauge railroad was built in 1895, Sumpter could be reached only by an old wagon road. The Powder River Gold Dredging Company worked the creek and recovered about $12 million in gold. The most important lode in this district and in Oregon, the North Pole-Columbia, was discovered in 1887. A 30-stamp mill was built in 1895 and operated until about 1908. The mine produced nearly $9 million in gold. After 1907 a total of about $39 million was recovered from the area. During 1917 a fire of unknown origin spread throughout the main street and destroyed more than 100 buildings. But the town was rebuilt and is still very much alive.

Dance Hall at Susanville, 1900 (Grant County Historical Museum, Canyon City, OR, #82-7-8).

Open today is a tour of the gold dredge, with explanations of how the precious metal was retrieved from the Sumpter Valley. Mining Heritage Days is held during July. For more information, call the Chamber of Commerce at (541) 894-2362.

SUSANVILLE (Grant) Gold strikes along Elk Creek brought about the founding of Susanville in 1864. The placer diggings were very rich, with one claim recovering more than $80,000 from a pocket only 70 feet deep. During 1865 the Badger Mine was located and a 20-stamp mill was erected to accommodate it, which processed ore valued at about $60,000. A Mr. Haskell brought in his hydraulic operation and recovered more than $160,000 in a single day, with a total output of almost $2 million. Haskell encountered a problem at one time when a Chinese night watchman was caught stealing. Someone saw the man with an unusually heavy, bulging suitcase walking down the road. When one of Haskell's employees asked the Chinese to open his satchel, it was loaded with gold nuggets, one of them found to be worth at least $1,000.

WALDO (Josephine) Sailors from the coast who were on their way to other claims found gold in Sailor Gulch during 1852. Early mining consisted of placer operations. Water was brought in from the Illinois River to the claims by the Sailors Diggings Water, Mining and Milling Company. This ditch was about 15 miles long and cost about $75,000 to build. When Waldo was established it became the largest settlement in the gulch and was named for William Waldo, candidate for California governor.

In 1856 there was talk that the local Indians might attack the miners. As a result, The *Crescent City Herald* printed a letter, "To Mr. Geo. Sam Rice:—I am requested by Maj. Bruce and Mr. Briggs to warn the people of Sailor Diggings that they are in imminent danger of being attacked by a band from two to three hundred Indians within the next 24 hours. You must get together within this day and do the best you can to prepare for a vigorous defense. Tell the merchants to secure their ammunition. P.S. If you are attacked let us know immediately, and all assistance possible will be rendered."[268] By the time the letter was printed, Indians had already attacked the diggings.

During 1857 a stage line established a route between the gulch and the town of Jacksonville. Waldo began its decline in the 1920s when the gold ran out, but had a slight revival with hydraulic operations. It was estimated that the total minimum production was $4 million.

WILLIAMS (Jackson) Gold found along Williams Creek caused this very small town to be founded about 1857. The stream was named for an Indian fighter named Captain Robert Williams. The camp itself didn't last long, but people still live there today and cling to a very tiny community.

One of the slabs from a marble quarry at Williams was cut for use in part of the Washington Monument at the nation's capital. In 1880 the four-ton block of marble was hauled by horses to the railroad in California. Its other claim to fame during the 1980s was a small restaurant (the only one in the area) that was known for its pigs. Everything in the building from wallpaper to plates, paper towel holders to salt shakers were pigs. During breakfast, one of the owner's sons would ride a live pig that wore a silly hat around the restaurant to the delight of the customers. Williams lost its unique status when the business moved to Grants Pass.

WOLF CREEK (Josephine) It's not known exactly when gold was found in the district, but placers along Graves and Wolf creeks probably began in 1860, and continued into the 1940s. During 1897 the Greenback Mine at Graves Creek was discovered by Len Browning and Edward Hannum. The little town of Wolf Creek was founded in 1872 and served as a supply center. Crushing the ore was first conducted using an arrastre, followed by a 40-stamp mill that produced more than $1 million before the Greenback shut down in 1906. The camp had a stage line that ran from the Wolf Creek Hotel to the mine, and a post office opened in 1905. Records of the area are sketchy for early production. The Daisy Mine at Jump Off Joe Creek yielded more than $200,000; the Livingstone, about $20,000, and the Silent Friend, a little more than $7,000. Placers recovered more than the lode mines after 1930, estimated at about $400,000.

During February of 2006, plans were made to relocate Howling Acres at Wolf Creek, which is a sanctuary for wolves. It is a nonprofit organization that takes in abused and abandoned animals raised by humans that are unable to live in the wild. The organization is expected to house a veterinary clinic, and educational center, and enough room for about 80 wolves or wolf hybrids.[269]

UTAH

ABAJO MOUNTAINS (San Juan) Small placer deposits were found at the foot of the mountains, which bear a Spanish word for "low." Discovered about 1893, the claims produced very little gold. The only known amount recovered was about 51 ounces, or a little more than $1,000.

ALTA (Salt Lake) Located at the headwaters of Little Cottonwood Creek, Alta was originally an Indian hunting ground and bears a Spanish name for "high." This present ski resort began its life as a mining camp on the slope of Mount Baldy about 1864 when silver was discovered. It was a notorious place comprised of a good 25 saloons, about 5,000 people, and became infamous for its more than 500 murders committed within a 30-year period. Two of the more disreputable saloons were the Bucket of Blood and Gold Miner's Daughter where many of the gunfights occurred.

The people who took up residence at Alta did not discover until much later that all "their" land was owned by a speculative company, which offered to sell the lots back to them. The local paper wrote: "We trust this matter will be amicably settled. It hangs at present like a pall over Alta, preventing improvements in building, and in other ways retarding growth of the city."[270] Rather than entering into lengthy litigations, the people paid the company.

Two versions are attributed to the discovery of silver in the area. General Patrick Conner from Fort Douglas and his men were on a surveying expedition when they discovered an ore outcrop in 1864. Another ascribes it to the wife of an Army surgeon who found a piece of silver and showed it to someone. In 1865 J.B. Woodman began working the Emma Mine, which brought in thousands of other prospectors. Unfortunately, Woodman and his partners almost starved during the first years of development. They tried to sell an interest in the property, but could find no takers. So they began to bring the ore down the canyon in green cowhides (called rawhiding, or tobogganing) to waiting wagons and hauled the silver by oxen to the railroad at Ogden, more than 50 miles away. They realized about $180 a ton. After learning about the mine, the Walker brothers and a Mr. Hussey purchased an interest. As a result it became a major producer and stimulated mining throughout the region. The Emma produced over $3 million. Between 1871 and 1877 two smelters were built to treat the ore and recovered more than $13 million. The demonetization of silver in 1893 was a devastating blow, and by 1895 Alta ceased to exist as a mining town.

Many of properties were located in the steep mountains that were prone to avalanches. One that occurred in 1874 took the lives of more than 60 men and buried Alta under tons of snow. The worst avalanche occurred in 1885 that not only killed many people and buried the town, but also caused a horrendous fire that just about destroyed Alta. But the town survived to become a premier ski resort.

BINGHAM (Salt Lake) This town was settled in Bingham Canyon and named for two Mormons, Sanford and Thomas Bingham, who were sent by Brigham Young to find a place to grow crops. They located some placer gold about 1848 and sent word back to Young, who was displeased. He ordered them to return home because "Instead of hunting gold, let every man go to work at raising wheat, oats, barley, corn and vegetables in abundance that there may be plenty in the land."[271] So they covered up their prospect hole, which would not be relocated for many years. The canyon was found to be rich in silver and gold when discovered by George B. Ogilvie about 1863. He was followed by General Patrick Conner (in charge of the Third California Infantry at Camp Douglas) and his men who also staked claims. That same year the Jordan Silver Mining Company was established. Development of the mines was slow due to lack of adequate transportation and supplies were quite expensive. But when placer gold was discovered at nearby Clay Bar, Bingham became very prosperous.

This region was historically the fourth largest producer of gold in the United States. Most of the early work utilized drift and hydraulic mining techniques. The diggings during 1868 yielded about $100,000, and by the end of 1873 it was estimated that about $2 million in gold had been realized. After the ore ran out, silver came into the picture. When the Bingham and Camp Floyd Railroad was built, it brought opportunity for cheap transportation, and hard rock mining began. Silver mining was conducted between 1880 and 1892 until the bottom dropped in prices during the Panic of 1893. Bingham managed to survive when copper was later discovered. Total production of the region was approximately $219 million in gold.

CASTLETON (Grand) Settled along Castle Creek during the 1870s, this town served as a supply and trading center for the miners in the region. But they didn't stay long because the ore soon pinched out. The land was later occupied by sheepherders who grazed their herds on the nearby La Sal Mountains. Castleton went on to become a small ranching community.

CORINNE (Box Elder) This site was named for the daughter of prominent resident J.A. Williamson, and served as a supply and freighting point for the mines located along the Montana Trail. Corinne was established northwest of Brigham in 1868 when Mark Gilmore and others founded the town. Their goal was to create a place that would not allow Mormons. Corinne experienced rapid growth with developers, freight forwarders, and miners seeking their fortunes in Idaho and Montana. A year later a smelter was built to process the ore, which was then transported across the Great Salt Lake in barges to its various destinations. Since the town banned Mormons, Brigham Young had the narrow-gauge Utah Northern Railroad built from Ogden to Franklin, Idaho, which bypassed the community. As a result, the freight wagons were no longer necessary and Corinne began its downward spiral. By 1903 it had lost its stature as a major supply camp.

CRESCENT CREEK (Garfield) Placer gold was discovered in the vicinity about 1888. A shortage of water made mining difficult and the disappointed prospectors were acquiring only between $1 and $10 a day. But one fellow, J.F. Wilson, found a placer that yielded him about $2,500.

DEER CREEK (Utah) Deer Creek was established about 1872 when the American Fork Railroad was built and the camp became its terminus. Ore from nearby mining camps was brought to Deer Creek where it was shipped out to treatment plants. By the late 1870s the ore began to run out, and so did Deer Creek.

DIVIDEND (Utah) Although mining began during 1869 when silver-bearing ore was found in the vicinity, growth was slow because of poor transportation until the 1870s when a railroad was built. Located a few miles east of Eureka, this camp was established about 1909 by Emil Raddatz after he found a silver outcropping. He sold all his assets in order to develop the mine, and when his money was just about gone he hit pay dirt. The camp was first named Standard for the Tintic Standard Mine which produced gold and silver from 1913 to 1953, and yielded more than $80,000. The name was changed to reflect the dividends stockholders received from the properties, which amounted to more than $1 million. By 1949 more than $19 million had been recovered.

EAGLE CITY (Wayne) In 1883 Cass Hite found placer gold in Glen Canyon and began the first gold rush in the state. He and others formed the Henry Mountain Mining District. About six years later Jack Butler and Jack Sumner discovered gold in the Henry Mountains on Mount Ellen and named their claim the Bromide Mine. This precipitated the establishment of Eagle City near the headwaters of Crescent Creek, named for its abundance of eagles. Since the area was so isolated water was hard to come by for the placer mines. At the turn of the century a dredge was brought in by the Hoskaninni Company. Nearby California Bar was reported to have recovered placer gold valued at about $10,000. But throughout the town's life, its mines never produced much and by 1900 the site was deserted.

EUREKA (Juab) Located in the Tintic Moun-

Dividend (courtesy Channing Lowe).

tains, Eureka was established in 1870 and given the Greek name for "I have found it." Its post office opened the same year. Paul Schettler, along with George and Gotlieb Beck, found gold and claimed what became the Eureka Mine. The ore in this region was overshadowed by the rich strikes at Alta and Park City until a few years later when the Sun Beam and Mammoth mines were located. Stamp mills were built to process ore that was initially valued between $1,000 and $1,500 a ton. During 1872 the gold was bringing more than $5,000 a ton. Eureka prospered with the arrival of the Utah Southern Railroad in 1878, enabling cost-effective transportation of the ore. By 1930 the mines had closed down.

One of the more interesting residents was a Mrs. Anna Marks, who moved to Eureka in the 1880s and was involved in numerous scrapes with land ownership. After acquiring quite a bit of real estate, she dared anyone to trespass on her land. When Mr. Cronin started to dig post holes to build a fence between the properties, Mrs. Marks went out and filled them up again, the last time jumping into one of the holes, whereupon Cronin continued to fill up the hole. She got of the hole, but before she could leave Cronin sat on her. She took him to court, but lost the case. The owner of the Tompkins Hotel later began to build his own fence, and Marks took potshots at him. Finally when the Rio Grande Railroad came through to build a spur across her property, they had to settle and pay her $500 in order to do so. Everyone breathed a sigh of relief when she passed away in 1912.[272]

FRISCO (Beaver) Located in the San Francisco Mountains, Frisco was settled in 1876, a year after silver was discovered. Its post office was established in 1877 and named for St. Francis of Assisi. A major claim was the Hornsilver Mine, so called because the silver curled like a ram's horn when thinly cut. The mine was discovered accidentally in 1875 by James Ryan and Samuel Hawkes of Pioche, and became the richest silver producer in the state. From a hole 900 feet long, 400 feet wide and 900 feet deep, more than $54 million was taken out in a ten-year period. The *Salt Lake Mining Review* wrote: "'This ledge looks kind of good,' said one as he drove his pick into a limestone outcrop. It was good. Beneath the weathered surface galena ore glistened in the sun. They agreed it was a 'bonanza' and scratched that name on a location notice."[273]

The rush was on, and Frisco became a rip-snorting town with more than 20 saloons, and shootings and killings the order of the day. After residents finally got tired of it they demanded, and got, a United States marshal named Pearson to restore order. Some of the roustabouts didn't think Pearson had what it took to tame them, but after killing five men the first day the town quieted down considerably.

The Utah Southern Railroad laid its tracks to Frisco in 1880, enabling shipment of ore that was cost-effective. With the fabulous wealth to be had, owners of the Hornsilver disregarded all mining tenets, working the men even as they dropped from 180-degree heat in the mine. The riches came to an end when a massive cave-in occurred, caused by

Frisco (courtesy Chad Carter).

badly timbered tunnels that had to bear the weight of snow and rain. Although a new shaft was sunk most of the miners left, taking even their houses with them. The Hornsilver was reopened for a time in 1928, but by that time not much was left of Frisco.

GLEN CANYON (Garfield) Placer gold was first discovered by Cass Hite about 1883. Although numerous claims were staked within a few years, they were not very rich because the gold was so fine it was difficult to recover, and the canyon was abandoned.

GOLD HILL (Tooele) The first mineral discoveries were made in 1858 by gold seekers heading to the California gold fields. But little work was done at that time because hostile Indians drove them away; their depredations continued until about 1869. Gold Hill was established in an isolated canyon a few miles from Ibapah. A crude smelter that was built in 1871 was operated using three bellows, but production was small until Colonel J.F. Woodman discovered two rich claims and built an amalgamation plant. His mines recovered about $300,000, but the other claims failed to produce as anticipated. Not much activity occurred until World War I, when a mill was built to treat the ore. By 1938 nothing was left and the town was deserted.

JACOB CITY (Tooele) This site was established during 1865 at the headwaters of Dry Canyon when gold was found nearby. Jacob City was built on the side of a mountain near the mines. As a result, bringing the ore down in carts hauled by mules was precarious at best, as described by an early pioneer: "Our trail led first to Dry Canyon, to arrive at which we passed through Jacob City. This city, not 'set upon a hill,' but hanging like a collection of crows nests on the side of a mountain, cannot be approached on wheels. Sure-footed horses and mules are rather doubtful of their foothold on its streets, paved with boulders and drained by the gully of a torrent. If heavy rains should swell the stream as they are liable to do, or an avalanche of snow, which every winter threatens, should descend, the flimsy structures of Jacob City would fly down into the abyss below like a pile of shingles before the storm."[274] A toll road was later carved out to the town of Stockton. One of the mines was averaging about $60,000 a month, but by the early 1900s the ore had run out and the place was abandoned.

Jacob City ruins (courtesy Channing Lowe).

KIMBERLY (Piute) Kimberly is located in the Tushar Mountains. During the 1890s gold and silver claims were staked, some of which Peter Kimberly later purchased. During its prosperous time the town had about 500 people. A mining district was organized around the Annie Laurie Mine in 1899, the most productive lode in the area. The gold was shipped on the Shepard Brothers Stageline to a railhead at Sevier. Kimberly saw its demise during the early 1900s when the ore ran out. There was a slight resurgence in the 1930s after a vein was relocated, but in a few years it, too, was gone. Total yields were estimated at about $2.7 million.

KNIGHTSVILLE (Juab) Jesse Knight was a mining man and financier who established the camp about 1897 after he realized the wealth of the newly discovered Humbug Mine. Knightsville was a typical company mining town, situated a few miles east of Eureka at the foot of Godiva Mountain. Its post office opened in 1909. A smelter was built at Knightsville during 1907 after rich silver ore was found, bringing more than 1,000 people there. After the mines played out in 1942 the site was abandoned. Knight later moved on to Silver City and built one of his mills there.

MAMMOTH (Juab) This site was established in the Tintic valley when silver outcroppings were discovered in 1870, and named Mammoth because of the large amount of ore that was located nearby. Its post office opened in 1890. Charles Crimson operated the Mammoth, but development was slow so Crimson sold his interests to Sam and William McIntyre in 1873. They were rewarded with a body of ore that produced for almost 75 years. In 1890 two mills were built to accommodate the mines. Total production exceeded $20 million. Decreasing values of ore in addition to high mining costs shut down all operations.

MANNING (Tooele) Located in Manning Canyon, the camp sprang up in the 1880s, named for L.S. Manning who promoted the use of cyanide in mining operations. It was established as a mill site for the claims at nearby Mercur. The Mercur Gold Mining Company used Manning's cyanide process, which enabled about a half ounce of gold to be extracted from each ton of ore. In 1895 the Salt Lake and Mercur Railroad was built and used in the transport of ore. When the town of Mercur constructed its own mill, Manning began its decline. Eventually the sites expanded to become one.

MERCUR (Tooele) General Patrick Connor and his men found silver in Mercur Canyon during the

Ore bin at Mercur (courtesy Chad Carter).

1860s, but the yields were disappointing. So the claims were sold to the Floyd Silver Mining Company, which invested $700,000 developing the property and the Camp Floyd District was organized. Unfortunately, the company recovered only $175,000. Some placer mining was attempted, but lack of water and low-grade ore brought operations to a halt. After Mack Gisborn realized more than $80,000 from his claim, miners hightailed it to Mercur. By the 1880s the boom was over, and after suffering from a fire during that time the place was not rebuilt.

Mercur took on new life when a prospector looking for gold discovered cinnabar with a high content of gold. But efforts to extract the gold were unsuccessful with amalgamation treatments used at that time, and the town finally succumbed, or so everyone assumed. However, in 1934 a side canyon called Horse Thief Gulch was found to be rich in gold, and within two years Mercur had risen from the ashes once again to become the second largest gold producing camp in Utah that lasted until about 1950. Total production from all the mines amounted to more than $20 million.

NEWHOUSE (Beaver) Newhouse was located in the San Francisco Range and developed in the 1870s after the first mine was discovered. Called the Cactus, it was worked unsuccessfully until Samuel Newhouse purchased the property which operated between 1905 and 1910. It was estimated the mine produced more than $3.5 million. First called Tent Town, the name was later changed. It was one of the few places that allowed no saloons, gambling, or houses of ill repute on orders from Newhouse.

OPHIR (Tooele) This camp was situated in Ophir Canyon south of Stockton. General Patrick Connor and his party came here during the 1860s after learning that the Indians were mining silver to make bullets, and found gold in addition to silver. A.W. Morris laid out the town about 1870. First called St. Louis, it was renamed for the mine that was located nearby. The town peaked about 1880 with a population of nearly 6,000. Mack Gisborn, who made a strike at Mercur, built a toll road from Ophir to Stockton over which ore was shipped to the railroads. Marcus Daly discovered the Zella claim, and later went to Montana where he became a copper king. The Kearsarge Mine at Ophir is said to have produced $1 million. One of the largest silver nuggets from the mine was exhibited at the St. Louis World's Fair in 1904. Since the ore beds were shallow, they were soon depleted. Early mines in the region recovered about $550,000. Efforts at smelting the ore initially met with small success, but the mines were abandoned by the end of 1865. Additional discoveries of silver ores during the early 1870s caused a brief period of prosperity that lasted until about 1890. Total production of gold in the district was about $2.1 million. During the 1990s there were still a few people living at Ophir.

PARK CITY (Summit) Park City never died and has a population today of more than 7,000, living its life as a ski resort, but it has preserved its old-day flavor. George G. Snyder named the town for Parley's Park located farther north which honored trapper Parley Pratt. Colonel Patrick Conner and his men came to the region in 1869 and stumbled upon gold and silver. A small stampede began during 1879 with the discovery of more ore in Walker and Webster gulches. When the famous Ontario vein was located, its bonanza ore bodies stimulated interest in the area and gave Park City early prominence in the up-and-down mining industry. The ore was deep in the ground and before long the workers reached the water table, flooding the mines. Additional capital was gathered and the properties were drained. Then in 1893 a decline in silver prices caused some of the claims to close down. During this hiatus, better methods of mining and more effective recovery methods were developed. In 1901 more rich discoveries were made. The Ontario mine produced more than $17 million in silver. Total production of gold amounted to about $16.3 million.

SAN JUAN RIVER (San Juan) Placers along the San Juan River were first located about 1879 and worked intermittently until 1891. Rumors of rich accumulations brought in investors who built mills and other equipment to recover the gold. But the ore was so fine and in too small amounts to make it worthwhile and the diggings were abandoned. The river was named for Spanish explorer Don Juan de Oñate, or Don Juan Maria de Rivera.

Old workings at Ophir (courtesy Channing Lowe).

SCRANTON (Tooele) Scranton was founded in the East Tintic Mountains about 1908 when the New Bullion Mine was located. A year later the post office was established. The site was deserted by 1910 when the mine played out. In 1971 a range fire occurred, burning what was left of the camp.

SHAUNTIE (Beaver) Shauntie was settled near the Star Range at the head of Shauntie Wash and honored a Paiute Indian, whose name was interpreted as "much." Starting out as a tent town in 1870, the camp served as a smelter center for the more than 1,500 claims. By 1875 Shauntie was on its way out after the smelter burned, followed by a fire that nearly destroyed the camp, and then the ore proved to be superficial.

SILVER CITY (Juab) Silver City had its beginnings in 1869 when George Rust found the remains of old mines with silver that had previously been worked by Indians. The camp was established in 1870 at the mouth of Dragon Canyon about

Wells Fargo building, Silver Reef (courtesy Chad Carter).

five miles south of Eureka. It became a trading center for the silver mines in the region, with a post office in 1871, and had a population of about 700. Someone once wrote that Silver City had "a billiard saloon, blacksmith shop, grog hole, some tents, several drunks, a free fight, water some miles off, a hole down 90 feet hunting a spring without success, and any number of rich or imaginary rich lodes in the neighborhood. The owners are all poor, and the poor men work for them. By next spring the poor will be poorer."[275] The mines shut down when water was encountered, and the cost of pumping it out proved too expensive. There was a small revival after William Hatfield located the Swansea Mine, which produced about $700,000 in silver. By the 1930s the place had been abandoned.

SILVER REEF (Washington) This camp was settled about 1866 near the Pine Valley Mountains. First called Bonanza Flats, the name was later changed to Silver Reef. Discovery of the silver was unusual in that it was found in a sandstone ledge by prospector John Kemple. But the amount of ore he saw was so small he left without further investigating the prospect. Tradition also says that William T. Barbee was credited with the discovery. During his travels he was welcome to stay overnight with a settler and, while warming himself by the fire, noticed something odd oozing from the sandstone fireplace, which turned out to be hornsilver. He wrote a letter to the *Salt Lake Tribune*: "We have an abundance of rich silver mines which your correspondent is now developing, and bringing to the surface their rich chlorides and horn silver—all in sandstone too.[276] A prospector and assayer from Pioche, Nevada, "Metalliferous" Murphy, was also thought to be the discoverer. When two miners brought in sandstone samples, Murphy saw the silver and later followed the men back to their quarry, taking credit for the discovery. Interest in the ore was generated during 1874 when Elijah Thomas sent his sample to Salt Lake City bankers. The most productive years were between 1877 and 1903 with recoveries amounting to over $10 million.

WEST DIP (Tooele) Established as an outgrowth of Mercur in 1895, West Dip served as a way station for travelers in addition to miners and freighters. An ore mill was built during 1898 to serve the nearby mines until 1917 when a fire destroyed it. The claims turned out to be marginal and the place was abandoned.

WASHINGTON

ADDY (Stevens) This small camp was established during the 1870s when a Mr. Fatzger claimed the land and built his grist mill. After the Spokane Falls & Northern Railroad came through in 1890, Addy changed from a small settlement to an outfitting and distribution center and acquired its post office the same year. The railroad shipped out hundreds of tons of rich silver ore from the area's mines. The site was named for Adeline Dudrey, wife of the first postmaster.

BARRON (Whatcom) First discovery of gold in 1893 is credited to Alex Barron, which brought in scores of prospectors. Located near the Cascade Summit, this camp supplied the miners after establishment of a trading post owned by Guy Waring (who earlier built one at Winthrop). The Slate Creek Mining District was formed and a small narrow-gauge railroad was built to the mines. In 1895 the Mammoth Mine was discovered and a 5-stamp mill was built to accommodate it. Later, the New Light and Azurite mines began operating and continued until about 1935. Eastern interests took over the Azurite in 1901, and had a wire-rope tramway built from the mine to a 10-stamp mill located at the junction of Slate and Bonita creeks. These two claims accounted for most of the lode gold production, estimated at nearly $1 million.

BLEWETT (Chelan) Gold was discovered in 1855 by a Captain Ingalls after an Indian showed him "color." Ingalls left for a period of time, then returned to the site, only to be killed. Further development did not occur until 1858 with the arrival of miners returning from the Fraser River strikes, at which time the camp was formed at the headwaters of the Swauk and Peshastin creeks. It was named for settler Edward Blewett. After a newspaper heralded the great news, storekeepers from Puget Sound were anxious to be the first to outfit the miners and headed to Blewett. Successful placer mining occurred first, followed in the 1870s by lode mining at nearby Culver Gulch. The difficulty in transporting ore was solved when a wagon road was built over the Wenatchee Divide. A 20-stamp mill was erected by the Chelan Mining Company, and a 4,000-foot cable tram took the ore from the mines to the mill. By 1910 operations had declined. Total production was about $17.5 million in gold.

BODIE (Okanogan) Located along Toroda Creek north of the town of Wauconda, this camp was established in 1900, although a post office was in place during 1899. Before the Colville Indian Reservation was opened to mining, Henry DeWitts made his strike less than a mile from the camp. He was unable to work his claim until the government separated that section off from the reservation. DeWitts staked several properties, one of which he sold to his brother, who found a rich pocket of ore worth about $80,000. The two later sold their interests to the Wrigley Brothers (famous chewing gum magnates). The mines were served by a stamp mill that processed the ore up until the 1930s. A little more than $150,000 was realized from the Bodie Mine.

BOLSTER (Okanogan) Bolster was settled in 1899 by J.W. McBride a few years after the discovery of gold in the area. Located a few miles from the Canadian border, the camp was named for mining promoter H. Bolster, and its post office opened in 1899. Serving as a trading center for the miners, the site lasted only until 1904.

Bodie (courtesy Dean LeMaster).

BOSSBURG (Stevens) This site was established along the Columbia River in 1888 with discovery of silver at the Young America and initially named for the mine. The camp was renamed in 1891 to honor Postmaster Chester S. Boss. Before mining occurred, the region was used as pasture grounds for the Hudson's Bay Company. After the silver discovery, a wagon road was built to a smelter established at the town of Colville. After demonetization of silver in 1893 the mine closed down. Bossburg was kept alive after a cable ferry was built across the Columbia River in order to ship out produce and lumber grown in this region.

CHESAW (Okanogan) Named for Chinese storekeeper Chee Saw, this town came into existence during the 1880s with a stampede of white miners after discovery of gold along May Ann Creek. When they arrived the men saw that many Chinese miners had already preceded them. Chee Saw was one of the most prominent residents, from whom the prospectors purchased most of their supplies. The area was not developed until the early 1900s with lode mines, but the assays did not carry their values and eventually the mines were abandoned. Only about $196,000 in gold was recovered. The region was later developed as an agricultural center.

CONCONULLY (Okanogan) First called Salmon City, the town's name was changed to Conconully which means "cloudy," "evil spirit," or "money hole." The year 1886 brought prospectors to the Salmon River when word got out about gold and silver. What began as the tents of miners turned into a settlement within a few years. The U.S. government cut off part of the Moses-Columbia Indian Reservation and opened it up for mining purposes. Ore from the claims had to be transported by wagon to the Columbia River, then to the railroad at Wenatchee, and finally on to the smelters. Conconully was town of disasters: a fire in 1892, the end of the gold and silver boom in 1893, and a flash flood the following year. The fire started in an empty store, and the blowing wind caused the town to go up in flames. The *Spokane Review* wrote, "The entire town was suddenly illuminated by a fierce, red glare ... a great mass of whirling sparks and cinders."[277] Conconully finally passed into oblivion.

EXCELSIOR (Whatcom) This short-lived camp was located along Wells Creek near the Nooksack River. It was home to the Excelsior Gold Mining Company, a hopeful name meaning "best," and the post office was established in 1902. The Great Excelsior Mine was discovered by W.H. Norton, who also built a stamp mill in 1900. From the beginning the mine contained little ore value and encountered problems in milling with a loss of over 50 percent. Norton sold his property in 1905 to an interest that rebuilt the mill around a gold cyanide process, but operations ceased about 1917. H.E. Barnes relocated the mine in 1922 because the former owners failed to do their annual assessment work. He worked his claim until 1960, but again very little gold was recovered.

GLADSTONE MOUNTAIN (Stevens) In 1915 J.E. Yoder found silver at the summit of the mountain, and a year later the property was developed by the Gladstone Mining Company. The first lode mine was the Electric Point Mine, so named because of the frequency of lightning strikes during thunderstorms. Then in 1917 the Gladstone was located. Both claims operated until about 1926 and recovered approximately $50,000.

GOLD BAR (Snohomish) Located near the mining town of Sultan, Gold Bar was established in 1899. Prospectors searching for ore in the gravels of the Skykomish River and its tributaries gave the place its name. But only traces of gold were found, and the camp went from mining camp to a construction camp for the Great Northern Railroad, and then later served as a logging headquarters.

GOLD HILL CITY (Okanogan) This was a very short-lived camp located at the foot of Palmer Mountain about five miles from Loomis. In 1892 John Reed and Dan McCauley discovered gold nearby. The property was later taken over by the Gold Hill Mining Company. Consideration was made to rebuild an old road that constantly washed out to transport the ore, but there was not enough gold so the idea was abandoned. The camp lasted until about 1910.

INDEX (Snohomish) Living its life as a trading and supply center, Index is located along the North Fork of the Skykomish River. Its name was taken from the mountain, for a pinnacle that looks like an index finger. Amos D. Gunn moved there in 1890, established the first tavern and hotel to serve the mines, and opened the post office in 1893. That same year a fire occurred when one of the children dropped a candle, and the conflagration demolished many of the buildings. When news arrived about the riches in the Klondike, most of the miners left. Index today is a bedroom community with a population of about 140.

LIBERTY (Kittitas) In 1868 settler Bent Goodwin went to Swauk Creek for cooking water and accidentally came across a gold nugget, attracting others who arrived to stake their claims, and Liberty was soon established along the stream. There were nuggets everywhere along all the small creeks in the area. So many, in fact, that the finer gold went unnoticed. One nugget was valued at more than $500, some miners realized about $1,000 from one pan of gold, and others averaged

Water-powered ore mill, Liberty (Library of Congress, Prints and Photograph Division, Historic American Buildings Survey, 19-LIB,1W-1).

$600 a day. Liberty declined after word arrived of the great Yukon gold rush.

Folklore says an old prospector was having his evening meal when a two-ton rock came flying down a hill and smashed through the cabin roof, hit the table and the fellow's meal went in all directions. He sat there for a minute checking out the damage, and hollered so all in the canyon could hear, "All right, dammit — who threw it?[278]

LOOMIS (Okanogan) Surviving as a crossroads community with a small store and restaurant, Loomis had its beginnings about 1870 as a winter cattle station. Lode gold discoveries the following year at nearby Chopaka Mountain changed all that. The town was named for store owner Julius A. Loomis, with a post office in 1889. Hiram F. Smith was thought to be the first discoverer of gold; he later became a member of the state legislature. News of gold brought in the miners from the overhyped Frazer River diggings in British Columbia. But upon their arrival here, one of the discouraged miners wrote a letter to his wife. "I have been humbugged. Gold is not as plentiful as we were led to believe, and it is ten times harder work to get it than anyone could have imagined."[279] And he would be further discouraged with the low-grade gold and the difficulty in hauling it to the mill because a road that had been crudely built was always washing out. Loomis began to decline in 1910. The 5-stamp mill yielded only about $113,000.

LOOP LOOP (Okanogan) This camp began in the 1890s and took its name from Loup Loup Creek, French for "wolf." Years earlier, Hudson's Bay trappers came to the region for the purpose of acquiring furs. There was some moderate mining, but falling ore prices caused a collapse in activity. Those who resided at the camp abandoned the site, taking anything of value that could be moved.

MILLER RIVER (King) First called Berlin, this camp was located near the Miller and Skykomish rivers. It was renamed for a local prospector, and spent its life as a supply center for the surrounding mines, with a post office opening in 1918.

MONTE CRISTO (Snohomish) Gold was discovered in 1889 on the west slope of the Cascade Range, and as a result the Silver Creek Mining District was formed. While searching for ore, F.W. Peabody and Joseph Pearsoll saw something shiny through their field glasses. The gold they located turned out to be quite rich, but transportation problems hampered them so they built a rough wagon road. The Rockefeller family later purchased the mine interest, had a concentrator built, and constructed a railroad in 1893 at a cost of $2 million. After the road washed out in 1896 and 1897, the property was sold to the Guggenheim smelter trust in 1903. Top producers were the Mystery/Pride and Justice mines. Gold from the district was valued at $6.1 million.

MOUNT BAKER (Whatcom) This site was named for the mountain, honoring Lieutenant Joseph Baker who was the first Englishman to see it. Gold deposits were worked on nearby Ruby Creek, but no real interest was aroused until richer deposits were discovered in 1897. R. Lambert, J. Post and L. Van Valkenburg located the Post-Lambert Mine. A year later L. Friede and H. Hahn from Oregon took ownership of the claim, paying $50,000 for it. A 10-stamp mill was brought in via horses and a skid-mounted steam engine. An aerial tramway was later built to bring the ore from the mines to the mill, which operated successfully until it burned down in 1907. Then the tram collapsed, and operations came to a halt. In 1915 there was a small resurgence when the Boundary Gold Company took over and rebuilt the mill and tram. But in 1924 a snow slide completely destroyed both, and that ended mining in the area. Gold values were estimated at about $1 million.

NIGHTHAWK (Okanogan) Named for the mine in recognition of the bird which is a relative of the whippoorwill family, the camp was established about 1903. Placer mining began along the Similkameen River, but came to an end temporarily when it was learned the area had been set aside for an Indian reservation. After the land was open, James M. Haggerty from Oregon came along and located three claims. Miners headed to Nighthawk after Haggerty got drunk one night and spilled the beans. In the bar were men from the mining camp of Loomis. Watching them through bleary eyes, Haggerty became belligerent and mouthed off, saying, "Well, it's a good thing Loomis can produce molasses. It'll never turn out any gold from that damn mine."[280] The Loomis fellows grabbed Haggerty, tarred and feathered him, and ran him out of town. No one ever saw him again. A spur line was run to Nighthawk by the Great Northern Railroad to ship out the gold. Total production before 1900 was about $1 million.

NORTHPORT (Stevens) Northport was a smelt-

Northport Smelting and Refining Co., Northport, ca. 1910 (University of Washington Libraries, Special Collections, UW4863).

ing town, formed in 1892 when the Union Smelting and Refining Company built its facility at a cost of $250,000. The equipment was built here because a smelter at British Columbia was so busy it could not handle the county's ore. Landowner and railroad man D.C. Corbin donated some of his property for the enterprise. He also had the Spokane Falls & Northern Railroad built in 1892 in order to ship out the gold. The post office was established the same year. Northport was located along the Columbia River just south of the Canadian border and somewhat isolated, since the only route there was a rough trail. But a little newspaper set up shop and in its first issue wrote: "It is already a town. Tomorrow — a few tomorrows hence, at any rate — it will be a city."[281] But that never happened. Northport suffered from fires, and then in 1894 it flooded, washing away most of the buildings. During 1897 another facility was built by the Northport Smelting and Refining Company, which operated until 1921.

OKANOGAN CITY (Okanogan) Located along the Similkameen River near the town of Oroville, this camp came into existence about 1871 when Hiram Smith discovered gold near Mount Chopaka. Okanogan is a Salish word meaning "rendezvous." Within one week more than 3,000 miners flocked to the site and Okanogan City thrived as a trading center for the area. While the miners were in the process of locating their claims, Chief Moses complained to authorities that the land was part of his reservation. As a result, a contingent of soldiers rode to Okanogan and forced the men to leave their claims, who moved on with the news of strikes elsewhere.

ORIENT (Ferry) Gold was discovered on First Thought Mountain about 1900, and the camp was established and platted by Alec Ireland. There, the Orient Improvement Company was formed. Located south of the Canadian border, the camp acquired a population of about 300. Numerous gold claims were located, and when the Great Northern Railroad completed its line through Orient, railroad workers joined the miners in their search for wealth. Settler George Temple established a dray line to haul machines and equipment for the mines in the vicinity. The First Thought Mine was the best producer of all the properties. Between 1904 and 1907 it yielded almost $685,000 in gold. But the escalating costs of recovering the gold exceeded ore payments, and the mines ceased operating.

First Thought Mint, Orient, ca. 1905 (Ferry County Historical Society, Republic, WA).

OROVILLE (Okanogan) Situated just south of the Canadian border, this camp was founded by Alexander McCauley in 1873 on what was once part of the land claimed by Chief Moses. With discovery of gold in the outlying areas, Oroville became populated with about 3,000 miners and 20 saloons, although the first store wasn't built until about 1891. It later served as a supply center and milling town. Ranchers took advantage of the situation and brought in their cattle to sell the meat to prospectors. Not much ore was found, however, and Oroville quietly faded away. It did serve for a time as a port of entry and customs and immigration station.

REPUBLIC (Ferry) Republic is situated near the San Poil River, just south of the Canadian border. Gold was first discovered along Eureka Creek in 1896 by John Welty. The same year Thomas Ryan and Philip Creaser located the Great Republic Mine, Creaser platted the town, and the post office was established in 1897. The camp flourished, as the *Republic Pioneer* reported: "Here is a little city that is moving right along. Large quantities of whiskey, flour and other necessities arrived during the week."[282] Along with the usual businesses, it also boasted more than 25 saloons. Serious prospecting began after part of the Colville Indian Reservation was opened in 1896. High gold content was shipped out until about 1901 when operations were temporarily halted. After 1903 the region was linked to smelters along the coast and activity resumed. High-grade ore was later found at the Republic Mine, bringing a slight revival. More than $17.5 million in gold was recovered.

ROGERSBURG (Asotin) First serving as a ferry waypoint in the 1860s, Rogersburg was established along the junction of the Snake and Grande Ronde rivers. Named for landowner G.A. Rogers, it later became a trading center with a post office in 1912. Tradition says gold was found on a bar of Shovel Creek by three prospectors about 1865. Needing supplies, they hid their gold and headed for Walla Walla to outfit themselves. Unfortunately, before any of them could return two of the men died; nobody knows what happened to the third. But before their demise, one of the men told someone of their find, and it wasn't too long before other prospectors followed their trail. But the gold didn't pan out and the miners left.

Mine near Republic, ca. 1916 (Ferry County Historical Society, Republic, WA).

RUBY (Okanogan) Once called the Babylon of Washington Territory, this camp was wild from the get-go. Cheap whiskey, gunpowder and rampant dishonesty characterized civic life. A mining engineer said in 1889 that the biggest liars and thieves he had met in other towns were honorable compared to the "beauts" who lived at Ruby City."[283] The first discoveries were located by Thomas D. Fuller and three partners; their claim would later become the Ruby Mine. The town was established by Fuller along the Salmon River near the mining camp of Conconully in 1886. Its post office opened two years later. Fuller later sold his interest to Jonathan Bourne, Jr., a senator from Portland, Oregon. D.C. Corbin, who owned the Spokane Falls & Northern Railroad, had planned on building a spur line to Ruby in the belief there were great quantities of silver. Hopes were high, as the *Ruby Miner* printed, "As Virginia City is to Nevada so is the town of Ruby.... It is out of debt and has money in the treasury."[284] But it was a flash in the pan, and the Panic of 1893 spelled doomed for Ruby. A flood, followed by a fire that consumed the camp, didn't help matters.

SILVERTON (Snohomish) Silverton served as a mine transportation center on the Everett & Monte Cristo Railroad. In 1891 both silver and gold were discovered near the Stillaquamish River, but the silver veins quickly petered out. By 1910 shipment of the ore had ceased. A total of about $3 million in gold was realized.

SPOKANE (Spokane) Acting for the North West Fur Company, Finan McDonald and Joco Finlay established a trading post called Spokane House in 1810 (later just Spokane). They took the name from the Spokane Indians, which signified "children of the sun," or "sun people." The chief called himself Illim-Spokane, "Chief of the sun people." Early residents near the falls were J.J. Downing and his family; in 1871 they homesteaded along the banks of the river. Encouraging settlement was the discovery of silver deposits in Idaho. Placer camps on the North Fork of the Coeur d'Alene River made Spokane a major supply point where heavily laden freight wagons and strings of pack animals transported goods to those mining camps.

SULTAN (Snohomish) In 1874 silver and gold were discovered along the Sultan River, followed by the establishment of the camp. Development was slow until about 1878 when mining companies became interested and brought in machinery and other equipment. The site was located at the confluence of the Sultan and Skykomish rivers,

Spokane when it was Spokane Falls, ca. 1887 (Northwest Room Spokane Public Library).

named for Snohomish chief Tseul-tud. Sultan became a supply center for the mines when the Great Northern Railroad built its line through town. About 1891 the 45 vein was discovered, soon to become one of the biggest strikes in the area. Its ore was transported via the Everett and Monte Cristo Railroad, but the tracks continually washed out, so operators of the 45 Mine built a plank road that went from the property to the Great Northern Railroad. Most of the ore production was sporadic, and there are no definitive amounts available, since its ore was combined with that of the Silver Creek District production amounts (see Monte Cristo).

TORODA (Ferry) Located about four miles from the Canadian border, Toroda was a supply depot during the 1890s for miners who raced to the region when they learned there were great riches to be taken. But rich they were not, and the discouraged men left.

TURK (Stevens) Turk was founded along Alder Creek to serve as a trading center, and the post office opened in 1905. It was originally formed by a mining company to simply house the workers.

TWISP (Okanogan) Twisp is a modification of t-wapsp, meaning "yellow jacket." Another derivation is from twistsp, an imitation of a yellow jacket's buzzing. It was also thought to be a Chinook word designating the forks of two rivers. H.C. Glover named the first site Gloversville and platted the town in 1897. When Amanda Burgar purchased land adjacent to Gloversville, she built a hotel and named that section Twisp. The town's economy was based on serving as a supply center during the mining era, and a number of mining districts were established. Gold discoveries were made along nearby Alder Creek during 1896. A major claim was the Alder that contained both gold and silver. In 1903 the Alder Group Mining and Smelter Company purchased the mine and further developed the property. The ore was shipped to Tacoma by railroad until 1941 when operations were closed by the U.S. government due to World War II. More than $300,000 in ore was recovered. Livestock, dairy farming and agriculture later supplanted mining.

WALLA WALLA (Walla Walla) With the discovery of gold in Idaho near the town of Orofino, miners flocked to Walla Walla which served as a major supply station. Wells Fargo also established a stage service to the mines. Someone wrote, "I got in sight of the City of Walla Walla — it was a lively little place of about two or three thousand inhabitants, mostly transient miners, teamsters, packers and land-seekers. There were three hotels and a lot of boarding houses ... one or two dance houses of evil repute, and one or two hurdy-gurdy dance houses of fair repute — taken altogether Walla Walla was a fast place, and a great outfitting place for the mines all over."[285] Prospectors also used the town as their winter headquarters. Those who did not had to make sure they stocked enough food and supplies at their diggings until spring.

Residents were astonished when they saw some of the pack trains enter town with camels. In 1850 the U.S. Army imported camels as pack animals to the Southwest desert. A few years later they sold them to freighters. The camels could each carry up to 500 pounds of supplies. As one old timer said, "They would be loaded with sacks of flour until you couldn't see anything of the camels except their heads.... They would go up and over the mountains in the roughest and steepest places and never refuse to keep moving along in their slow, deliberate way."[286]

WENATCHEE (Chelan) Wenatchee sits at the foothills of the Cascade Mountains, a favorite spot for many of the Northwest tribes who once used it as a council ground and camp. The name was interpreted from a Yakima word for "robe of the rainbow," "good place," or "boiling waters." Two men named Ingraham and McBride opened a trading post to sell supplies to the Indians. In 1855 the Gold King Mine was discovered nearby, but little mining occurred until 1894 when the property recovered about $1,600 in gold. Construction of the Highline Canal in 1903 brought water to the Wenatchee region. Large-scale mining began during the 1940s after the Lovitt Mining Company purchased the Gold King. Total production of the district came to about $4 million. After the gold was gone, farmers moved in and cultivated apple orchards. By the 1920s Wenatchee became one of the important apple-producing regions in the world, and is still a prosperous town.

WINTHROP (Okanogan) Winthrop spent its life as a distribution point for miners heading to the placer diggings in the Slate Creek District. A trading post was also established in 1891 by Guy Waring (who later built one at Barron). Located at the confluence of the Chewack and Methow rivers, the camp was named for author Theodore Winthrop. The town is still active and the highway through town heads into the North Cascades National Park.

WYOMING

ATLANTIC CITY (Fremont) Before any settlement occurred, this was the site where a party of Mormons died in a horrendous blizzard during 1857. Atlantic City was established in 1868 with the discovery of gold. It was named either because the town was on the Atlantic side of the Continental Divide, or for the city in New Jersey. Mills were built to treat the ore from nearby claims, and the town grew to about 2,000 people. Along with the usual businesses established, Atlantic City also boasted an opera house and the first beer garden and brewery in the state. By 1875 most of the ore had been exhausted. A milling company moved in about 1904 and built a cyanide plant, but with so little gold left the project was abandoned. The town saw a slight resurgence during the Depression when another company moved in and put a dredge into operation, recovering more than $700,000 in gold. During the 1960s iron ore was found in the mountains; U.S. Steel came in and developed an open mine, bringing new life to the town. Atlantic City was so close to South Pass, some of its gold values are included there.

KERWIN (Park) Kerwin was a short-lived camp formed by William Kerwin shortly after he discovered gold in 1885, but it would be about ten years before any serious development. The Shoshone River Mining Company was established, and the ore it recovered had to be shipped out by mule. Kerwin never had a large population, probably no more than 200 miners. In 1905 one of the men was killed in a mine explosion, and two years later an avalanche descended on the camp, killing a number of people. After it was all over everyone abandoned Kerwin. Gold production records are unknown. The Forest Service later purchased the land.

MOORE'S GULCH (Albany) Prospectors found gold during 1868 in the gulch. Not much is known of the place, but some of the claims brought out about $8,000. Total production of the placers was estimated at close to $40,000.

SOUTH PASS CITY (Fremont) During 1842 a trapper found gold along the Sweetwater River, but he was killed before anyone learned of his discovery. Mining was sporadic for a number of years because of Indian attacks. During 1866 prospectors stopped at the river to make camp for the night and found the gleaming slivers of gold. But the rush didn't really begin until the next year when the Cariso Mine was discovered. It would yield between $1 and $2 million. Eager miners rushed to new discoveries. Someone reported: "I do not deem it necessary at this time to speak of the extent or richness of our mineral veins. That we have them of extraordinary value both in gold and silver is now, I believe, a well established fact.... This city will soon take rank among the magic cities of the West.[287] A sawmill was built so miners could construct their shelters, and South Pass was soon established. The town also had Pony Express and Overland Stage stations. Stocks and leases were manipulated by speculators who tried to get control of all the claims, but the rich veins soon played out and by the 1870s most had left the scene. Approximately $6 million in gold was taken from the mines.

Prominent resident Esther Hobart Morris was responsible for the first women's suffrage in the state. She invited legislative candidates to a tea

Remnants of a saloon at South Pass (Library of Congress, Prints and Photographs Division, Historic American Buildings Survey, 7-SOPAC).

party in 1868 and convinced them to vote for the bill. It became law, which was the first in the U.S. for women's equal rights. Mrs. Morris went on to become the first woman justice of the peace in the world.

Glossary

Alluvial— Pertaining to material carried or laid down by moving water along a stream bed.

Amalgamation— Extraction of precious metals by alloying them with mercury.

Arrastre— A Spanish device used for the free milling of gold or silver; a circular pit whose bottom is lined tightly with rock facing, over which heavy stones were dragged by horse or mule power, to grind ore. Arrastre is derived from Spanish meaning "to drag along the ground."

Assay— Term used to determine value of ore in ounces per ton.

Auriferous— Dirt that contains particles of gold, or gold in combination with some other mineral.

Bar— Accumulation of gravel along the banks of a stream forming a ridge, which when worked by the miners for gold are called bar diggings.

Batea— Apparatus used by Mexican miners, which was a wide, shallow vessel, usually of wood, used for panning ore.

Blue Lead— The auriferous gravel and cement in ancient river channels that was usually a dark blue. Best known are in Sierra, Nevada and Placer counties.

Bullion— Gold or silver that was milled and smelted to a form approaching purity, then molded into bars.

Carbonates— Ores containing a considerable proportion of carbonate of lead.

Chimney—An ore shoot; a large, rich mineral in a vein.

Chloride— A compound of chlorine with another element or radical. A salt of hydrochloric acid.

Claim— The boundaries that encompass legal ownership of a piece of ground presumed to contain a mineral vein or placer deposit.

Color— Term used to describe minute specks of gold in the gravel.

Concentrator— A facility where the ore is separated into ore values and rejected material.

Crosscut— A short, lateral tunnel in a deep mine used for exploratory, ventilation or communication purposes.

District— A section of country usually designated by name and described or understood as being confined within certain natural boundaries, in which gold or silver may be found in paying quantities.

Dredge— A shallow draft barge used in the efficient method of placer mining; crawling over the ground that was soaked with water, and washed a steady stream of gravel, depositing the waste behind it.

Drift—A deep mining tunnel that "drifts" with the course of the vein.

Dry Diggings— Literally picking out particles of gold with a knife from a stream, or digging gold out by knife or other implement in rock crevices.

Dry Washing— Where there was not sufficient water, a pan full of gravel was simply tossed into the air and caught again, with the gold staying in the pan while the dirt blew away.

Float— Pieces of gold or silver washed out from a hidden vein.

Flume— Inclined waterway to transport water to the hydraulics.

Free Gold— Loose gold not in a matrix of other material.

Glory Hole— A small, but unusually rich deposit of gold.

Gold Dust— Very fine particles of gold.

Grubstake— Supplies were furnished to prospectors with the promise the grubstaker would get a share of a claim.

Gulch— Any ravine, chasm or canyon where gold was found.

Hard Rock— Ore that had to be blasted out as opposed to ore soft enough to be worked with hand tools.

High-Grading— Clandestinely acquiring of exceptionally rich pieces of ore from a mine and selling it to assayers.

Hydraulic Mining— A form of placer mining that involved the use of powerful jets of water to wash away hillsides where alluvial deposits of gold occurred. This method of mining was outlawed because it destroyed the landscape.

Jewelry Rock— Very rich ore.

Lode— A vein of larger-than-normal proportions; a mine as opposed to placers which are taken from the rivers and dry riverbeds.

Old Channels— Ancient dry riverbeds left when, due to natural forces, the course of a river changed, taking an alternate route. The old channels had never been touched by early miners.

Placer Mining— Extraction of free gold from alluvial deposits using the methods of panning, rocking, or sluicing.

Quartz Ore–A matrix (gangue) of crystallized silicon dioxide containing varying amounts of gold or silver. Often referred to by miners simply as quartz.

Refractory— Exceptionally hard ore not yielding to mill treatment.

Rocker— An extension of the pan method of washing gold, the cradle was rocked steadily while a stream of dirt and water was funneled through it to separate the gold.

Seam Diggings or Mining— Decomposed bedrock filled with irregular seams of gold-bearing quartz.

Sluice— Either a ditch or a long, inclined wooden device used to wash gold in large quantities in placer mining.

Smelting— The process of melting ore in order to release the desired minerals from worthless minerals.

Stamp Mill— A device that crushed ore by means of a series of pestles (stamps), that descended, crushing the rock.

Tailings— The waste material deposited by mills and smelters after most of the valuable ore had been separated.

Tram— A conveyance for moving ore suspended by pulleys that run down cables, such as ore baskets from mines in the mountains down to the mill on a valley floor.

Vein— A clearly defined body of ore lying between walls of natural rock.

Appendix: The Miner's Ten Commandments

Placerville Herald, June 4, 1853

A man spake these words and said: I am a miner, who wandered "from down east and came to sojourn in a strange land and 'see the elephant.'" And behold I saw him and bear witness, that from the key of his trunk to the end of his tail, his whole body has passed before me: and I followed him until his huge feet stood still before a clabboard shanty: then, with his trunk extended, he pointed to a candle-card tacked upon a shingle as though he would say read, and I read

THE MINER'S TEN COMMANDMENTS

I. Thou shalt have no other claim than one.

II. Thou shall not make unto thyself any false claim, nor any likeness to a mean man by jumping one: whatever thou findest on the top above, or on the rock beneath, or in a crevice underneath the rock — for I am a jealous dog and will visit the miners round with my presence to invite them on my side: and when they decide against thee thou shalt have to take thy pick and thy pan, thy shovel and thy blankets with all thou hast and go "prospecting" both north and south to seek good diggings: and thou shalt find none. Then when thou has returned in sorrow thou shalt find that thine own claim is worked out, and no pile made thee, to hide it in the ground, or in an old boot beneath thy bunk, or in a buckskin or bottle underneath the cabin, but has paid all that was in thy purse away, worn out thy boots and thy garments so that there is nothing good about them but the pickets, and thy patience be likened unto the garments: and at last thou shalt hire thy body out to make thy board and save thy bacon.

III. Thou shalt not go prospecting before thy claim gives out. Thou shalt not take thy money, nor thy gold dust, nor thy good name, to the gambling table in vain: for monte, twenty-one, roulette, faro, lansquenet and poker will prove to thee, that the more thou puttest down, the less thou shalt take up: and when thou thinkest of thy wife and children, thou shalt not hold thyself guiltless, but insane.

IV. Thou shalt not remember what thy friends do at home on the Sabbath day, lest the rememberance may not compare favorably with what thou doest. Six days thou mayest dig or pick all that thy body can stand under; but the other day is Sunday, when thou shalt wash all the dirty shirts, darn all thy stockings, tap all thy boots, mend all thy clothing, chop the whole week's fire-wood, make up and bake thy bread and boil thy pork and beans, that thou wait not when thou returnest from thy long tour, weary. For in six days' labor only thou canst not work enough to wear out the body in two years; but if thou workest hard on Sunday also, thou canst do it in six months and thou, and thy son, and thy daughter, thy male friend and thy female friend, thy morals and thy conscience be none the better for it: but reproach thee shouldst thou ever return with thy worn-out body to thy mother's fireside, and thou strive to justify thyself, because the trader and the blacksmith, the carpenter and the merchant, the tailors, Jews, and bucaneers defy God and civilization, by keeping not the Sabbath day, and wish not for a day of rest, such as memory, youth and home made hallowed.

V. Think more of all thy gold and how canst make it fastest, than how thou wilt enjoy it, after thou hast ridden, rough-shod, over thy good old parents' precepts and examples, that thou mayest have something to reproach and sting thee, when thou art left *alone* in the land where thy father's blessing and thy mother's love sent thee.

VI. Thou shalt not kill thine own body by working in the rain, even though thou shalt make enough to buy physic and attendance with — neither shalt thou kill thy neighbor's body by shooting him, except he give thee offence — then upon principle of honor; without principle; thou mayest, even though by "keeping cool" thou hadst saved his live and thy conscience.

VII. Thou shalt not grow discouraged, and think of going home before thou hast made thy "pile," because thou hast not "struck a lead," or found a "rich crevice," nor sunk a hole upon a "pocket," lest in going home thou shalt leave four dollars a day, and go to work, ashamed, at fifty cents, and serve the right: for here, by staying, thou mightest strike a lead and fifty dollars a day, and thy manly self-respect and then go home with enough to make thyself and others happy.

VIII. Thou shalt not pick out specimens from the company pan, and put them into thy mouth or in thy purse. Neither shalt thou take from thy cabinmate his gold dust to add to thine, lest he find thee out, and straightaway call his fellow-miners together, and they hang thee, or give thee fifty lashes and two hours to leave the country, or brand thee like a horse thief with R upon thy cheek, to be "known and read of all men" Californians in particular. And if thou steal a shovel, or a pick, or a pan, from thy toiling fellow-miner, hanging will be too good for thee, and thou ask to be kicked and cow hided for thy pains: and forever hang down thy head.

IX. Thou shalt not tell any false tales about "good diggings in the mountains" to thy neighbor, that thou mayest benefit a friend who hath mules, and provisions, and tools, and blankets he cannot sell — lest in deceiving thy neighbor, when he returneth through the snow, with aught save his rifle, he present thee with contents thereof, and like a dog, thou shalt fall down and die.

X. Thou shalt not covet thy neighbor's gold nor his claim, nor undermine his bank in following a lead, nor move his stake, nor wash the tailings from his sluice's mouth, nor throw dirt upon his bank, And if thy neighbor have his family here, and thou love and covet his daughter's hand in marriage, thou shalt lose no time in seeking her affection; and when thou hast obtained it, thou shalt "pop the question" like a man, lest another more manly than thou art, should step in before thee, and thou covet her in vain, and in the anguish of disappointment, thou shalt quote the language of the great and say, "Let her rip!" and thy future be that of a poor, lonely, despised and comfortless bachelor.

The end.

A LITTLE ONE THROWN IN — Thou shalt not dig up a public road, unless thou canst afford to fix it again as good as before, otherwise thou injurest the teamster to benefit thyself, and he curse thee every time he passeth. Amen.

Notes

1. William R. Hunt, *North of 53: The Wild Days of the Alaska-Yukon Mining Frontier, 1870–1914* (New York: Macmillan, 1974), p. 238.
2. David B. Wharton, *The Alaska Gold Rush* (Bloomington: Indiana University Press, 1972), p. 156.
3. Federal Writers' Project, *Alaska: Last American Frontier* (New York: Macmillan, 1943), p. 204.
4. *Alaska's Seward Peninsula*, (Anchorage: Alaska Geographic, 1987), pp. 49–53.
5. John Muir, *Travels in Alaska* (New York: Penguin Books, 1993), p. 208.
6. "Discovery and Value." *The Fairbanks Miner* 1, no. 1 (May 1903): 1–2.
7. Wharton, p. 36.
8. Ibid., p. 118.
9. Jim Halloran, "Alaska's Gold Rushes." *California Mining Journal* 67, no. 12 (August 1998): 25.
10. Hunt, p. 229.
11. Ibid., p. 112.
12. Wharton, p. 262.
13. "The True Story of Skaguay." *The Skaguay News*. December 31, 1897, p. 1.
14. *The Ferndale* (CA) *Enterprise*, April 29, 1898.
15. Lambert Florin, *Alaska, the Yukon and British Columbia Ghost Towns*, (Seattle: Superior, 1971), p. 71.
16. Matt Volz, "Gold Rush Days." *Mail Tribune*. September 18, 2005, p. 5C.
17. Muir, pp. 21–22.
18. Philip Varney, *Arizona's Best Ghost Towns* (Flagstaff, AZ: Northland Press. 1980), p. 103.
19. Arizona Bureau of Mines, *Gold Placers and Placering in Arizona*. Bulletin No. 168 (Tucson: University of Arizona, 1961), p. 56.
20. Muriel S. Wolle, *The Bonanza Trail* (Bloomington: Indiana University Press, 1953), p. 100.
21. James E. Sherman and Barbara H. Sherman, *Ghost Towns of Arizona* (Norman: University of Oklahoma Press, 1969), p. 38.
22. Federal Writers' Project, *Arizona: A State Guide* (New York: Hastings House, 1940), p. 192.
23. Will C. Barnes, *Arizona Place Names* (Tucson: University of Arizona Press, 1988), p. 141.
24. Philip Varney, *Arizona Ghost Towns and Mining Camps* (Phoenix: Arizona Highways, 1994), p. 124.
25. Frank Love, *Mining Camps and Ghost Towns* (Los Angeles: Westernlore Press, 1974), p. 21.
26. Arizona Bureau of Mines, *Gold Placers and Placering in Arizona*, p. 19.
27. Ibid., p. 18.
28. Love, p. 99.
29. Byrd H. Granger, *Arizona's Names (X Marks the Place)* (Tucson: Falconer, 1983), p.316.
30. Arizona Bureau of Mines. *Gold Placers and Placering in Arizona*. p. 26.
31. Ibid., p. 82.
32. Ibid., p. 29.
33. Sherman and Sherman, p. 128.
34. Varney, *Arizona's Best Ghost Towns*, p. 113.
35. Barnes, p. 562.
36. *Mariposa* (CA) *Gazette*. April 1, 1862.
37. T.H. Watkins, *Gold and Silver in the West* (Palo Alto, CA: American West, 1971), p. 275.
38. "Alleghany Lighted by Electricity This Evening." *The Daily Transcript*, December 23, 1910.
39. *History of Amador County, California* (Oakland, CA: Thompson & West, 1881), p. 221.
40. *History of Placer County* (Oakland, CA: Thompson & West, 1882), p. 189.
41. Phil T. Hanna. *Dictionary of California Land Names*. Los Angeles: Automobile Club of Southern California, 1946, p. 18.
42. Francis P. Farquhar (ed.), *Up and Down California in 1860–1864* (Berkeley: University of California Press, 1966), p. 537
43. Evelyn Wells and Harry C. Peterson, *The '49ers*, (Garden City, NY: Doubleday, 1949), p. 93.
44. Irene Paden and Margaret Schlightman, *The Big Oak Flat Road to Yosemite* (Yosemite, CA: Yosemite Natural History Association, 1959), p. 133.
45. Gary D. Stumpf, *Gold Mining in Siskiyou County, 1850–1900* (Yreka, CA: Siskiyou County Historical Society, 1979), p. 43.
46. Farquhar, p. 416.
47. Wolle, p. 131.
48. Chuck Lawliss, *Ghost Towns, Gamblers & Gold* (New York: Gallery Books, 1985), p. 63.
49. Joseph H. Jackson, *Anybody's Gold: The Story of California's Mining Towns* (San Francisco: Chronicle Books, 1970), p. 206.
50. Hubert H. Bancroft. *California Pioneer Register and Index, 1542–1848*. Baltimore: Regional, 1964, p. 91.
51. *Calico Ghost Town* (Buena Park, CA: Knott's Berry Farm, 1952), p. 14.

52. Ibid, pp. 23–24.
53. Erwin G. Gudde, *California Gold Camps* (Berkeley, CA: University of California Press, 1975), p. 59.
54. G. B. Glasscock, *A Golden Highway* (Indianapolis: Bobbs-Merrill, 1934), p. 276
55. Jackson, p. 220.
56. Ibid., p. 222.
57. *History of Placer County*, p. 52.
58. Ibid., p. 54.
59. Ibid., p. 57.
60. Charles H. Shinn, *Mining Camps: A Study in American Frontier Government*. Reprint of 1884 edition. (New York: Harper Torchbooks, 1965), p. 245.
61. Wolle p. 124.
62. G. Ezra Dane, *Ghost Town* (New York: Tudor, 1941), p. 286.
63. *Randsburg Miner*, November 8, 1906.
64. Vardis Fisher and Opal L. Holmes, *Gold Rushes and Mining Camps of the Early American West* (Caldwell, ID: Caxton Printers, 1968), p. 351.
65. Federal Writers' Project, *California: A Guide to the Golden State*. Fourth printing. (New York: Hastings House, 1945), p. 476.
66. Laurence I. Seidman, *The Fools of '49* (New York: Knopf, 1976), p. 22.
67. Erwin G. Gudde, *California Place Names* (Los Angeles: University of California Press, 1962), p. 102.
68. A.H. Koschmann and M.H. Bergendahl, *Principal Gold-Producing Districts of the United States*. Geological Survey Professional Paper 610. (Washington, DC: U.S. Government Printing Office, 1968), p. 58.
69. Federal Writers' Project. *California: A Guide to the Golden State*, p. 554.
70. Philip Johnson, *Lost and Living Cities of the California Gold Rush* (Los Angeles: Automobile Club of Southern California, 1948), pp. 31–32.
71. Glasscock, p. 133.
72. Ibid., p. 138.
73. *Morning Union*, April 20, 1922.
74. *History of Nevada County*. First printed 1971. (Oakland, CA: Thompson & West, 1880), p. 65.
75. Hanna, p. 119.
76. Richard Silva, "First Gold Discovery." *The Siskiyou Pioneer* 8, no. 3, 2004, p. 23.
77. Wells and Peterson, pp. 249–252.
78. *Placer Herald*, January 14, 1854.
79. "Fire at Hornitos," *San Joaquin Valley Argus*, June 14, 1873.
80. Glasscock, p. 305
81. Doyce B. Nunis (ed.), *The Golden Frontier* (Austin: University of Texas Press, 1962), p. 29.
82. Koschmann and Bergendahl, p. 78.
83. *History of Placer County*, p. 391.
84. Federal Writers' Project. *California: A Guide to the Golden State*, p. 489.
85. *History of Amador County*, p. 168.
86. Ruth Pittman, *Roadside History of California* (Missoula, MT: Mountain Press, 1995), p. 86.
87. *Daily Alta*, January 1, 1862.
88. Peter Browning, *Yosemite Place Names* (Lafayette, CA: Great West Books, 1988), p. 82.
89. Robert I. Alotta, *Signposts & Settlers: History of Place Names West of the Rockies* (Chicago: Bonus Books, 1994), p. 335.
90. Pittman, p. 104.
91. Katherine White (comp.), *A Yankee Trader in the Gold Rush* (Boston: Houghton Mifflin, 1930), p. 274.
92. Koschmann and Bergendahl, p. 68.
93. Richard H. Peterson, "Women in the California Gold Rush," *California Mining Journal* 7, no. 11 (July 1998): 37.
94. Johnson, p. 18.
95. Ibid., p. 26.
96. Charles H. Shinns, *Mining Camps: A Study in American Frontier Government*. First published 1884. (New York: Harper Torchbooks, 1965), p. 115.
97. Bancroft, p. 68.
98. *San Andreas Independent*, August 27, 1859.
99. John Muir, *The Mountains of California* (San Francisco: Sierra Club, 1989), pp. 247–248.
100. *History of Nevada County*, p. 83.
101. Ibid., p. 79
102. Mary K. Pakui, *Place Names of Hawaii* (Honolulu: University of Hawaii Press, 1974), p. 262.
103. Glasscock, p. 186.
104. *San Joaquin Valley Argus*, August 26, 1876.
105. Pittman, pp. 47–48.
106. Seidman, p. 145.
107. "Gone." *The Daily Humboldt Standard*, Monday, September 6, 1886.
108. Elisabeth Margo, *Taming the Forty-Niner* (New York: Rinehart, 1955), p. 18.
109. Helen S. Giffen (ed.), *The Diaries of Peter Decker* (Georgetown, CA: Talisman Press, 1966), p. 296.
110. *History of Siskiyou County* (Oakland, CA: Stewart, 1881), p. 118.
111. "Bailey Bagged," *The Daily Humboldt Standard*, Thursday, September 21, 1899.
112. Carol Barrett, *As It Was, Stories from the History of Southern Oregon and Northern California* (Ashland, OR: Jefferson Public Radio, 1998), p. 54.
113. Shinn, p. 223.
114. Chauncey L. Canfield (ed.), *The Diary of a Forty-Niner* (New York: Houghton Mifflin, 1920), p. 47.
115. Johnson, p. 45.
116. Dane, p. 2
117. *History of Placer County*, p. 188.
118. Wells and Peterson, p. 140.
119. Ibid., p. 206.
120. Ibid., p. 61.
121. Pittman, p. 33.
122. Judy Barras, *Tehachapi: The Formative Years* (Reed, 1973), p. 75.
123. Giffen, p. 18.
124. Lambert Florin, *Western Ghost Town Shadows* (Seattle: Superior, 1964), p. 35.
125. Dane, p. 31
126. Glasscock, p. 110.
127. *History of Nevada County*, p. 71.
128. *History of Siskiyou County*, p. 53.
129. Ibid., p. 198.
130. Peter J. Delay, *History of Yuba and Sutter Counties* (Los Angeles: Historic Record, 1924), p. 268.
131. Caroline Bancroft, *Colorado's Lost Gold Mines and Buried Treasure* (Boulder, CO: Johnson Books, 1983), pp. 9–11.
132. Duane A. Smith, *Rocky Mountain Mining Camps* (Lincoln: University of Nebraska Press, 1967), p. 49.
133. Caroline Bancroft, pp. 11–13.
134. Florin, *Western Ghost Town Shadows*, p. 186.
135. Maxine Benson, *1001 Colorado Place Names* (Lawrence: University Press of Kansas, 1994), p. 31.
136. Federal Writers' Project, *Ghost Towns of Colorado* (New York: Hastings House, 1947), p. 8.
137. Ibid., p. 33.
138. Rick Steber, *Miners: Tales of the West*, Vol. 9 (Prineville, OR: Bonanza, 1990), p. 8.

139. Federal Writers' Project, *Ghost Towns of Colorado*, p. 81.
140. Muriel S. Wolle, *Stampede to Timberline* (Denver: Sage Books, 1965), p. 223.
141. Benson, p. 84.
142. James McTighe, *Roadside History of Colorado* (Boulder, CO: Johnson Books, 1989), p. 161.
143. Benson, p. 107.
144. Rodman W. Paul, *Mining Frontiers of the Far West: 1848–1880* (New York: Holt. Rinehart and Winston, 1963), p. 117.
145. David Dary, *Seeking Pleasure in the West* (New York: Knopf, 1995), p. 219.
146. Watkins, p. 96.
147. Perry Eberhart, *Guide to the Colorado Ghost Towns and Mining Camps* (Denver: Sage Books, 1959), p. 285.
148. Wolle, *The Bonanza Trail*, p. 426.
149. Wolle, *Stampede to Timberline*, p. 290.
150. Eberhart, p. 139.
151. Lawliss, p. 164.
152. *La Plata Miner*, May 6, 1876.
153. Federal Writers' Project, *Ghost Towns of Colorado*, pp. 60–61.
154. Benson, p. 202.
155. Wolle, *Stampede to Timberlin*, p. 174.
156. Leonard J. Arrington, *History of Idaho*, Vol. 1 (Moscow: University of Idaho Press, 1994), pp. 241–242.
157. Nunis, pp. 219–221.
158. Todd Webb, *Gold Strikes and Ghost Towns* (New York: Doubleday, 1961), p. 129.
159. Norman D. Weis, *Ghost Towns of the Northwest* (Caldwell, ID: Caxton Printers, 1981), p. 172.
160. Steber, p. 35.
161. Lambert Florin, *Ghost Town Album* (New York: Bonanza Books, 1962), p. 124.
162. Betty Derig, *Roadside History of Idaho* (Missoula, MT: Mountain Press, 1996), p. 146.
163. Wolle, *The Bonanza Trail*, p. 254.
164. Derig, p. 287.
165. Arrington, pp. 373–74.
166. "Early History of Idaho," Teacher Resource Material (Boise: Idaho Superintendent of Public Instruction, 1909), p. 12.
167. Federal Writers' Project, *Idaho: A Guide in Word and Picture* (Caldwell, ID: Caxton Printers, 1937), p. 389.
168. Wolle, *The Bonanza Trail*, p. 208.
169. Muriel S. Wolle, *Montana Paydirt* (Denver: Sage Books, 1963), p. 114.
170. Lambert Florin, *Ghost Town Trails* (Seattle: Superior, 1963), p. 80.
171. Wolle, *Montana Paydirt*, p. 387.
172. Roberta C. Cheney, *Names on the Face of Montana* (Missoula, MT: University of Montana, 1971), p. 101.
173. Ibid., p. 105.
174. Wolle, *Montana Paydirt*, p. 326.
175. Lambert Florin, *Ghost Town Treasures* (Seattle: Superior, 1965), p. 80.
176. Wolle, *Montana Paydirt*, p. 298.
177. Ibid., p. 221.
178. Ibid., p. 208.
179. Robert Silverberg, *Ghost Towns of the American West* (Athens: Ohio University Press, 1994), p. 219.
180. Smith, p. 49.
181. Cheney, p. 247.
182. Florin, *Ghost Town Treasures*, p. 88.
183. Don Ashbaugh, *Nevada's Turbulent Yesterday: A Study in Ghost Towns* (Los Angeles: Westernlore Press, 1963), p. 177.
184. Browning, p. 74.
185. Wolle, *The Bonanza Trail*, p. 343.
186. Ashbaugh, p. 327.
187. Stanley Paher, *Nevada Ghost Towns & Mining Camps* (Berkeley: Howell-North Books, 1970), p. 166.
188. Nell Murbarger, *Ghosts of the Glory Trail* (Palm Desert, CA: Desert Magazine Press, 1956), p. 43.
189. Harold Weight and Lucile Weight, *Rhyolite, The Ghost City of Dreams* (Twentynine Palms, CA: Calico Press, 1953), pp.12–13.
190. Shawn Hall, *A Guide to the Ghost Towns and Mining Camps of Nye County, Nevada* (New York: Dodd, Mead, 1981), p. 26.
191. Murbarger, p. 111.
192. Ashbaugh, p. 288.
193. Federal Writers' Project, *Nevada: A Guide to the Silver State* (Portland: Binfords & Mort, 1940), p. 119.
194. Maureen Johnson, *Placer Gold Deposits of Nevada*. Geological Survey Bulletin 1356. (Baldwin Park, CA: Gem Guides, 1972), p. 14.
195. R.W. Raymond, *Old Mines of California and Nevada*. First published 1869. (Toyahville, TX: Frontier Book, 1964), p. 82.
196. Maureen Johnson, p. 1.
197. Ashbaugh, p. 185.
198. Paher, p. 141.
199. White, pp. 221–222.
200. Watkins, p. 165.
201. Koschmann and Bergendahl, p. 192.
202. Raymond, p. 47.
203. Helen S. Carlson, *Nevada Place Names: A Geographical Dictionary* (Reno: University of Nevada Press, 1974), p. 125.
204. Federal Writers' Project. *Nevada: A Guide to the Silver State*, p. 169.
205. Wolle, *The Bonanza Trail*, p. 351.
206. Fisher and Holmes, p. 173.
207. Lambert Florin, *Nevada Ghost Towns* (Seattle: Superior, 1971), p. 34.
208. Murbarger, p. 149.
209. Ibid., p. 102.
210. W. Turrentine Jackson, *Treasure Hill* (Tucson: University of Arizona Press, 1963), p. 25–26.
211. Carlson, p. 131.
212. Florin, *Western Ghost Town Shadows*, p. 54.
213. Paher, p. 54.
214. Murbarger, p. 84.
215. Paherp. 84.
216. Ibid., p. 332.
217. Paher, p. 54.
218. White, p. 235.
219. Florin, *Nevada Ghost Towns*, p. 67.
220. Ibid., p. 66
221. Federal Writers' Project, *Nevada: A Guide to the Silver State*, p. 221.
222. Murbarger, p. 211.
223. Weight and Weight, p. 26.
224. Federal Writers' Project, *Nevada: A Guide to the Silver State*, p. 264.
225. Ashbaugh, p. 168.
226. W. Turrentine Jackson, p. 139.
227. Richard Moreno, *Roadside History of Nevada* (Missoula, MT: Mountain Press, 2000), p. 99.
228. Carlson, p. 215.
229. Ashbaugh, p. 198.
230. Paher, p. 136.
231. Paul, p. 108.
232. Paher, p. 326.

233. W. Turrentine Jackson, p. 49.
234. Raymond, p. 87.
235. Shawn R. Hall, *Old Heart of Nevada: Ghost Towns and Mining Camps of Elko County* (Reno: University of Nevada Press, 1998), p. 68.
236. *Tybo Sun*. Saturday, November 3, 1877.
237. Federal Writers' Project, *Nevada: A Guide to the Silver State*, p. 134.
238. Robert N. Johnson, *California-Nevada Ghost Town Atlas* (Susanville: Johnson, 1967), p. 7.
239. Alotta, p. 481.
240. Hubert Howe Bancroft, *The Works of Hubert Howe Bancroft. Volume XXV, History of Nevada, Colorado, and Wyoming* (San Francisco: History, 1890), p. 99.
241. Warren Hinckle and Fredric Hobbs, *The Richest Place on Earth: The Story of Virginia City and the Heyday of the Comstock Lode* (Boston: Houghton Mifflin, 1978), p. 132.
242. Smith, p. 96.
243. Carlsonp. 245.
244. Sherman, James E., and Barbara H. Sherman. *Ghost Towns and Mining Camps of New Mexico*. Norman: Univ. of OK Press, 1977, p. 42.
245. Julyan, Robert. *The Place Names of New Mexico*. Albuquerque: Univ. of NM, 1998, p. 111.
246. Smith, Duane A. *Rocky Mountain Mining Camps*. Lincoln: University of Nebraska Press, 1967, p. 252.
247. Florin, Lambert. *Ghost Town Trails*. Seattle: Superior, 1963, p. 133.
248. Sherman and Sherman, p. 85.
249. Hubert Howe Bancroft, *The Works of Hubert Howe Bancroft, Volume XVII: History of Arizona and New Mexico* (San Francisco: History, 1889), p. 755.
250. Sherman and Sherman, p. 162.
251. Ibid., p. 164.
252. Florin, *Ghost Town Album*, p. 37.
253. Sherman and Sherman, p. 174.
254. Wolle, *The Bonanza Trail*, p. 17.
255. Florin, *Ghost Town Album*, p. 51.
256. Donald C. Miller, *Ghost Towns of Washington and Oregon* (Boulder, CO: Pruett, 1977), p. 65.
257. Willard Street and Elsie Street, *Sailors' Diggings* (Wilderville: Wilderville Press, 1973), p. 7.
258. Miles F. Potter, *Oregon's Golden Years* (Caldwell, ID: Caxton Printers, 1987), p. 39.
259. Connie Fowler and J.B. Roberts, *Buncom: Crossroads Station* (Jacksonville, OR: Buncom Historical Society, 1995), p. xiii.
260. Paul Fattig, "Buncom's Back ... Not That it Actually Ever Went Away," Medford: *Mail Tribune*, Sunday, January 8, 2006, 4A.
261. Miller, p. 75.
262. Potter, p. 155.
263. Miller, p. 81.
264. Barrett, p. 61.
265. "Oregon's Hardrock Mine Cleanup Problem" (Portland: State of Oregon Department of Environmental Quality, 2003).
266. Bert Webber and Margie Webber, *Jacksonville, Oregon* (Fairfield, WA: Ye Galleon Press, 1982), p. 107.
267. Miller, p. 104.
268. Street and Street, p. 21.
269. Sanne Specht, "Wolf Habitat Welcomed by Soon-To-Be New Home," *Mail Tribune*, February 5, 2006, 4B.
270. Smith, p. 73.
271. Florin, *Ghost Town Album*, p. 75.
272. Federal Writers' Project, *Utah: A Guide to the State* (New York: Hastings House, 1941), p. 415.
273. Ibid., p. 324.
274. Fayette Lapham, "The Mormons," *Historical Magazine* 8, no. 5, 1870, p. 322.
275. Federal Writers' Project, *Utah: A Guide to the State*, p. 413.
276. Wolle, *The Bonanza Trail*, p. 187.
277. Bruce Wilson, *Late Frontier: A History of Okanogan County, Washington* (Okanogan, WA: Okanogan County Historical Society, 1990), p. 160.
278. Weis, p. 83.
279. Ruth Kirk and Carmela Alexander, *Exploring Washington's Past* (Seattle: University Washington Press, 1990), p. 61.
280. Florin, *Ghost Town Trails*, p. 51.
281. Florin, *Ghost Town Album*, p. 138.
282. Ibid., p. 136.
283. Kirk and Alexander, p. 69.
284. Wolle, *The Bonanza Trail*, p. 278
285. Nunis, p. 186
286. Kirk and Alexander, p. 182.
287. Smith, p. 46.

Bibliography

Alaska

Alaska's Seward Peninsula. Anchorage: Alaska Geographic, 1987.

Albanese, M. D., et al. *Livengood Studies Resumed: A Summary of Recent Mineral-Resource Investigations of the Alaska Division of Geological and Geophysical Surveys.* Fairbanks: Alaska Division of Geological and Geophysical Surveys, 1985.

Bundtzen, T. K., et al. *Review of Alaska's Mineral Resources.* N.p.: State of Alaska Department of Commerce and Economic Development, 1982.

Cavagnol, Joseph J. *Postmarked Alaska.* Holton, KS: The Gossip Printery, 1957.

Cook, Michael W. "BLM Lake Valley Trail." *New Mexico Magazine* 79, no. 9 (September 2001): 24.

Couch, James S. *Philately Below Zero: A Postal History of Alaska.* State College, PA: American Philatelic Society, 1953.

Dickerson, Ora B. *120 Years of Alaska Postmasters, 1867–1987.* Scotts, MI: Cammarata, 1989.

Federal Writers' Project. *Alaska: Last American Frontier.* New York: Macmillan, 1943.

Florin, Lambert. *Alaska, the Yukon and British Columbia Ghost Towns.* Seattle: Superior, 1971.

Halloran, Jim. "Alaska's Gold Rushes." *California Mining Journal* 67, no. 12 (August 1998): 23–27.

Haycox, Stephen. *Alaska, An American Colony.* Seattle: University of Washington Press, 2002.

Holzheimer, F. W. "Placer Operations Innoko District, 1926." Alaska Territorial Department of Mines, 1926.

Hunt, William R. *North of 53: The Wild Days of the Alaska-Yukon Mining Frontier, 1870–1914.* New York: Macmillan, 1974.

_____. *Alaska, A Bicentennial History.* New York: Norton, 1976.

Jasper, Martin W. "Trip to Copper River Region." Memo to Department of Mines. October 9, 1956.

Muir, John. *Travels in Alaska.* New York: Penguin Books, 1993.

Naske, Claus M., and Herman E. Slotnick. *Alaska: A History of the 49th State.* Grand Rapids, MI: Eerdmans, 1979.

Orth, Donald J. *Dictionary of Alaska Place Names.* Washington, D.C.: U.S. Government Printing Office, 1971.

Phillips, James W. *Alaska-Yukon Place Names.* Seattle: University of Washington Press, 1973.

Reed, John C., and Robert R. Coats. *Geology and Ore Deposits of the Chichagof Mining District, Alaska.* Bulletin 929. Washington, D.C.: U.S. Government Printing Office, 1941.

Robinson, M. S., and T. K. Bundtzen. "Historic Gold Production in Alaska." *Mines & Geology Bulletin* 28, no. 3 (September 1979).

Smith, Phillip S. *Mineral Industry of Alaska in 1938.* Washington, D.C.: U.S. Government Printing Office, 1939.

Stewart, R. L. *Prospecting in Alaska.* Juneau: Department of Mines, 1949.

United States Coast Pilot: Alaska, Part I, Dixon Entrance to Yakutat Bay. Washington, D.C.: U.S. Government Printing Office, 1932.

Wharton, David B. *The Alaska Gold Rush.* Bloomington: Indiana University Press, 1972.

Wimmler, Norman L. *Placer Mining in Alaska in 1924 and 1925.* N.p.: Alaska Territorial Department of Mines, 1925.

Arizona

Arizona Bureau of Mines. *Gold Placers and Placering in Arizona.* Bulletin No. 168. Tucson: University of Arizona, 1961.

Barnes, Will C. *Arizona Place Names.* Tucson: University of Arizona Press, 1988.

Dillon, Richard. *Arizona's Amazing Towns.* Tempe: Four Peaks Press, 1992.

Farish, Thomas E. *History of Arizona.* Phoenix: Thomas E. Farish, 1915.

Federal Writers' Project. *Arizona: A State Guide*. New York: Hastings House, 1940.
Granger, Byrd H. *Arizona's Names (X Marks the Place)*. Tucson: Falconer, 1983.
Hamilton, Patrick. *A Manual of Reliable Information Concerning the Territory*. Prescott, AZ, 1881.
Heylmun, Dr. Edgar B. "Greaterville Placers, Arizona." *California Mining Journal*. May, 1997:54.
_____. "Oro Blanco Mining District, Arizona." *California Mining Journal*. May, 1998:7.
_____. "Yellow Jacket Mine, Arizona." *California Mining Journal*. February, 1997:38.
Hill, J. M., et al. *Advance Chapter from Contributions to Economic Geology*. Bulletin 430-A. Washington, D.C.: U.S. Government Printing Office, 1910.
Hodge, Hiram C. *Arizona As It Is*. Cambridge: Riverside Press, 1877.
Love, Frank. *Mining Camps and Ghost Towns*. Los Angeles: Westernlore Press, 1974.
Miner, Carrie M. "Vulture Mine." *Arizona Highways*. September, 1999:18.
Newton, Charles H. *Place Names in Arizona*. Phoenix: Primer, 1954.
Patera, Alan H., and John S. Gallagher. *Arizona Post Offices*. Lake Grove, OR: The Depot, 1988.
Preston, Douglas. *Cities of Gold*. New York: Simon & Schuster, 1992.
Schrader, F. C. *Mineral Deposits of the Cerbat Range, Black Mountains and Grand Wash Cliffs, Mohave County*. Arizona: USGS Bulletin 397, 1909.
Sheridan, Thomas E. *Arizona, A History*. Tucson: University of Arizona Press, 1995.
Sherman, James E., and Barbara H. Sherman. *Ghost Towns of Arizona*. Norman: University of Oklahoma Press, 1969.
Trimble, Marshall. *Roadside History of Arizona*. Missoula: Mountain Press, 1986.
Varney, Philip. *Arizona's Best Ghost Towns*. Flagstaff: Northland Press. 1980.
_____. *Arizona Ghost Towns and Mining Camps*. Phoenix: Arizona Highways, 1994.

California

As We Were Told. Coarsegold, CA: Coarsegold Historical Society, 1990.
Axon, Gordon V. *The California Gold Rush*. New York: Mason/Charter, 1976.
Bancroft, Hubert H. *California Pioneer Register and Index, 1542–1848*. Baltimore: Regional, 1964.
Birkett, Charles V. "The Ready Relief Mine." *San Diego Historical Society Quarterly* 9, no. 4 (October 1963).
Bowen, Oliver E. "Mines Near Yosemite." *Mineral Information Service* 15, no. 3 (March 1962): 1.
Browning, Peter. *Place Names of the Sierra Nevada*. Berkeley, CA: Wilderness Press, 1986.
_____. *Yosemite Place Names*. Lafayette, CA: Great West Books, 1988.
Calico. Ghost Town. Buena Park, CA: Knott's Berry Farm, 1952.
California State Mining Bureau. *State Mineralogist, Report XIV, of the State Mineralogist, Mines and Mineral Resources of Siskiyou County, of Portions of California. Chapters of State Mineralogist's Report Biennial, Period 1913–1914*. San Francisco: California State Printing Office, 1916.
Canfield, Chauncey L., ed. *The Diary of a Forty-Niner*. New York: Houghton Mifflin, 1920.
Chalfant, W. A. *Gold, Guns & Ghost Towns*. Stanford, CA: Stanford University Press, 1947.
_____. *The Story of Inyo*. Bishop, CA: Pinon Books, 1964.
Clark, W. B. "Gold Districts of California." *California Division of Mines and Geology*. Bulletin 193, 1976.
_____. "Gold Mining at Alleghany." *Mineral Information Service* 16, no. 6 (June 1963): 12.
Coburn, Jesse L. *Letters of Gold: California Postal History Through 1869*. New York: U.S. Philatelic Classics Society, 1984.
Cronise, Titus F. *Natural Wealth of California*. San Francisco: Bancroft, 1868.
Dane, G. Ezra. *Ghost Town*. New York: Tudor, 1948.
Darling, Curtis. *Kern County Place Names*. Fresno: Pioneer, 1988.
Delay, Peer J. *History of Yuba and Sutter Counties, California with Biographical Sketches, etc*. Los Angeles: Historic Record, 1924.
Detzer, Jordan E. *Exploring Behind San Diego: Bibles. Bullets & Bullion on the Border*. San Diego: Jordan E. Detzer, 1980.
Egenhoff, Elisabeth L. "H. H. Bancroft." *Mineral Information Service* 17, no. 7 (July 1964): 127.
Farquhar, Francis P. *History of the Sierra Nevada*. Berkeley: University of California Press, 1966.
_____, ed. *Up and Down California in 1860–1864. The Journal of William H. Brewer*. Berkeley: University of California Press, 1966.
Federal Writers' Project. *California: A Guide to the Golden State*. Fourth printing. New York: Hastings House, 1945.
_____. *Death Valley: A Guide*. Riverside: Riverside Press, 1939.
Frickstad, Walter N. *A Century of California Post Offices, 1848 to 1954*. Oakland: Philatelic Research Society, 1955.
Giffen, Helen S. ed. *The Diaries of Peter Decker: Overland to California in 1849 and Life in the Mines, 1850–1851*. Georgetown, CA: Talisman Press, 1966.
Glasscock, G. B. *A Golden Highway: Scenes of History's Greatest Gold Rush Yesterday and Today*. Indianapolis: Bobbs-Merrill, 1934
Gudde, Erwin G. *California Gold Camps: A Geographical and Historical Dictionary of Camps, Towns, and Localities Where Gold Was Found and Mined; Wayside Stations and Trading Centers*. Berkeley: University of California Press, 1975.

_____. *California Place Names*. Berkeley: University of California Press, 1962.

Guide to the Mother Lode Country. Los Angeles: Automobile Club of Southern California, 1971.

Hafen, LeRoy R. *Journals of Forty-Niners*. Glendale: Clarke, 1954.

Hanna, Phil T. *Dictionary of California Land Names*. Los Angeles: Automobile Club of Southern California, 1946.

Hill, Mary R. "Silver." *Mineral Information Service* 16, no. 6 (June 1963): 1.

History of Amador County, California, with Illustrations and Biographical Sketches of Its Prominent Men and Pioneers. Oakland, CA: Thompson & West, 1881.

History of Nevada County, California, with Illustrations Descriptive of Its Scenery, Residences, Public Buildings, Fine Blocks, and Manufactories. Oakland, CA: Thompson & West, 1880.

History of Placer County, California, with Illustrations and Biographical Sketches of Its Prominent Men and Pioneers. Oakland, CA: Thompson & West, 1882.

History of Siskiyou County, California, Illustrated with Views of Residences, Business Buildings and Natural Scenery. Oakland, CA: Stewart, 1881.

History of Tuolumne County, California. San Francisco: Alley, 1882.

Holliday, J. S. *The World Rushed In: The California Gold Rush Experience*. New York: Simon and Schuster, 1983.

Jackson, Donald D. *Gold Dust*. New York: Knopf, 1980.

Jackson, Joseph H. *Anybody's Gold: The Story of California's Mining Towns*. San Francisco: Chronicle Books, 1970.

Johnson, Philip. *Lost and Living Cities of the California Gold Rush*. Los Angeles: Automobile Club of Southern California, 1948.

Johnson, Robert N. *California-Nevada Ghost Town Atlas*. Susanville, CA: Johnson, 1967.

Leroy, Frank. "125 Years of Mining History in the Panamints." *International California Mining Journal* 69, no. 12 (August 2000): 37

Mallicoat, Mark. "Hardrock in Downieville." *International California Mining Journal* 70, no. 3 (November 2000): 39.

Margo, Elisabeth. *Taming the Forty-Niners*. New York: Rinehart, 1955.

Martin, Don W., and Betty W. Martin. *The Best of the Gold Country*. Walnut Creek, CA: Pine Cone Press, 1987.

Mother Lode of California. Los Angeles: Automobile Club of Southern California, 1959.

Muir, John. *The Mountains of California*. San Francisco: Sierra Club, 1989.

Paden, Irene, and Margaret Schlightmann. *The Big Oak Flat Road to Yosemite*. Second printing. Yosemite, CA: Yosemite Natural History Association, 1959.

Pakui, Mary K. *Place Names of Hawaii*. Honolulu: University of Hawaii Press, 1974.

Perlot, Jean-Nicolas. *Gold Seeker: Adventures of a Belgian Argonaut During the Gold Rush Years*. New Haven, CT: Yale University Press, 1985.

Pittman, Ruth. *Roadside History of California*. Missoula, MT: Mountain Press, 1995.

"Place Names of the California Gold Rush." *American Geographic*. October 1945.

Raymond, R. W. *Old Mines of California and Nevada*. Toyahville, TX: Frontier Book, 1964. First published 1889.

Reisenberger, Felix, Jr. *The Golden Road*. New York: McGraw Hill, 1962.

Sanchez, Nellie Van de Grift. *Spanish and Indian Place Names of California*. San Francisco: Robertson, 1930.

Seidman, Laurence I. *The Fools of '49*. New York: Knopf, 1976.

Settle, Glen, and Kathy Hansen. *Mojave—A Rich History of Rails, Mines and Flight*. Bakersfield, CA: Kern-Antelope Historical Society, 1996.

Shinn, Charles H. *Mining Camps, A Study in American Frontier Government*. Gloucester, MA: Peter Smith, 1970. First printed 1884.

Silva, Richard. "First Gold Discovery." *The Siskiyou Pioneer* 8, no. 3. 2004: 22–32.

Stein, Lou. *San Diego County Place Names*. Leucadia, CA: Tofua Press, 1988.

Stong, Phil. *Gold in Them Hills*. New York: Doubleday, 1957.

Stumpf, Gary D. *Gold Mining in Siskiyou County, 1850–1900*. Yreka, CA: Siskiyou County Historical Society, 1979.

Thompson, Robert A. *Historical and Descriptive Sketch of Sonoma County, California*. Philadelphia: Everts, 1877.

Vandor, Paul E. *History of Fresno County, California*. Los Angeles: Historic Record, 1919.

Van Wormer, Stephen R., and James D. Newland. "The History of Hedges and the Cargo Muchacho Mining District, Part II." *Journal of San Diego History* 42, no. 3 (Summer 1996).

Ver Planck, William E. "History of Mining in Northeastern San Bernardino County." *Mineral Information Service* 14, no. 9 (September 1961): 1.

Visiting the Mother Lode of California. Los Angeles: Automobile Club of Southern California, n.d.

Wells, Evelyn, & Harry C. Peterson. *The '49ers*. Garden City, NY: Doubleday, 1949.

White, Katherine, comp. *A Yankee Trader in the Gold Rush: The Letters of Franklin A. Buck*. Boston: Houghton Mifflin, 1930.

Zauner, Phyllis, and Lou Zauner. *California Gold: Story of the Rush to Riches*. Sonoma, CA: Zanel, 1980.

Colorado

Arps, Louisa W., and Elinor E. Kingery. *High Country Names*. Boulder, CO: Johnson Books, 1994.

Bancroft, Caroline. *Colorado's Lost Gold Mines and Buried Treasure*. Boulder, CO: Johnson Books, 1983.

Bauer, William H., James L. Ozment, and John H. Willard. *Colorado Postal History: The Post Offices*. Crete, NE: J-B, 1971.

Benson, Maxine. *1001 Colorado Place Names*. Lawrence: University Press of Kansas, 1994.

Black, Robert C., III. *Island in the Rockies: The History of Grand County, Colorado, to 1930*. Boulder, CO: Pruett, 1969.

Bright, William. *Colorado Place Names*. Boulder, CO: Johnson Books, 1993.

Eberhart, Perry. *Guide to the Colorado Ghost Towns and Mining* Camps. Denver: Sage Books, 1959.

___. *Ghosts of the Colorado Plains*. Athens: Ohio University Press, 1986.

Eichler, George R. *Colorado Place Names*. Boulder, CO: Johnson Books, 1977.

Federal Writers' Project. *Colorado: A Guide to the Highest State*. Fourth printing. New York: Hastings House, 1946.

_____. *Ghost Towns of Colorado*. New York: Hastings House, 1947.

Greever, William S. *Bonanza West: The Story of the Western Mining Rushes, 1848–1900*. Norman: University of Oklahoma Press, 1963.

Heylmum, Dr. Edgar B. "Placer Gold in Colorado." *California Mining Journal*. July 1997:35.

Klucas, Gillian. *Leadville, The Struggle to Revive an American Town*. Washington, D.C.: Island Press, 2004.

McTighe, James. *Roadside History of Colorado*. Boulder, CO: Johnson Books, 1989.

Renz, Dolores C. "Paradox Valley of Colorado. A Study in Enigmas." *The Brand Book*. November 1949.

Ubbelohde, Carl, Maxine Benson, and Duane A. Smith. *A Colorado History*. Boulder, CO: Pruett, 1965.

Vandenbusche, Duane. *The Gunnison Country*. Fourth printing. Gunnison, CO: B & B, 1992.

Wolle, Muriel S. *Stampede to Timberline, The Ghost Towns and Mining Camps of Colorado*. Denver: Sage Books, 1965.

Idaho

Arrington, Leonard J. *History of Idaho. Volume 1*. Moscow: University of Idaho Press, 1994.

Barber, Floyd R., and Dan W. Martin. *Idaho in the Pacific Northwest*. Caldwell, ID: Caxton Printers, 1956.

Boone, Lalila. *Idaho Place Names*. Moscow: University of Idaho Press, 1988.

Conley, Cort. *Idaho for the Curious*. Riggins, ID: Backeddy Books, 1982.

Derig, Betty. *Roadside History of Idaho*. Missoula, MT: Mountain Press, 1996.

Federal Writers' Project. *Idaho: A Guide in Word and Pictures*. Caldwell, ID: Caxton Printers, 1937.

_____. *The Idaho Encyclopedia*. Caldwell, ID: Caxton Printers, 1983.

Friedman, Ralph. *Tracking Down Idaho*. Caldwell, ID: Caxton Printers, 1978.

Greever, William S. *Bonanza West: The Story of the Western Mining Rushes, 1848–1900*. Moscow: University of Idaho Press, 1963.

Hart, Arthur A. *Basin of Gold*. Boise: Idaho City Historical Foundation, 1986.

Heylmun, Dr. Edgar B. "Gold at Yankee Fork, Idaho." *California Mining Journal* 68, no. 4 (December 1998): 27–29.

Morgan, Clay, and Steve Mitchell. *Idaho unBound*. Ketchum, ID: West Bound Books, 1995.

Sparling, Wayne. *Southern Idaho Ghost Towns*. Caldwell, ID: Caxton Printers, 1981.

Montana

Cheney, Roberta C. *Names on the Face of Montana: The Story of Montana's Place Names*. Missoula: University of Montana, 1971.

Federal Writers' Project. *Montana, A State Guide Book*. New York: Viking Press, 1939.

Miller, Donald C. *Ghost Towns of Montana*. Boulder, CO: Pruett, 1974.

Noyes, Al. *History of Southern Montana*: Helena, MT: State, 1915.

Reflections of the Past. Bozeman, MT: Montana Ghost Town Preservation Society, 1983.

Paul, Rodman W. *Mining Frontiers of the Far West: 1848–1880*. New York: Holt, Rinehart and Winston, 1963.

Wolle, Muriel S. *Montana Paydirt*. Denver: Sage Books, 1963.

Wright, John B. Montana Places: *Exploring Big Sky Country*. Mesilla, NM: New Mexico Geographical Society, 2000.

Nevada

Ashbaugh, Don. *Nevada's Turbulent Yesterday: A Study in Ghost Towns*. Los Angeles: Westernlore Press, 1963.

Bowers, Michael L. *The Sagebrush State*. Reno: University of Nevada Press, 1996.

Carlson, Helen S. *Nevada Place Names: A Geographical Dictionary*. Reno: University of Nevada Press, 1974.

Couch, Bertrand F., and Jay A. Carpenter. *Nevada's Metal and Mineral Production (1859–1940, Inclusive)*. University of Nevada Bulletin 37, no. 4, 1943.

Elliott, Russell R. *History of Nevada*. Reno: University of Nevada Press, 1987.

Federal Writers' Project. *Nevada: A Guide to the Silver State*. Portland, OR: Binfords & Mort, 1940.

_____. *Origin of Place Names: Nevada*. Reno: Nevada

State Department of Highways and State Department of Education, 1941.
Florin, Lambert. *Nevada Ghost Towns.* Seattle: Superior, 1971.
Gamett, James, and Stanley W. Paher. *Nevada Post Offices, An Illustrated History.* Las Vegas: Nevada, 1983.
Hall, Shawn R. *Old Heart of Nevada. Ghost Towns and Mining Camps of Elk County.* Reno: University of Nevada Press, 1998.
_____. *A Guide to the Ghost Towns and Mining Camps of Nye County, Nevada.* New York: Dodd, Mead, 1981.
_____. *Romancing Nevada's Past, Ghost Towns and Historic Sites of Eureka, Lander, and White Pine Counties.* Reno: University of Nevada Press, 1994.
_____. "Austin, Nevada." *Northeastern Nevada Historical Quarterly.* Fourth Quarter, 1992.
Harris, Robert P. *Nevada Postal History.* Santa Cruz: Bonanza Press, 1973.
Hinckle, Warren, and Fredric Hobbs. *The Richest Place on Earth: The Story of Virginia City and the Heyday of the Comstock Lode.* Boston: Houghton Mifflin, 1978.
Jackson, W. Turrentine. *Treasure Hill: Portrait of a Silver Mining Camp.* Tucson: University of Arizona Press, 1963.
Johnson, Robert N. *California-Nevada Ghost Town Atlas.* Susanville, CA: Johnson, 1967.
Johnson, Maureen G. *Placer Gold Deposits of Nevada.* Geological Survcy Bulletin 1356. Baldwin Park, CA: Gem Guides, 1972.
Leigh, Rufus W. *Nevada Place Names: Their Origin and Significance.* Salt Lake City: Deseret Press, 1964.
Moreno, Richard. *Roadside History of Nevada.* Missoula, MT: Mountain Press, 2000.
Murbarger, Nell. *Ghosts of the Glory Trail.* Palm Desert, CA: Desert Magazine Press, 1956.
Nash, Thomas. *Historic Mills and Mill Tailings as Potential Sources of Contamination in and Near the Humboldt River Basin, Northern Nevada.* USGS Bulletin 2210-D. Washington, DC: Department of the Interior, 2003.
Paher, Stanley. *Nevada Ghost Towns & Mining Camps.* Berkeley, CA: Howell-North Books, 1970.
Patterson, Edna. *Who Named It? History of Elko County Place Names.* Elko, NV: Warren & Mary Monroe, 1964.
Raymond, R. W. *Old Mines of California and Nevada.* Toyahville, TX: Frontier Book, 1964 (1869 reprint).
Reproduction of Thompson and West's History of Humboldt County, 1881, With Illustrations and Biographical Sketches of Its Prominent Men and Pioneers. Berkeley, CA: Howell-North Books, 1958.
Thompson, David. *In Nevada.* New York: Knopf, 1999.
Toll, David W. *The Complete Nevada Traveler: A Guide to the State.* Reno: University of Nevada Press, 1976.
Townley, John. "Missing Gold in Tuscarora." *Treasure World.* March 1972: 34–37.
Weight, Harold, and Lucile Weight. *Rhyolite, The Ghost City of Golden Dreams.* Twentynine Palms, CA: Calico Press, 1953.
Williams, Brad, and Choral Pepper. *Lost Treasures of the West.* New York: Holt, Rinehart and Winston, 1975.

New Mexico

Barbara, James E., and H. Sherman. *Ghost Towns and Mining Camps of New Mexico.* Norman: University of Oklahoma Press, 1977.
Calentine, D. D. "Gold and the Sleepy-Headed Lady." *Frontier Times.* May 1975:12.
Casey, Robert L. *Journey to the High Southwest.* Seattle: Pacific Search Press, 1983.
Clark, Mary Grooms. *A History of New Mexico: A Mark of Time.* Canyon, TX: Staked Plains Press, 1983.
Federal Writers' Project. *New Mexico: A Guide to the Colorful State.* New York: Hastings House, 1940.
Fugate, Francis L., and Roberta B. Fugate. *Roadside History of New Mexico.* Missoula, MT: Mountain Press, 1989.
Gregg, Andrew. *New Mexico in the 19th Century.* Albuquerque: University of New Mexico, 1968.
Helbock, Richard W. *Post Offices of New Mexico.* Las Cruces, NM: Richard W. Helbock, 1981.
Julyan, Robert. *The Place Names of New Mexico.* Fourth printing. Albuquerque: University of New Mexico Press, 2003.
Pearce, T. M. *New Mexico Place Names: A Geographical Dictionary.* Albuquerque: University of New Mexico Press, 1965.

Oregon

Andrews, Wesley. "Baker City in the Eighties: Boyhood Memories." *Oregon Historical Quarterly* 50, no. 2, 1949: 83–97.
Barklow, Irene. *From Trails to Rails.* Enterprise, OR: Enchantments, 1987.
Barrett, Carol. *As It Was, Stories from the History of Southern Oregon and Northern California.* Ashland, OR: Jefferson Public Radio, 1998.
Battaile, Connie H. *The Oregon Book.* Missoula, MT: Mountain Press, 1998.
Burch, Albert. "Development of Metal Mining in Oregon. *Oregon Historical Quarterly* 43, 1942: 107–119.
"East Fork of the Illinois River Watershed Analysis." *Bureau of Land Management.* July 2000.
Federal Writers' Project. *Oregon: End of the Trail.* Portland, OR: Binfords & Mort, 1940.
Florin, Lambert. *Oregon Ghost Towns.* Seattle: Superior, 1970.
Fowler, Connie, and J.B. Roberts. *Buncom: Crossroads*

Station. Jacksonville, OR: Buncom Historical Society, 1995.
Gaston, Joseph. *Centennial History of Oregon*. Chicago: Clark, 1912.
Guest, Anna L. "The Historical Development of Southern Oregon, 1825–1852." Master's thesis, University of California, 1924.
Gullick, Bill. *Roadside History of Oregon*. Missoula, MT: Mountain Press, 1991.
Helbock, Richard. *Oregon Post Offices, 1847–1982*. Lake Oswego, OR: Raven Press, 1985.
McArthur, Lewis L. *Oregon Geographic Names*. Portland: Oregon Historical Society, 1974.
Miller, Donald C. *Ghost Towns of Washington and Oregon*. Boulder, CO: Pruett, 1977.
Oregon Metal Mines Handbook: Josephine County. Bulletin 14-C. Portland: Department of Geology and Mineral Industries, 1942.
Oregon Metal Mines Handbook: Northwestern Oregon. Bulletin 14-D. Portland: Department of Geology and Mineral Industries, 1951.
"Oregon's Hardrock Mine Cleanup Problem." Portland: State of Oregon Department of Environmental Quality, 2003.
Potter, Miles F. *Oregon's Golden Years*. Caldwell, ID: Caxton Printers, 1987.
Powers, Alfred. *Little Known Tales from Oregon History*. Vol. 1. Geoff Hill, ed. Bend, OR: Sun, 1988.
_____. "Development of Metal Mining in Oregon." *Oregon Historical Quarterly* 43, no. 2 (June 1941).
Smith, Donald T., comp. *Catalog of Northwest Railway Post Offices*. Eugene, OR: Northwest RPO Study Group, 1977.
Street, Willard, and Elsie Street. *Sailors' Diggings*. Wilderville, OR: Wilderville Press, 1973.
U.S. Forest Service. *East Fork of the Illinois River Watershed Analysis*. Grants Pass, OR: Bureau of Land Management, 2000.
Webber, Bert, and Margie Webber. *Jacksonville, Oregon*. Fairfield, WA: Ye Galleon Press, 1982.

Utah

Bancroft, Hubert H. *History of Utah*. San Francisco: History, 1890.
Bennett, Cynthia L. *Roadside History of Utah*. Missoula, MT: Mountain Press, 1999.
Carr, Stephen. *The Historical Guide to Utah Ghost Towns*. Salt Lake City: Western Epics, 1972.
Federal Writers' Project. *Utah: A Guide to the State*. New York: Hastings House, 1941.
Gallagher, John S. *Post Offices of Utah*. Burtonsville, MD: The Depot, 1977.
Johnson, Maureen G. *Placer Gold Deposits of Utah. Geological Survey*. Bulletin 1357. Gem Guides Book, 1992. First published 1973.
Lapham, Fayette. "The Mormons." *Historical Magazine* 8, no. 5 (May 1870): 322.
Powell, Allan K. *The Utah Guide*. Boulder, CT: Fulcrum, 1998.
Shaw, Lloyd. W. *Utah Post Offices*. Bountiful, UT: Lloyd W. Shaw, 1998.
Thompson, George A. *Some Dreams Die: Utah's Ghost Towns and Lost Treasures*. Salt Lake City: Dream Garden Press, 1982.
Van Cott, John W. *Utah Place Names*. Salt Lake City: University of Utah Press, 1990.

Washington

Boardman, Tim, and Richard W. Helbock. *Washington Post Offices*. Lake Oswego, OR: La Posta, 1986.
Davenport, Marge. *Best of the Old Northwest*. Tigard, OR: Paddlewheel Press, 1980.
Dow, Edson. *Passes to the North*. Wenatchee, WA: Wenatchee Bindery & Printing, 1963.
Edwards, Jonathan. *History of Spokane County*. San Francisco: Lever, 1900.
Federal Writers' Project. *Washington: A Guide to the Evergreen State*. Portland, OR: Binfords & Mort, 1941.
Heylmum, Dr. Edgar B. "Blewett District, Washington." *California Mining Journal* 67, no. 8 (April 1998): 13
Hitchman, Robert. *Place Names of Washington*. Tacoma: Washington State Historical Society, 1985.
Hull, Lindley. *A History of Central Washington*. Spokane: Shaw & Borden, 1929.
Kirk, Ruth, and Carmela Alexander. *Exploring Washington's Past*. Seattle: University Washington Press, 1990.
Lockley, Fred. *History of the Columbia River Valley: From the Dalles to the Sea*. Vol. 1. Chicago: Clark, 1928.
Meany, Edmond S. *Origin of Washington Geographic Names*. Seattle: University Washington Press, 1923.
Miller, Donald C. *Ghost Towns of Washington and Oregon*. Boulder: Pruett, 1977.
Phillips, James W. *Washington State Place Names*. Seattle: University Washington Press, 1971.
Phipps, Richard T., et al. "Inactive and Abandoned Mine Lands: Spada Lake and Cecile Creek Watershed Analysis Units, Snohomish and Okanogan Counties, Washington." Washington State Department of Natural Resources. Open File Report 2003. February 2003.
Steele, Richard F. *An Illustrated History of Stevens, Ferry, Okanogan and Chelen Counties, State of Washington*. Spokane: Western Heritage, 1904.
Wilson, Bruce. *Late Frontier: A History of Okanogan County, Washington*. Okanogan, WA: Okanogan County Historical Society, 1990.
Wolff, Fritz E., et al. "Inactive and Abandoned Mine Lands. Lone Jack Mine, Mount Baker Mining District, Whatcom County, Washington." Washington State Department of Natural Resources. Information Circular 98. March 2005.

Wyoming

Federal Writers' Project. *Wyoming: A Guide to Its History, Highways, and People.* New York: Oxford University Press, 1948.

Moulton, Candy. *Roadside History of Wyoming.* Missoula, MT: Mountain Press, 1995.

Urbank, Mae. *Wyoming Place Names.* Missoula, MT: Mountain Press, 1988.

General

Alotta, Robert I. *Signposts & Settlers, The History of Place Names West of the Rockies.* Chicago: Bonus Books, 1994.

Bancroft, Hubert Howe. *The Works of Hubert Howe Bancroft, Volume XVII: History of Arizona and New Mexico.* San Francisco: History, 1889.

_____. *The Works of Hubert Howe Bancroft, Volume XXV: History of Nevada, Colorado, and Wyoming.* San Francisco: History, 1890.

Beebe, Lucius, and Charles Clegg. *The American West.* New York: Dutton, 1955.

Blevens, Don. *Peculiar, Uncertain, and Two Egg.* Nashville, TN: Cumberland House, 2000.

Carter, William. *Ghost Towns of the West.* Menlo Park, CA: Lane Magazine & Book, 1971.

Dary, David. *Seeking Pleasure in the Old West.* New York: Knopf, 1995.

Emmons, William H. *Gold Deposits of the World.* New York: McGraw-Hill, 1937.

Fisher, Vardis, and Opal L. Holmes. *Gold Rushes and Mining Camps of the Early American West.* Caldwell, ID: Caxton Printers, 1968.

Florin, Lambert. *Ghost Town Treasures.* Seattle: Superior, 1965.

_____. *Western Ghost Town Shadows.* Seattle: Superior, 1964.

_____. *Ghost Towns of the Southwest.* New York: Promontory Press, 1970.

_____. *Ghost Town Trails.* Seattle: Superior, 1963.

_____. *Ghost Town Album.* New York: Bonanza Books, 1962.

Greever, William S. *Bonanza West: The Story of Western Mining Rushes, 1848–1900.* Moscow: University of Idaho Press, 1963.

Joyce, P.W. *Pocket Guide to Irish Place Names.* Belfast: Appletree Press, 1984.

Koschmann, A. H., and M. H. Bergendahl. *Principal Gold-Producing Districts of the United States.* Geological Survey Professional Paper 610. Washington, DC: U.S. Government Printing Office, 1968.

Lawliss, Chuck. *Ghost Towns, Gamblers and Gold.* New York: Gallery Books, 1985.

Linford, Laurance D. *Navajo Places: History, Legend, Landscape.* Salt Lake City: University of Utah Press, 2000.

Mills, A.D. *A Dictionary of English Place Names.* New York: Oxford University Press, 1991.

Nunis, Doyce B. (Ed.). *The Golden Frontier—The Recollections of Herman Francis Reinhart, 1851–1896.* Austin: University of Texas Press, 1962.

Nestor, Sandy. *Our Native American Legacy.* Caldwell, ID: Caxton Printers, 2001.

_____. *Indian Placenames in America, Vol. 1, Cities, Towns and Villages.* Jefferson, NC: McFarland, 2003.

_____. *Indian Placenames in America, Vol. 2, Mountains, Canyons, Rivers, etc.* Jefferson, NC: McFarland, 2004.

Rascoe, Jesse E. *Western Treasures, Lost and Found.* Fort Davis, TX: Frontier, 1971.

Room, Adrian. *Place-Names of the World.* Totowa, NJ: Rowman and Littlefield, 1974.

Silverberg, Robert. *Ghost Towns of the American West.* Athens: Ohio University Press, 1994.

Smith, Duane A. *Rocky Mountain Mining Camps.* Lincoln: University of Nebraska Press, 1967.

Steber, Rick. *Miners: Tales of the West, Vol. 9.* Prineville, OR: Bonanza, 1990.

Vokac, David. *The Great Towns of the West.* San Diego: West Press, 1985.

Watkins, T. H. *Gold and Silver in the West.* Palo Alto, CA: American West, 1971.

Webb, Todd. *Gold Strikes and Ghost Towns.* New York: Doubleday, 1961.

Weis, Norman D. *Ghost Towns of the Northwest.* Caldwell, ID: Caxton Printers, 1981.

Williams, Brad, and Choral Pepper. *Lost Legends of the West.* New York: Holt, Rinehart & Winston, 1970.

Wolle, Muriel S. *The Bonanza Trail: Ghost Towns and Mining Camps of the West.* Bloomington: Indiana University Press, 1953.

Young, John P. *Journalism in California.* San Francisco: Chronicle, 1915.

Index